The Sheriff Clerk
Sheriff Court House,
Hope Street,
LANARK.

SCTS
Library Services
Withdrawn from Stock

CONNELL

on

**THE AGRICULTURAL HOLDINGS
(SCOTLAND) ACTS**

CONNELL

on

The Agricultural Holdings (Scotland) Acts

Sixth Edition

by

CHARLES H. JOHNSTON, Q.C., M.A., LL.B.
Sheriff Substitute of Lanarkshire at Glasgow

and

KENNETH M. CAMPBELL, M.A., LL.B., W.S.

EDINBURGH   GLASGOW
WILLIAM HODGE & COMPANY LIMITED

COPYRIGHT © C. H. JOHNSTON AND K. M. CAMPBELL, 1970

First Published	- -	1923
Second Edition	- -	1928
Third Edition -	- -	1938
Fourth Edition	- -	1951
Fifth Edition -	- -	1961
Sixth Edition -	- -	1970

Printed in Great Britain by William Hodge & Company, Limited
ISBN 0 85279 003 1

PREFATORY NOTE

THIS Sixth Edition is the first for which no member of the Connell family bears responsibility, but we wish to acknowledge the debt we owe not only to the work of the late Sir Isaac Connell, S.S.C., but also to the experience and care of his son, Sir Charles G. Connell, W.S., who was concerned with several recent editions. Any errors or omissions—we hope there are none—are our responsibility.

The permission of the Controller of H.M. Stationery Office has been obtained for the reproduction of the various Acts and Statutory Instruments to which we refer wholly or in part. We have altered the text of the 1949 Act to conform to what we believe is the effect of amending legislation, and we hope that we have shewn in each case the source of the alterations. Readers are, however, invited to look for themselves at the official texts of the various amending statutes detailed in paragraph (2) of the Introduction.

Our thanks are due to Mr. J. A. Batters, LL.B., who has been good enough to prepare the index.

The law is stated as at 1st May, 1970.

C. H. J.

K. M. C.

CONTENTS

	PAGE
TABLE OF CASES AND ARBITRATIONS CITED	x
INTRODUCTION	1
The Object of the Acts	1
Commencement and Dates of Operation of the Agricultural Holdings (Scotland) Acts	2
Landlord and Tenant	3
Holdings Affected	4
Minimum Term of a Lease	4
Provisions for Written Leases	5
Variation of Rent	6
Absolute Right to Increase Rent	10
Variation of Terms of Tenancy as to Permanent Pasture	10
Continuance of Lease Not Affected by Variation of Term	11
MISCELLANEOUS PROVISIONS RELATING TO LANDLORD AND TENANT	11
Freedom of Cropping	11
Tenant's Right to Remove Fixtures	12
Some Fixtures at Common Law	13
Penal Rent and/or Liquidated Damages	14
Record of Holding	14
Removing for Non-Payment of Rent	15
Bequest of Lease	16
Succession to Holdings	17
Notice to Quit	19
Provisions as to Security of Tenure	21
Restriction on Operation of Notices to Quit	21
Provisions as to Consents for Purposes of Sec. 25	22
Application for Certificate of Bad Husbandry	23
Notice to Quit Part of Holding	23
Provisions Regarding Compensation for Disturbance	23
Amount of Compensation for Disturbance	26
Sum for re-organisation of Tenant's Affairs	26
Compensation for Early Resumption	28
Compensation for Improvements	29
Improvements for which the Act Allows Compensation	31
Permanent Improvements (Part I of each of Schedules I, II, and III)	32
Drainage, etc. (Part II)	33
Temporary Improvements (Part III)	35
Artificial Manures and Feeding Stuffs	37
Temporary Pasture	39
High Farming	42

CONTENTS

	PAGE
Ascertaining the Value of Improvements	43
"Substituted" Compensation for Improvements	45
Improvements made under Repealed Statutes	46
Compensation for Damage by Game—*i.e.,* Deer, Pheasants, Partridges, Grouse and Black Game	46
Notices of Intention to Claim Compensation and Lodging of Particulars	48
Contracting out	49
Charging the Estate with Compensation	51
Special Provisions as to Market Gardens	51
ARBITRATION UNDER THE ACT	54
Claims and Questions which are, or may be, referred to Arbitration, and conditions applicable thereto	54
Claims by the Tenant	55
Claims by Landlord	63
Particulars of Claims	64
Relevancy and Competency of Claims	64
Appointment of Arbiter	66
Disqualification and Removal of Arbiter	69
Minute of Submission	70
Procedure in Arbitration	70
Appointment of Clerk	71
Receiving Claims and Objections	71
Hearing and Inspection	72
Evidence	73
Stated Case	76
Note of Proposed Findings	79
Expenses	80
The Award	82
Reduction of Award	85
ARBITRATIONS AND VALUATIONS (OUTSIDE THE ACT) GENERALLY BETWEEN AWAY-GOING AND INCOMING TENANTS	89
The Subjects of Common Law Arbitration or Valuation	90
Appointment of Arbiters and Oversman	93
Procedure and Principles of Valuation	94
Valuation of Sheep Stocks	95
Hill Farming Act, 1946	99
The Duties of Arbiters and Oversman	101
Procedure in Making Valuations	102
Away-going Crop Valuations	102
TEXT OF THE AGRICULTURAL HOLDINGS (SCOTLAND) ACT, 1949, AND SCHEDULES WITH ANNOTATIONS	105

APPENDICES

		PAGE
1.	Forms - - - - - - - - - - - -	224
2.	The Agricultural Holdings (Specification of Forms) (Scotland) Instrument, 1960 - - - - - - -	277
3.	Sheep Stocks Valuation (Scotland) Act, 1937 - - -	285
4.	Excerpts from Hill Farming Act, 1946, relating to Sheep Stock Valuations - - - - - - - -	288
5.	The Agricultural Records (Scotland) Regulations, 1948 -	296
6.	Excerpts from the Agriculture Act, 1958, relating to Scotland	302
7.	Excerpt from the Succession (Scotland) Act, 1964 - -	316
8.	Excerpts from the Agriculture (Miscellaneous Provisions) Act, 1968 - - - - - - - - - - -	322
Index - - - - - - - - - - - - -		343

TABLE OF CASES AND ARBITRATIONS CITED

PAGE

Adam v. Smythe - - - - 217
Adams v. Gt. North of Scotland
    Railway Co. - - - 70, 85
Addie & Sons v. Henderson &
    Anor. - - - - - 69
Admiralty v. Burns - - - 137
Alexander's Trs. v. Dymock's
    Trs. - - - - - 101
Allan v. Thomson - - - 121
Allan's Trs. v. Allan & Son - 141
Alston v. Chappell - - - 87
Alston's Trs. v. Muir - - 122, 135
Anderson v. McCall - - - 121
Anderson v. Tod - - - - 121
Armstrong & Co. v. McGregor
    & Co. - - - - - 121
Auckland (Lady) v. Dowie 48, 125

Ballantyne v. Brechin - - - 129
Barbour v. M'Douall
                  35, 122, 153, 159
Barr v. Strang & Gardner - - 198
Barrow Green Estate Co. v.
    Walker's Executors - - 172
Baxter v. Macarthur - - - 79
Bebington v. Wildman - 136, 175
Bell v. Graham 45, 90, 160, 161, 218
Bellshill & Mossend Co-operative Society Ltd. v. Dalziel
    Co-operative Society Ltd. - 88
Benington-Wood v. Mackay - 143
Benington-Wood's Trustee v.
    Mackay - - - - 143
Bennett v. Stone - - - - 202
Bennetts v. Bennet - - - 54
Bennie v. Mack - - - - 154
Bennion and National Provincial Bank's Arbitration - 217
Bernays v. Prosser - - 106, 197
Berry v. Allen - - - - 92
Berwick v. Baird - - - - 199
Bevan v. Chambers - - - 201
Beveridge & Ors. v. McAdam - 152
Bickerdike v. Lucy - - 179, 199
Bignall v. Gale - - - - 73
Black v. Clay - - - - 200
Black (Alex.) & Sons v. Paterson
                  113, 199

PAGE

Black v. Williams & Co.
    (Wishaw) Ltd. - - - 86
Blaikie v. Aberdeen Railway Co. 76
Blair v. Meikle - - - - 161
Blay v. Dadswell - - - - 147
Boyd v. Wilton - - - - 172
Bradshaw v. Bird - - - 202
Brand's Trs. v. Brand's Trs. - 124
Breadalbane, Marquis of, v.
    Robertson - - - - 154
Breadalbane, Marquis of, v.
    Stewart - - - - 137, 201
Breslin v. Barr & Thornton - 81
Brodie v. Ker, McCallum v.
    Macnair 65, 67, 106, 129, 186
Brodie-Innes v. Brown - 66, 170
Brown v. College of St. Andrews 91
Brown v. Mitchell - - 38, 44
Broxburn Oil Co. Ltd. v. Earl of
    Buchan - - - - - 218
Brunskill v. Atkinson - - - 210
Bryson v. Mitchell - - - 102
Buccleuch, Duke of, v. Tod's Trs. 13
Buchanan v. Smith - - - 74
Buchanan v. Taylor - - - 46
Budge v. Hicks - - - - 140
Budge v. Mackenzie - - - 187
Buerger & Co. v. Barnett - - 78
Burnett v. Henry - - - - 74
Burnett v. Gordon - - - 143

Caird - - - - - - 76
Caledonian Railway Co. v.
    Lockhart - - - 74, 102
Callander v. Smith - - 154, 178
Cameron v. Ferrier - - - 195
Cameron v. Mackay - - - 84
Cameron v. Menzies - - - 85
Cameron v. Nicol 54, 89, 101
Campbell v. McHolm - - 73
Carnegie v. Davidson - - - 143
Carron Co. v. Donaldson - - 122
Carruthers v. Hall - - - 86
Cathcart v. Chalmers - - 36, 49
Cave v. Page - - - 137, 155
Chalmers Property Investment
    Co. Ltd. v. MacColl
                67, 184, 217, 254

TABLE OF CASES

Chalmers Property Investment Co. Ltd. v. Bowman - 77, 86
Chalmers' Tr. v. Dick's Tr. - 137
Chapman v. Dept. of Agriculture - 38
Chapman v. Lockhart - 288
Christison's Trs. v. Callender Brodie - 67, 200
Christopher Brown Ltd. v. Genossensschaft, etc. - 66, 70
Clark v. Hume - 154
Coates v. Diment - 177
Collett v. Deely - 217, 252
Commissioners for Crown Lands v. Grant - 189, 193
Cooper v. Muirden - 143
Cooper v. Pearce - 199
Coutts v. Barclay-Harvey - 184, 201
Coutts v. Smith's Trs. - 68
Cowan v. Wrayford - 140
Cowane's Hospital, Patrons of, v. Rennie - 143
Cowdray v. Ferries - 65, 187
Cowe v. Millar - 113, 135
Craighall Cast-Stone Co. v. Wood Bros. - 137
Crawford v. McKinlay - 143
Crichton v. North British Railway Co. - 76
Crichton-Stuart v. Ogilvie - 137
Crossley v. Clay - 85
Crown Estate Commissioners v. Gunn - 9, 113
Cunningham v. Fife County Council - 199
Cushnie v. Thomson - 135
Dale v. Hatfield Chase Corporation - 153, 202
Dalgleish v. Livingston - 199
Davidson v. Chiskan Estate Co. Ltd. - 187
Davidson v. Hunter - 38, 44
Davidson v. Logan 42, 72, 89, 101, 102
Davies ex parte - 143
Dean v. Secretary of State for War - 153
Department of Agriculture for Scotland v. Goodfellow - 136
Derby, Lord and Fergusson's Contract re. - 199, 202
Digby and Penny, In re. - 25, 140
Disraeli Agreement re. - 137

Donald v. Sheill & Anor. - 86
Donaldson v. Regent Photo Playhouse - 135
Donaldson's Hospital v. Esselmont - 65, 188
Dow Agrochemicals v. Lane - 143
Drew v. Drew - 69, 73
Drummond v. Thomson - 199
Drysdale v. Jameson - 46
Duff v. Keith - 200
Duguid v. Muirhead - 20, 107, 135
Dunbar's Trs. v. Bruce - 199
Dundas v. Hogg and Anor. - 86
Dundas v. Morison - 200
Dundee Corporation v. Guthrie 68, 113, 217
Dunlop & Co. v. Meiklem - 199
Dunlop v. Mundell - 287
Dunmore, Earl of, v. McInturner 73
Dunn v. Fidoe - 105
Dyke, ex parte, in re Morrish - 200

Eastern Angus Properties Ltd. v. Chivers & Sons Ltd. - 143
Edell v. Dulieu - 122, 135
Edinburgh & Glasgow Railway Co. v. Miller - 68
Edinburgh & Glasgow Railway Co. v. Marshall & Miller - 101
Edinburgh Corporation v. Gray 137
Edinburgh Magistrates v. The Lord Advocate - 154
Edmunds v. Woollacott - 141
Egerton v. Rutter - 139, 200
Elibank, Lord, v. Hay - 129
Ellis v. Lewin - 88, 218
Elphinstone, Lord, v. Monkland Iron Co. - 200
Elwes v. Maw - 124
Erskine's Trs. v. Crombie - 95
Euman v. Dalziel & Co. - 77
Evans v. Glamorgan County Council - 153
Evans v. Jones - 172
Evans v. Roper - 143
Evans (F.R.) (Leeds) v. Webster 217, 252, 254
Ewing & Co. v. Dewar - 101

Farrer v. Nelson - 126
Farrow v. Orttewell - 153
Faviell v. Gaskoin - 29
Fenton v. Howie - 143

TABLE OF CASES

Ferguson v. Norman - - - 218
Ferrier v. Alison - - - - 102
Findlay v. Munro
    30, 40, 41, 43, 114, 170
Fleming v. Middle Ward of
  Lanark - - - - - 149
Fletcher v. Fletcher - - - 129
Fletcher v. Robertson - 74, 90
Forbes v. Lady Saltoun's Exrs. - 135
Forbes v. Pratt - - - - 152
Forbes v. Underwood - 68, 101
Forbes v. Ure - - - - 200
Forbes-Sempill's Trs. v. Brown 141
Forsyth-Grant v. Salmon - - 218
Frankland v. Capstick - - 184
Fraser v. Murray's Trs. - - 131
Fraser's Trs. v. Maule & Son - 201
French v. Elliott - 22, 140, 227, 232

Galbraith & Anor. v. Ardna-
  cross Farming Co. - - 187
Galbraith's Trs. v. Eglinton Iron
  Co. - - - - - 149
Gale v. Bates - - - - 121
Galloway, Earl of, v. Elliot - 152
Galloway, Earl of, v. McClelland
  (1915 S.C.)
    30, 38, 39, 40, 43, 164, 210
  (1917 Sh.Ct.) - - - 78
Gardiner v. Abercromby - 160, 201
Gates v. Blair - - - - 198
Gibson v. Farie - - - - 179
Gibson v. Fotheringham - 89, 102
Gibson v. McKechnie - - 143
Gibson v. Sherret - - - 156
Gillies v. Fairlie - - - - 137
Gilmour v. Osborne's Trs. - 145
Gladstone v. Bower - - - 106
Glasgow, Yoker and Clydebank
  Railway Co. v. Lidgerwood 54
Glencruitten Trs. v. Love - 137
Glendinning v. Board of Agri-
  culture - - - - - 83
Goldsack v. Shore 99, 106, 187, 207
Gordon v. Abernethy - - - 102
Gordon (Lindsay's Trs.) v.
  Welsh - - - - - 200
Graham v. Gardner - - 7, 113
Graham v. Stirling - 100, 107, 135
Graham v. Wilson-Clarke - - 145
Grant v. Girdwood - - - 87
Grant v. Murray - - - - 143
Gray v. Ashburton - - - 81

Gray v. Low - - - - 200
Grewar v. Moncur's Curator
  Bonis - - - - 179, 199
Grieve & Sons v. Barr - - 107
Griffiths v. Morris - - - 81
Gulliver v. Catt - - - 172, 187
Guthe v. Broatch - - 7, 113

Haden and Ors. (Scott's Exrs.)
  v. Hepburn - - - - 200
Halliday v. William Fergusson
  & Sons and others - - 141
Halliday v. Semple - - 86, 217
Hamilton's Trs., Duke of, v.
  Fleming - - - - 154, 157
Hamilton v. Duke of Hamilton's
  Trs. - - - - - 199
Hamilton v. Lorimer - 21, 23, 137
Hamilton v. Peeke - - - 84
Hamilton v. Reid's Trs. - - 163
Hamilton Ogilvy v. Elliot 54, 65, 159
Hammon v. Fairbrother - - 140
Hammond, ex parte - - - 199
Harrison-Broadley v. Smith - 50
Hart v. Cameron - - - 141
Harvey and Mann's Arbitration 153
Harvey v. Gibsons - - - 76
Harvey v. Skelton - - - 85
Heaven & Kesterton v. Etablisse-
  ments Francois Albiac et
  Cie - - - - - - 81
Hemington v. Walter - - 139, 202
Henderson v. Corporation of
  Glasgow - - - - 77
Hendry v. Walker - 137, 152, 155
Hewson v. Matthews - - - 139
Highland Railway Co. v. Mit-
  chell - - - - - 76
Hollings v. Swindell - - - 105
Holman v. Peruvian Nitrate Co. 121
Holme v. Brunskill - - - 135
Holmes Oil Co. Ltd. v. Pump-
  herston Oil Co. Ltd. - 73, 89
Hopper - - - - - 85, 89
Horswell v. Alliance Assurance
  Co. - - - - - 80
Hoth v. Cowan - - - 65, 136
Houison-Crauford's Trs. v.
  Davies - - - 99, 106, 186
Howie v. David Lowe & Sons
  Ltd. - - - - - 131
Howkins v. Jardine - - - 105
Hunter v. Barron's Trs. - 20, 201

TABLE OF CASES

	PAGE
Hunter v. Miller	122
Imrie's Tr. v. Calder	200
Inchiquin v. Lyons	135
Ingham v. Fenton	202
Inglis v. Moir's Tutors	126
Inland Revenue v. Assessor for Lanarkshire	199
Irving v. Church of Scotland General Trustees	131
Iveagh (Earl) v. Ministry of Housing and Local Government	80
Jackson v. Galloway	82
James v. Dean	201
Jamieson v. Clark	217, 252
Johnston v. Cheape	69
Johnston v. Glasgow Corporation	77
Johnston v. Malcolm	153
Johnstone v. Hughan	42
Jones v. Gates	140
Kedwell v. Flint & Co.	180
Kennedy v. Johnstone	49, 131, 200
Kent v. Conniff	172, 207
Kestell and Anor., v. Langmaid	153
Keswick v. Wright	153
Kidd v. Byrne	126
Kilmarnock Estates Ltd. v. Barr	113
Kininmonth v. British Aluminium Co. Ltd.	137
Kirkcaldy v. Dalgairns	94
Kirkland v. Gibson	200
Kyd v. Paterson	87
Lancaster & Macnamara	199
Lawrence v. Bristol & Somerset Railway Co.	86
Lean v. Inland Revenue	199
Ledingham v. Elphinstone	72, 73
Lendon v. Keen	82
Lendrum v. Ayr Steam Shipping Co.	78
Lennox v. Reid	129
Leschallas v. Woolf	124
Lewis v. Haverford-West R.D.C.	81, 87
Lindsay v. Bell	95
Lindsay's Trs. v. Welsh	200
Linton's Trs. v. Wemyss Landed Estates Co. Ltd.	131

	PAGE
Livingstone v. Beattie	199
Logan v. Leadbetter	72, 89, 95
Longair v. Reid	143
Lothian's Trs., Marquis of, v. Johnston	131
Lower v. Sorrell	51
Lowther v. Clifford	179
Lyons v. Anderson	113
McBay v. Birse	143
McCallum v. Macnair (See Brodie v. Ker)	
McCallum v. Robertson	79
MacCormick v. Lord Advocate	65
Macdonald v. Macdonald	73, 74
McDouall's Trs. v. MacLeod	129, 137
M'Farlane v. Mitchell	135
McGavin v. Sturrock's Tr.	200
McGhie and Anor. v. Lang and Anor.	105
Macgregor v. Board of Agriculture	153
McGregor v. Stevenson	72, 89, 102
McIntyre v. Anderson	91
McIntyre v. Board of Agriculture	149
McIntyre v. Forbes and Fraser	96, 97
McIver v. McIver	200
Mackay & Son v. Leven Police Commissioners	102
Mackenzie v. Cameron	131
Mackenzie v. Laird	4, 106
Mackenzie v. McGillivray	30, 163, 164, 210
Mackenzie v. Tait	143
McKenzie v. Buchan	198
McKenzie v. Clark	69
McKenzie v. Hill	72
Mackessack & Son v. Molleson	200
McKessock v. Drew	83
MacKinnon v. Martin	131
Mackinnon v. Secretary of State for Scotland	101
Mackintosh v. Lord Lovat	199
McKinley v. Hutchison's Trs.	137
McLaren v. Aikman	84
McLaren v. Lawrie	143
McLean's Trustee v. McLean	200
Macleod v. Urquhart	201
MacMaster v. Esson	156
Macnab v. Willison (1955)	42
Macnab of Macnab v. Willison (1960)	137, 218

TABLE OF CASES

	PAGE
Macnabb v. Anderson (1955)	140
Macnabb v. Anderson (1957)	140
McNair's Trs. v. Roxburgh	72
Macpherson v. Secretary of State	295
McQuaker v. Phoenix Assurance Co.	218
McQuater v. Fergusson	30, 82, 163, 218
Malcolm v. McDougall	198
Masters v. Duveen	179
Mears v. Callender	124, 157
Meggeson v. Groves	121, 122, 124
Methven v. Burn	188
Meux v. Cobley	124
Miller v. Millar	85
Miller v. Muirhead	124
Miller & Son v. Oliver & Boyd	86
Miller's Trs. v. Berwickshire Assessor	154
Mills v. Rose	137, 156
Miln v. Earl of Dalhousie	163
Ministry of Agriculture v. Dean	200
Mitchell v. Cable	73, 85
Mitchell-Gill v. Buchan	69, 79
Moncrieffe v. Ferguson	200
Montgomerie v. Wilson	100, 107, 135
Morison v. Thomson's Trs.	79
Morris & Anor. v. Muirhead Buchanan & Macpherson & Ors.	141
Morrison-Low v. Howison	141, 232
Morse v. Dixon	179
Morton (Earl) Trs. v. Macdougall	49
Moseley v. Simpson	85, 87
Mountford v. Hodkinson	139
Mowbray v. Dickson	73, 87
Murray v. Nisbet	141
Napier v. Wood	86, 87
Nelson v. Allan Brothers & Co. (U.K.) Ltd.	77
Nivison v. Howat	72, 89, 102
North and South-Western Junction Railway Co. v. Brantford Union	78
North British Railway Co. v. Barr	87
North British Railway Co. v. Wilson	74
O'Connor and Whitelaw's Arbitration, In re.	73, 88

	PAGE
Officer v. Nicolson	121
Osler v. Lansdowne	201
Palmer & Co. v. Hoskin & Co.	78
Paterson v. Glasgow Corporation	74
Paynter v. Rutherford and Ors.	287
Pearson and I'Anson	157, 178
Pendreigh's Tr. v. Dewar	201
Pentland v. Hart	141
Perrens & Harrison v. Borron	87
Perry v. Stopher	81
Piggott v. Robson	137
Pollich v. Heatley	86
Pott's Jud. Factor v. Glendinning and Ors.	97, 99, 287
Poyser and Mills' Arbitration, re	80, 136, 195
Premier Dairies v. Garlick	123
Price v. Romilly	25, 141
Purser v. Worthing Local Board	199
Purves v. Rutherford	91
R. v. Agricultural Land Tribunal ex p. Palmer	105
Rae & Cooper v. Davidson	135
Ramsay v. McLaren and Anor.	67, 217
Ramsden & Co. v. Jacobs	88
Reid v. Dawson	106
Reid v. Duffus Estate Ltd.	16, 131
Reid's Executors v. Reid	38
Rennie v. Patrons of Cowane's Hospital	89
Richards v. Pryse	199
Richardson v. Worsley	80
Riddel's Exrs. v. Milligan's Exrs.	199
Roberts v. Magor	124, 152
Robertson v. Cheyne	101
Robertson v. Ross & Co.	149
Robertson's Trs. v. Cunningham	184, 217
Robinson v. Rowbottom	199
Rochester and Chatham Joint Sewerage Board v. Clinch	136, 147
Roddan v. McCowan	47
Roger v. Hutcheson and Ors.	65, 82, 91, 122
Ross v. Macdonald	187
Ross v. Monteith	200
Russell v. Freen	201

TABLE OF CASES

Russell & Harding's Arbitration,
In re. - - - - 198, 199
Rutherford v. Maurer 4, 105, 106

Salmon v. Eastern Regional Hospital Board - - - 198
Sangster v. Noy - - - - 136
Saunders-Jacobs v. Yates - - 179
Schofield v. Hincks - - - 200
Sclater v. Horton - - - 113
Scott v. Livingstone - - - 136
Scott v. Ritchie - - - - 95
Scott v. Scott - - - 48, 136
Secretary of State v. Anderson - 295
Secretary of State v. Davidson - 113
Secretary of State v. Fraser - 139
Secretary of State v. John Jaffray and Ors. - - - 189
Secretary of State v. Prentice 136, 148
Secretary of State v. Sinclair (1960) - - - - 110, 276
Secretary of State v. Sinclair (1962) - - - - 9, 113
Secretary of State v. Young and Ors. - - - 8, 72, 73, 113
Seggie v. Haggart - - - 136
Sellar v. Highland Railway Co. 69
Selleck v. Hellens - - - 140
Service v. Duke of Argyll - - 131
Shaw v. Porter - - - - 201
Shepherd & Anor. v. Lomas - 141
Sheriff v. Christie and Anor. 68, 69
Sherwood (Baron) v. Moody - 51
Sim v. McConnell and Anor. - 86
Simpson & Henderson - - 217
Simson's Trs. v. Carnegie - - 91
Sinclair v. Clyne's Tr. - 65, 157, 200
Sinclair v. Fraser - - - 68, 102
Sloss v. Agnew - - - - 131
Smeaton Hanscomb & Co., v. Sassoon I. Setty Son & Co. 81
Smith v. Callander - - - 25
Smith v. Grayton Estates Ltd. 51, 107
Smith v. Liverpool and London and Globe Insurance Co. - 67
Smith v. Richmond - - - 124
Stark v. Edmondstone - - 120
Steele v. McIntosh Bros. - - 77
Steele v. Steele - - - - 86
Stewart v. Brims - 72, 141, 217, 232
Stewart v. Maclaine - - - 156
Stewart v. Moir - - - - 136

Stevens v. Sedgeman - - - 105
Stirrat & Anor. v. Whyte 20, 50, 105, 106, 136, 137, 140, 198
Stormonth-Darling v. Young - 199
Strachan v. Hunter - - 135, 136
Strang v. Stuart - - 200, 201, 207
Stringer and Riley Brothers' Arbitration - - - - 84
Stuart v. Smith - - - - 102
Sutherland & Co. v. Hannevig Brothers - - - - 84
Swinburne v. Andrews - - 201

Tabernacle Permanent Building Society v. Knight - - 77
Tayleur v. Wildin - - 135, 140
Taylor v. Burnett's Trs. - - 110
Taylor v. Earl of Moray - - 198
Taylor v. Fordyce - - - 199
Taylor v. Neilson- - - - 83
Taylor v. Steel Maitland - 25, 178
Taylor's Tr. v. Paul - - - 200
Tebb's & Sons v. Edwards 139, 199
Teignmouth U.D.C. v. Elliott - 140
Thallon v. Wemyss - - - 86
Thomas v. Jennings - - - 124
Thomson v. Earl of Galloway - 125
Thomson v. Lyall - - - 131
Tidball v. Marshall - - - 152
Todd v. Bowie - - - - 200
Tombs v. Turvey - - - 156, 202
Trotter v. Torrance - - - 137
Turnbull v. Millar - 32, 157, 244
Turner v. Hutchison - - - 157
Turner v. Wilson - - - 137, 143
Turton v. Turnbull - - - 140
Tweeddale, Marquis of, v. Brown - - - - - 122

University of Edinburgh v. Craik 143

Van Grutten v. Trevenen - - 136

Waddell v. Howat 156, 199, 201, 202
Waldie v. Mungall - - - 200
Walford Baker & Co. v. Macfie & Sons - - - - - 87
Walker v. Hendry - - 107, 137
Walker's Trs. v. Manson 106, 113, 157
Walker, &c. v. McKnights - 200
Wallis, In re.- - - - - 199
Ward v. Scott - - - - 139

TABLE OF CASES

	PAGE
Wardell v. Usher	179
Ware v. Davies	209
Watson v. Robertson	68, 101
Watters v. Hunter	136, 179, 199, 201
Webster & Co. v. Cramond Iron Co.	152
Welch v. Jackson	86
Westlake v. Page	152
Weston v. Duke of Devonshire	175
Whatley v. Morland	78
Whittaker v. Barker	201
Wight v. Earl of Hopetoun	100, 107, 135
Wight v. Marquis of Lothian's Trs.	131
Wilbraham v. Colclough and Ors.	139, 194
Williams v. Lewis	121
Williams v. Wallis & Core	87
Williamson v. Stewart	96, 97
Willson v. Love	127
Wilson v. Stewart	200
Wilson-Clarke v. Graham	141
Windsor R.D.C. v. Otterway & Try	78
Woodhouse ex parte	80
Yool v. Shepherd	199
Young v. Oswald	46
Young v. Steven	143
Yuill v. Semple	143

THE AGRICULTURAL HOLDINGS (SCOTLAND) ACT, 1949

AS AMENDED

INTRODUCTION

This introduction consists of a general account of the provisions of the Agricultural Holdings (Scotland) Act, 1949, as amended by *inter alia* the Agriculture Act, 1958, and the Succession (Scotland) Act, 1964 and as affected by the provisions of the Agriculture (Miscellaneous Provisions) Act, 1968 with such references as have been thought advisable to the Agricultural Holdings (Scotland) Acts, 1923-1948, which the 1949 Act consolidated and repealed, and a summary of the Law and Practice of Agricultural Arbitration and Valuation in Scotland.

(1) THE OBJECT OF THE ACTS

The object of the Acts is to encourage the tenant to farm well and to make necessary improvements to his holding. This has been done by giving him substantial security of tenure and rights to compensation for: (i) improvements, (ii) disturbance and (iii) damage by game. The Acts make important provisions with regard to the terms of leases and the rights and obligations of landlords and tenants in regard to the maintenance and repair of buildings and equipment on farms. There are also subordinate provisions relating to removing for non-payment of rent, the fixing of rent by arbitration, notice of termination of tenancy, bequest of lease, compensation for fixtures, freedom of cropping and sale of produce, and market gardens.

The Acts up to and including the 1949 Act progressively extended the tenant's rights in relation to security of tenure and compensation. The Agriculture Act, 1958, was, to some extent, an attempt to redress the balance in favour of the landlord. It entitled the landlord to obtain an economic rent for his land.

It enabled him, subject to certain limited rights open to heirs and legatees of the tenant, to terminate the lease at the tenant's death or at the expiry of the lease where the tenant died during its stipulated endurance, but these provisions have been largely restricted by the 1968 Act mentioned below. It amended the provisions restricting the operation of notices to quit. It transferred to the Land Court certain powers which belonged under the previous Acts to the Secretary of State, and, except in certain minor matters, it abolished the functions of the Agricultural Executive Committees.

The Succession (Scotland) Act, 1964 having abolished primogeniture provided a new method of transferring a deceased tenant's lease to one of his heirs and introduced a new Section 21 of the 1949 Act to establish a procedure for intimation by the acquirer of a lease and for objection by the landlord.

The Agriculture (Miscellaneous Provisions) Act, 1968 represents a swing in favour of the tenant. Part II of the Act gives a tenant who is displaced for development, forestry or any other non-agricultural purpose the right to claim a sum to assist in the reorganisation of his affairs. Part III limits the effect of Section 6 (3) of the 1958 Act by taking away the landlord's right to give an incontestable notice to quit to a 'near relative' successor. The landlord can still give an incontestable notice to quit to a successor who is not a near relative of the deceased tenant.

(2) Commencement and Dates of Operation of the Agricultural Holdings (Scotland) Acts

The 1923 Act came into operation on 7th July, 1923, but some of its provisions affected contracts entered into prior to its commencement. The 1931 Act came into operation on 31st July, 1931, but did not affect improvements made or begun before that date. The 1948 Act came into operation on 1st November, 1948. (We are here concerned with Part I only of the Agriculture (Scotland) Act, 1948). The 1949 Act came into operation on 24th November, 1949. In certain cases it is necessary to refer to, and found on, the repealed Acts; in some cases the Act applies to contracts entered into after specified dates, and in others to contracts entered into after the commencement. The 1958 Act received the Royal Assent on 1st August, 1958, but all of its

INTRODUCTION

provisions did not immediately come into effect. The effect of Section 6 (which relates to succession to holdings) was postponed to 1st September, 1958. Section 3 (2) (which relates to notices to quit) came into force on 1st November, 1958 (S.I. 1958 No. 1705). The Agriculture (Miscellaneous Provisions) Act, 1963 received the Royal Assent and came into effect on 15th May, 1963. The Succession (Scotland) Act, 1964 received the Royal Assent on 10th June, 1964 and came into effect on 10th September, 1964. The Agriculture (Miscellaneous Provisions) Act, 1968 received the Royal Assent and came into effect on 3rd July, 1968.

1. *Contracted-for improvements* attract compensation only where the contract was entered into on or after 1st January, 1921 (the date of coming into operation of the Act of 1920). Sec. 37

2. *Substituted compensation* can be claimed only for old improvements specified in Part III of the Second or the Third Schedules (except as regards market garden improvements embraced in the Fourth Schedule) under agreements entered into before 1st January, 1921. Secs. 42 & 67

3. *Compensation for the continuous adoption of a special standard or system of farming* is not payable in respect of any such improvement made before the date of a record; nor can a landlord claim for dilapidations under sec. 57 or sec. 58 unless there is a record. Sec. 56 Sec. 59

4. *Additional payments under the 1968 Act* are not payable unless notice to quit was served after 1st November, 1967, or resumption took place after 3rd July, 1968. 1968 Act Sec. 11

(3) LANDLORD AND TENANT

Landlord means 'any person for the time being entitled to receive the rents and profits or to take possession of any agricultural holding'. A tenant means 'the holder of land under a lease' and lease means generally 'a letting for a term of years, or for lives, or for lives and years, or from year to year'. In the case of sub-tenants, as of others, the period of let must be not less than one year. Even a let for a shorter period than from year to year (unless it be approved by the Secretary of State before being Sec. 93

entered into) may give rise to a claim under the Act as if it were a lease of the land from year to year (sec. 2). If the principal tenant and sub-tenant are in the relation of landlord and tenant as defined in the Act, the latter would appear to have right to such claims as are competent to a tenant.

'Landlord' and 'tenant' are defined as including the executor, administrator, assignee, heir-at-law, legatee, disponee, next-of-kin, guardian, *curator bonis,* or trustee in bankruptcy of the landlord or tenant. This definition does not overrule the express terms of the lease providing for the exclusion of legatees, executors or assignees.

(4) Holdings Affected

Sec. 1

In the 1949 Act an 'agricultural holding' is defined as meaning 'the aggregate of the agricultural land comprised in a lease not being a lease under which the said land is let to the tenant during his continuance in any office, appointment or employment held under the landlord'. 'Agricultural land' is defined as 'land used for agriculture which is so used for the purposes of a trade or business and includes any other land which, by virtue of a designation of the Secretary of State under sub-sec. (1) of sec. 86 of the Agriculture (Scotland) Act, 1948, is agricultural land within the meaning of that Act'. 'Trade or business' is not limited to agricultural trade or business.[1] The definition of 'agriculture' in sec. 93 (1) of the 1949 Act seems designed to cover every arable, pastoral and livestock-raising activity. The effect of the definition is to include all areas of agricultural land comprised in a lease.

(5) Minimum Term of a Lease

Secs. 2 & 3

A lease of an agricultural holding may not now be created for a period shorter than from year to year without the prior approval of the Secretary of State except in grazing and mowing lets for a specified period[2] of less than a year or where the lease is in respect of land which the grantor holds on a lease for a lesser period than from year to year. Section 3 deals with tacit relocation.

[1] *Rutherford* v. *Maurer* [1962] 1 Q.B. 16.
[2] See *Mackenzie* v. *Laird*, 1959 S.C. 266.

(6) Provisions for Written Leases

Where there is no written lease embodying the terms of a tenancy, the landlord or the tenant is entitled to request the other to enter into an agreement. If no such agreement is made, the matter can be referred to arbitration. A similar request and reference to arbitration can be made where a subsisting written lease held on tacit relocation contains no provisions for one or more of the matters specified in the Fifth Schedule to the Act, or contains a provision inconsistent with that schedule or with sec. 5 which deals with the liability for maintenance of fixed equipment. The agreement which is requested by the one party from the other, or is referred to arbitration, will be one containing provision for all these matters. [Sec. 4]

On any such reference the arbiter must specify the terms of the existing tenancy and make provision for the matters specified in the Fifth Schedule in so far as the terms of the tenancy, as agreed between the landlord and the tenant, make no provision for them. These matters include names of parties, particulars of the holding referable to a map or plan, the term or terms for which the holding is let, the rent with the dates on which it is payable, and certain undertakings by landlord and tenant against the outbreak of fire.

In every new lease there is to be deemed to be incorporated (a) an undertaking by the landlord to put the fixed equipment into repair as soon as possible, to provide all fixed equipment for the maintenance of efficient production and to effect during the tenancy necessary replacements or renewals and (b) a provision limiting the liability of the tenant as to maintenance of fixed equipment to a standard of repair (natural decay and fair wear and tear excepted) similar to that immediately after the property was put in repair. It is, however, open to either party after the lease has been entered into to undertake to execute on behalf of the other party any work which the lease obliges that party to do. Provisions in leases entered into on or after 1st November, 1948, requiring a tenant to pay any part of the fire insurance policy premium over fixed equipment are null and void. Should a dispute arise it is to be submitted to arbitration. [Sec. 5]

Where the liability for the maintenance or repair of fixed equipment has been transferred under sec. 4 from the tenant to the landlord the landlord may require arbitration as to the amount of [Sec. 6] compensation which would have been payable in respect of the previous failure by the tenant to discharge his liability if the tenant had quitted the holding on the termination of his tenancy at the date from which the transfer takes effect. A similar provision regulates the case where the liability as above has been transferred from landlord to tenant.

The arbiter may vary the rent where he considers such variation equitable on a reference under secs. 4 or 5 (5) of the Act.

(7) Variation of Rent

Sec. 7

Under the 1949 Act, either landlord or tenant may serve a written demand on the other for a reference to arbitration as to rent. This provision affects all leases whether entered into before or after the passing of the 1948 Act. The arbitration is to decide the question—

> What rent should be payable in respect of the holding as from the next ensuing day on which the tenancy could have been determined by notice to quit given at the date of demanding the reference.

The method of serving a demand for a reference is found in sec. 90.

Sec. 7 (3)

A reference to arbitration on this subject cannot lawfully be demanded if the consequential increase or reduction of the rent would take effect as from a date earlier than five years from certain specified dates which are as follows:—

(a) the commencement of the tenancy, or
(b) the date as from which there took effect a previous increase or reduction of rent (whether under sec. 7 or otherwise), or
(c) the date as from which there took effect a previous direction under sec. 7 that the rent should continue unchanged.

There are important exceptions to the above stated general rule. The following variations are to be disregarded for the purposes of sec. 7 (3):—

(a) any variation under sec. 6 (3).
(b) an increase of rent under sec. 8 (1) or (2).
(c) a reduction of rent under sec. 34.

INTRODUCTION

The arbiter must be appointed and accept office before the date when the alteration of rent is to take effect.[1] Sec. 7 (2)

The arbiter, in fixing the new amount of the rent, is not entitled to take into account any increase in the rental value of the holding which is due to improvements executed wholly or partly at the expense of the tenant without any equivalent allowance or benefit from the landlord in consideration of them. If, however, the improvement was executed under an obligation imposed on the tenant by the lease, then the effect of the improvement on the rental value can be taken into account. The continuous adoption of a special system of farming is for this purpose an improvement executed at the tenant's expense. Sec. 7 (4)

Moreover, the arbiter may not take into account any improvement which was executed by the landlord in so far as he has received, or will receive, a state grant in respect of the improvement. He must not fix the rent at a lower amount because of any dilapidation or deterioration of, or damage to buildings or land, which has been caused or permitted by the tenant. Sec. 7 (2)

The duty of an arbiter appointed to fix a new rent for an agricultural holding is a difficult one. He must examine the lease or, if there is none, ascertain the terms of the tenancy by reference to correspondence or otherwise. He will then require to make an extensive examination of the farm, as exhaustive as if he were himself to be an offerer for the tenancy. He should ascertain the nature of the soil and size of fields over the whole area of the farm; carefully inspect the buildings and cottages to see how far they satisfy the requirements of the farm and if they will last the period of the tenancy; ascertain the yield of the crops, the general state of cultivation and stock carried and the state of the roads, drains, ditches, hedges, embankments and fences. The situation of the farm in relation to transport facilities and markets is also a relevant consideration. He should make inquiry to ascertain if the land is subject to flooding or the stock to disease. He is usually accompanied on his inspection by both parties or their representatives, from whom he can naturally obtain much information, and in issuing his proposed findings and award, it is usual to add an explanatory note stating the grounds on which his decision to increase or reduce the rent has been arrived at. An arbiter appointed (otherwise than by agreement) under sec. 77

---

[1] *Graham* v. *Gardner,* 1966 S.L.T. (Land Ct.) 12.

of, or the Sixth Schedule to, the 1949 Act must furnish a statement, written or oral, of the reasons for the decision if requested to do so, on or before the giving of notification of the decision. See sec. 12 of the Tribunals and Inquiries Act, 1958, and paragraph 24 of Part II to the First Schedule to that Act.

1958 Act
Sec. 2

On what basis is a new rent to be fixed? The 1958 Act amends sec. 7 of the 1949 Act and lays down as the criterion for the arbiter: 'the rent at which having regard to the terms of the tenancy . . . the holding might reasonably be expected to be let in the open market by a willing landlord to a willing tenant *there being disregarded . . . any effect on rent of the fact that the tenant who is a party to the arbitration is in occupation of the holding'*. The criterion laid down prior to the passing of the 1958 Act in *Guthe* v. *Broatch,* 1956 S.L.T. 170, was the proper rent to be paid for the particular holding by the particular tenant to the particular landlord. This test, no longer applies. The rent to be fixed is such a rent as a farmer would pay for the farm in its present condition if it were put on the market. The arbiter should not take into account what the particular tenant actually in the farm can pay, but what rent would be paid by a new tenant with reasonably adequate means and capable of farming to good advantage. In fixing what may be called the 'economic rent' as opposed to a fair or equitable rent, the arbiter will have regard to the general state of agriculture and, the rent being fixed for a five-year period at least (unless the tenancy be sooner terminated), the general trend of prices in one direction or another. The arbiter's discretion is, however, limited as was indicated by the Land Court in *Secretary of State for Scotland* v. *Young and Others*[1] where it was stated that a member of the Court was now 'less in the position of relying upon his own expert knowledge and more in the position of assessing evidence of external circumstances'. In discussing open market value the note to that case states that ' . . . it is now important that in rent cases parties should produce evidence tending to show what the open market value of a lease of comparable subjects would be'. The note also contains the following passage, 'In the *Encyclopaedia of the Laws of Scotland* Sheriff Dickson says: "In accordance with those commercial principles which generally obtain in connection with contracts, whether relating to land or other matters, the fair value of an article is the

[1] *Secretary of State for Scotland* v. *Young and Others,* 1960 S.L.C.R. 31.

highest price which it will bring on free exposure in the open market".' This might suggest that the rent must be fixed at a rate equivalent to the highest offer proved to have been made for a holding which could be shown to be comparable but it is considered that two qualifications ought to be made. Firstly although a seller will normally accept the highest price offered it is commonly found that a landlord, since he has a continuing interest in the subjects let, does not accept the highest rent offered and as the Land Court has pointed out he may on a long term view be wiser not to do so but to prefer a tenant at a lower rent if he is more likely to maintain the fertility and productivity of the holding.[1] Thus the reference to what may be reasonably expected would justify an arbiter in disregarding 'freak' offers especially if they were not in fact accepted. Secondly although an arbiter must have regard to evidence on rent of comparable holdings he may have reservations regarding the evidence unless he is in a position to inspect the comparable holdings himself as has been done in some cases[2] and it is recommended that he should make such inspections whenever possible. He will also require to be satisfied as to the terms on which allegedly comparable holdings are held.

The arbiter must, of course, have regard to the general level of rents being obtained in the area and whilst he is entitled to rely upon his own knowledge evidence should be led by the parties and the best evidence is 'evidence of values at which transactions have recently been concluded by offer and acceptance in the open market after advertisement of the subjects'.[1]

In the case *Rennie* v. *The Patrons of Cowane's Hospital,* 1959 S.L.T. (Notes) 76, the tenant, at a public roup, deliberately offered a rent in excess of what he believed the subjects to be worth. At the five-year break he sought to have the rent reduced, and the First Division of the Court of Session held that the Land Court (upon an application under sec. 78) was entitled to reduce the rent and rejected the argument that rent could only be altered by arbitration when a change of circumstances was established. The 1958 Act did not apply to this case.

In a reference to the Land Court by the same parties in 1967 the rent was raised but not to a figure as high as that originally offered. (1967 Stirling R.N. 69.)

---

[1] *Crown Estate Commissioners* v. *Gunn,* 1961 S.L.C.R.App. 173
[2] See, e.g., *Secretary of State for Scotland* v. *Sinclair,* 1962 S.L.C.R. 6.

(8) Absolute Right to Increase Rent

Sec. 8

Under certain conditions, the landlord is absolutely entitled to have the rent increased. This right arises where he has carried out any improvement of the nature specified in sec. 8 (1) of the 1949 Act.

In respect of any such improvement the landlord can serve a written notice on the tenant within six months from the completion of the improvement. If he does so, the rent of the holding is increased by an amount equal to the increase in the rental value of the holding which is attributable to the carrying out of the improvement. The date as from which the increase operates is the date of the completion of the improvement, unless the improvement was completed before 1st November, 1948, when that is the operative date.

If the landlord obtained a State grant in respect of the improvement, that fact will not affect his right to an increase of rent under sec. 8. The arbiter will still have to determine the amount of the increase in the rental value of the holding, which is attributable to the carrying out of the improvement, but when that amount has been ascertained, it has to be 'reduced proportionately' to the amount of the State grant. The simplest interpretation of these words is that the amount by which the rent would be increased is to be reduced by a sum representing the portion of the increase which is due to the grant.

(9) Variation of Terms of Tenancy as to Permanent Pasture

Sec. 9

Landlord or tenant can demand a reference to arbitration on the question whether the amount of land required to be maintained as permanent pasture in terms of the lease should be reduced in the interests of efficient farming.

The arbiter may order a reduction in the area which the lease requires to be in permanent pasture but may stipulate that at the termination of the tenancy the tenant will leave in pasture an area not exceeding what the lease originally required. Section 63 (1) excludes compensation for restoration of pasture in accordance with such a stipulation and requires averaging of the value of pasture (to prevent inferior pasture being treated as that for which compensation is excluded).

(10) CONTINUANCE OF LEASE NOT AFFECTED BY VARIATION OF TERM

The addition of new terms or the variation of existing terms under the provisions of the Act noticed above do not have the effect of creating a new lease. ^{Sec. 10}

Miscellaneous Provisions Relating to Landlord and Tenant

(11) FREEDOM OF CROPPING

No conditions can be imposed on a tenant, by lease or otherwise, to prevent him from practising any system of cropping of the arable land (which does not include land in grass which by the lease is to be retained in the same condition throughout the tenancy), or from disposing of the produce of the holding. These are statutory rights, and the tenant cannot be subjected to any penalty, forfeiture, or liability for exercising them. ^{Sec. 12} ^{Sub-sec. (5)}

There are, however, important conditions attending the exercise of these rights. The tenant must previously have made, or as soon as may be, make suitable and adequate provision to protect the holding from injury or deterioration, such provision, in the case of disposal of produce, consisting in the return to the holding of the full equivalent manurial value of all crops sold off or removed from the holding in contravention of custom, lease, or agreement. ^{Sub-sec. (1)}

These rights of freedom of cropping and disposal of produce do not apply— ^{Sub-sec. (4)}

(a) in the case of a tenancy from year to year, as respects the year before the tenant quits the holding, or any period after he has given or received notice to quit which results in his quitting the holding; or

(b) in any other case, as respects the year before the expiration of the lease.

It will be observed that the tenant is not bound to follow the rotation of cropping prescribed in his lease, nor, for that part, the rotation customary in the district, nor is he bound to consume on the farm the produce grown on it. A general obligation to follow the rules of good husbandry still holds good, but a tenant cannot be bound guilty of a breach of these rules simply because he acts contrary to long-practised methods or to the express terms of his contract.

What is the meaning of 'full equivalent manurial value to the holding' of all crops sold or removed in contravention of custom, contract, or agreement? It has no necessary relation to cost. What is intended is retoration of fertility, leaving the land in the condition in which it would have been had the crops not been removed. Assuming that the tenant is taken bound to have his turnips consumed by sheep, he can nevertheless sell and remove the turnips, but he must bring back the equivalent of their manurial value to the land.

Sec. 12 (2)

If the tenant exercises these rights in such a manner as to injure or deteriorate the holding, the landlord may recover damages and obtain interdict restraining the exercise of the rights in that manner. The question whether a tenant is exercising, or has exercised, his rights in such a manner as to injure or deteriorate the holding, or as to be likely to do so, falls to be determined by arbitration. A certificate by the arbiter as to his determination of the question will, in any proceedings, including arbitrations, under sec. 12, be conclusive proof of the facts stated in the certificate.

(12) Tenant's Right to Remove Fixtures

Sec. 14

A tenant who, without being taken bound to do so, has erected any engine, machinery, fencing or other fixture or any building (other than a building in respect of which he is entitled to compensation under the Act or otherwise) which was not erected or affixed in place of some fixture or building belonging to the landlord, is entitled to remove the same before, or within six months or such other time as may be agreed, after the termination of the tenancy. Before the removal, however, he must pay all rent owing, and perform or satisfy all other obligations to the landlord; he must do no avoidable damage to any other building or other part of the holding, and must make good any unavoidable damage. He must also give the landlord notice one month before both the exercise of the right and the termination of the tenancy of his intention to remove the fixtures and others, and any time within that month the landlord may, by notice in writing to the tenant, elect to purchase them at a value to be fixed, failing agreement, by a single arbiter the basis of value being 'the value to an incoming tenant'.

It appears to be evident that the engine, machinery, and fencing here referred to, must be of the nature of fixtures. The words 'or other fixture' indicate this. Ordinary moveable machinery would not, it is thought, be affected. Again, 'any building' would probably not apply to a mere temporary wooden erection.

Ordinary wood and wire fencing and wooden structures not built into the ground would not seem to be 'fixtures' within the meaning of the Act.[1] 'Fixtures' have frequently been defined in the Courts.[2] 'Whatever moves, or is capable of being moved, from place to place without injury, or change of nature in itself, or in the subject with which it is connected, is moveable'.[3]

Some Fixtures at Common Law

It may be useful here to enumerate some accepted fixtures, viz.—
1. Things planted in the ground with the intention that they should grow there, such as trees, shrubs, turf, and vines in a vinery.
2. Gates, dykes, walls, and railings, intended to be more or less permanent.
3. Any building with foundations sunk in the ground, even a vinery or conservatory.
4. Doors and windows.
5. Statues, pictures, and mirrors, which form part of an architectural design and are not easy to replace.
6. Ordinary mantlepieces.
7. Machinery built in, or specially adapted to particular premises.[4]
8. Leathern belt necessary for fixed machinery.
9. The keys of a door and the millstone of a mill.

The following are Not Fixtures at Common Law
1. Temporary fencing.
2. Buildings resting on, not built into, the ground.

[1] *Duke of Buccleuch* v. *Tod's Trustees*, 1871 9 M. 1014.
[2] Rankine on Leases, 3rd edn. p. 286, and Land Ownership, 4th edn. 117.
[3] Bell's Prin., 1472. As to moveable things which become heritable by accession, see Bell's Prin., 1473.
[4] It is doubtful if machinery attached to buildings by bolts or screws is a fixture.

3. Furniture, pictures, andmirrors.
4. Grates and gas fittings.
5. Ornamental mantlepieces.
6. Machinery merely resting on the ground, unless of exceptional size and weight.

Sec. 16

(13) PENAL RENT AND/OR LIQUIDATED DAMAGES

Since the year 1900 penal rents and liquidated damages have been placed in the same position as ordinary damages, i.e., they can be enforced only up to the amount of the actual loss sustained. This applies to penal rents or liquidated damages for breach or non-fulfilment of any of the terms or conditions in the lease. This does not prevent the landlord from enforcing such penalties as are not stated in sterling. In certain events he might stipulate for power to withdraw certain rights or privileges, or to irritate the lease, and conditions of irritancy have been strictly interpreted by the Courts.

(14) RECORD OF HOLDING

Section 5 of the 1949 Act makes it obligatory to have a record of fixed equipment (defined in sec. 93) made in the case of all new leases. In addition the landlord or tenant is entitled, at any time, to require a record to be made of the condition of the fixed equipment on, and of the cultivation of, the holding. The tenant may also require that the record shall include any existing improvements executed by him or in respect of the carrying out of which he has, with the written consent of the landlord, paid compensation to an outgoing tenant. Any fixtures or buildings which under the Act he is entitled to remove, he may also require to be included. The record falls to be made by a person to be appointed by the Secretary of State, and is to be in such form as may be prescribed. Farm records, exhaustive and well compiled, offer considerable advantages.

Sec. 17
Sub-sec. (1)

Sub-sec. (2)

Such a record must be made (1) before a tenant can claim compensation for the adoption of a special standard or system of cultivation (section 56) and the improvement must be made after the date of the record, and (2) before a landlord can make a claim for dilapidations under secs. 57 and 58. But see sec. 57 (3) as to compensation in respect of matters in accordance with a written lease.

A record to be of permanent value must be exhaustive and, more than that, compiled and written so that it may convey to a reader years later a true picture of the state of the holding when it was written. Vagueness of description and general expressions capable of wide interpretation are to be deprecated. To say that a fence is 'not in very good repair' or that a certain building is 'inadequate' conveys very little useful information. A plan is useful; all fields, fences or hedges, and buildings should be numbered for identification and the most useful references are to an up to date Ordnance Survey plan; a schedule of cropping should be provided with particulars of temporary pasture; and details of construction of bridges, mill-lades, roads, embankments, fences, drains, dippers, or fixtures, with repairs necessary, should be exactly given. Each field may be dealt with separately, with observations on its water supply, drains, state of cultivation, and soil. Any further information, such as particulars of stock, its condition and quality, and the quality of crops would be useful when eventually a comparison comes to be made with the state of the farm at a later date. Any consideration or allowances _{Sub-sec. (4)} made by the landlord to the tenant, or vice versa, should be shown in the record.

It may sometimes occur that the landlord and tenant are not _{Sec. 17 (6)} agreed on points arising in connection with the preparation of such a record, particularly with reference to 'improvements' for which the tenant is entitled to claim compensation, and 'fixtures or buildings' which the tenant is entitled to remove. Either party may apply to the Land Court to determine any question or _{Sub-sec. (3)} difference arising. The cost of making the record is to be borne equally by the landlord and the tenant, unless they agree otherwise.

(15) Removing for Non-Payment of Rent

When a tenant is six months in arrear with his rent, the landlord _{Sec. 19} may raise an action in the Sheriff Court for his removal at the next ensuing term of Whitsunday or Martinmas. The only answer open to the tenant is to pay up or find security to the satisfaction of the Sheriff for the arrears and for one year's additional rent.

The tenant who is removed under this section has all the rights of an away-going tenant as if his lease had naturally expired at the quitting term. Accordingly, he will be entitled to all the rights and privileges specified in the lease; and it is thought that he may also be entitled to compensation for improvements, but not for disturbance.

(16) Bequest of Lease

By sec. 20 of the 1949 Act the tenant could bequeath his holding to any person unless the lease contained an express exclusion of legatees. Sec. 6 of the 1958 Act restricted the tenant's right of bequest to 'any member of his family', but this has been extended by the Succession (Scotland) Act, 1964 to the tenant's son-in-law or daughter-in-law or any person 'who would be or would in any circumstances have been entitled to succeed to the estate on intestacy' by virtue of that Act. This right is subject to certain conditions, namely:—

The legatee must intimate the bequest to the landlord within twenty-one days after the tenant's death, unless he is prevented, by some unavoidable cause, from doing so within that time, in which event notice must be given as soon as possible thereafter. Such notice imports acceptance by the legatee of the bequest.

Within a month after receipt of the notice, the landlord may give a counter-notice refusing to receive the legatee as tenant. Failing the giving of such a counter-notice the lease is to be held as binding on landlord and legatee as from the date of the death of the tenant. If the landlord gives a counter-notice, the legatee may apply to the Land Court to declare that he is tenant under the lease. If the landlord establishes a reasonable ground of objection, the legatee will fail. Pending the Court proceedings, the legatee may have possession of the holding unless, on cause shown, the Land Court otherwise directs. The statute does not suggest any particular ground of objection to a legatee, and it is left to the Land Court to decide whether any ground stated by the landlord is reasonable. It might be a good reason that the legatee had not sufficient capital, or had no experience of practical agriculture, or that he was a person of dissolute habits. Objections considered by the Land Court have usually been personal to the legatee. See for example the majority decision in *Reid* v. *Duffus Estate, Ltd.*, 1955 S.L.C.R. 13.

If the legatee does not accept the bequest or if the bequest is Sec. 21
declared null and void by the Land Court, the lease is treated as
intestate estate of the tenant in accordance with Part I of the
Succession (Scotland) Act, 1964. The lease vests in the executor
who may within one year transfer the deceased's interest in the
lease to one of the persons entitled to succeed to or to claim legal
rights out of the estate. The acquirer of the lease must within
21 days of the acquisition give notice to the landlord who may
within one month serve counter-notice in the same way as with a
legatee. The procedure then differs however as it is for the
landlord within one month of serving counter-notice to apply to
the Land Court to terminate the acquirer's interest. A termination of the acquirer's interest in this way is to be treated, for the
purposes of the statutory provisions as to compensation, as the
termination of his tenancy; but he is not entitled to compensation
for disturbance.

SUCCESSION TO HOLDINGS

Irrespective of his rights of objection under secs. 20 and 21 of the
1949 Act, the landlord may, under sec. 6 (3) of the 1958 Act,
terminate the tenancy after the death of a tenant but only if the
successor is not a near relative of the deceased tenant. 'Near
relative' means husband, wife, son or daughter or adopted son or
daughter. The provisions of the sub-section may be summarised
thus:

(a) *Successor not a near relative*

If a notice to quit is given to a legatee or acquirer who has
acquired right to a tenancy in terms of sec. 20 or 21 of the 1949
Act but is not a near relative then the provisions of sec 25 (1) of
that Act are not to apply.

If the lease has expired or has less than two years to run of its
stipulated endurance the landlord can terminate the tenancy at the
term (being the term of outgoing stipulated in the lease or the
corresponding term in any succeeding year) following the first
anniversary of the tenant's acquiring right as legatee or acquirer.

If the lease has more than two years to run of its stipulated
endurance the landlord can terminate the tenancy at the stipulated
term of outgoing but must wait until it is competent to serve the

notice to quit. (Under sec. 24 of the 1949 Act notice to quit cannot be given more than two years before the termination of the tenancy.)

1958 Act, sec. 6 (4)

The notice to quit must bear to be given in pursuance of sub-sec. 3 of sec. 6. If it does not state that the tenancy is being terminated for the purpose of using the land for agriculture only an additional payment will be due to the tenant under sec. 9 of the 1968 Act.

(b) Near relative successor

Part 3 of the Agriculture (Miscellaneous Provisions) Act, 1968 removes the landlord's former right to serve an incontestible notice to quit under section 6 (3) of the 1958 Act on a 'near relative' as above defined. Section 25 (1) applies to a notice to quit served on a near relative and he thus has the right to serve a counter-notice. There are, however, limitations on the near relative's security of tenure and these are contained in section 18 (2) of the 1968 Act which provides that the Land Court shall consent to a notice to quit served on a near relative successor if they are satisfied:

(a) that the near relative successor has neither sufficient training in agriculture nor sufficient experience of farming to enable him to farm the holding with reasonable efficiency; or

(b) that the holding to which (or to part of which) the notice to quit relates is not capable of providing full-time employment for an occupier and at least one other man and that the notice is given to enable the landlord to effect an amalgamation within two years of the termination of the tenancy; or

(c) that the near relative successor occupies as owner or tenant agricultural land other than the holding (or where the holding is part of a unit other than the unit) and has done so since before the death of the tenant whom he succeeded and the other unit is capable of providing full-time employment for an occupier and for at least one other man. The reference to a holding forming part of a unit would cover the case of the owner occupier who tenants the adjoining farm and works the two farms together. He could not be removed under paragraph

(c) (unless he had yet another 'two man' farm). There may be room for argument where non-contiguous subjects are alleged to form an agricultural unit.

In deciding under para. (b) and (c) whether a holding is capable of providing employment for more than one man it is to be assumed that the holding or unit is farmed under reasonably skilled management, that a system of husbandry suitable for the district is followed and that the greater part of the feeding stuffs required by any livestock kept on the holding or unit is grown there.

A notice to quit served under Section 18 (2) must state the grounds on which it is given and in the case of proposed amalgamation must specify the other land.

The sub-section has a proviso (the same as that to Section 26 (1) of the 1949 Act) that even if satisfied on one of the grounds mentioned the Court must withhold consent to the notice to quit if it appears to them that a fair and reasonable landlord would not insist on possession.

In consenting to the operation of a notice to quit for the purpose of an amalgamation the Court must impose such conditions as they think necessary for securing that the holding will be amalgamated with the other land specified and within two years after the termination of the tenancy.

(17) NOTICE TO QUIT

The Removal Terms (Scotland) Act of 1886, provides that where Sec. 24 under a lease of a house (which includes a dwellinghouse or building let along with land for agricultural purposes), the tenant's entry or removal be one or other of the terms of Whitsunday or Martinmas, the tenant shall, in the absence of express stipulation to the contrary, enter to or remove from the house (any custom or usage to the contrary notwithstanding) at noon on the 28th May, if the term be Whitsunday, or at noon on 28th November, if the term be Martinmas, or on the following day at the same hour, where the said terms fall on a Sunday. But, nevertheless, forty days warning to quit continues to be required prior to 15th May and 11th November respectively.

The quitting terms of Whitsunday and Martinmas from farmhouse and farmsteading cottages, &c., are, accordingly, 28th May in respect of Whitsunday, and 28th November in respect of Martinmas, unless there be express stipulation to the contrary. The statute which so regulates the dates of removal from buildings does not affect land let for agricultural purposes, and, when the entry to or ish from such land is Whitsunday or Martinmas, 15th May and 11th November are the applicable days, in the absence of facts and circumstances clearly showing that the parties understood and intended otherwise. It has been held that the parties may put their own interpretation on 'Whitsunday' and 'Martinmas'.[1] For example, if a tenant got entry only at 28th November, and Martinmas was stated in the lease as the term of entry, it would be very difficult, if not impossible, for the landlord to establish that the ish was 11th November. So far as the 1949 Act is concerned the definition clause (sec. 93) provides that Whitsunday and Martinmas shall mean 28th May and 28th November respectively. See below however as to notice to quit.

It may be suggested that, in every case in which Whitsunday and Martinmas are referred to in a lease, the particular month and day should also be stated thus—'Whitsunday (28th May)', 'Martinmas (28th November)'.

Sec. 24

Written notice of intention to terminate a tenancy requires to be given not less than one year, and not more than two years, before the termination of the lease.

It has been decided that contracting out of the provisions of the 1923 Act with regard to notices to quit is incompetent.[2]

Failing such notice, the lease is renewed by tacit relocation for another year, and thereafter from year to year, and, in the case of any lease so renewed, the period of notice required to terminate the tenancy is to be not less than one year and not more than two years. A tenancy for a year renewed on tacit relocation thus requires the same notice to quit as one for two years or more.

Sec. 24 (3)

It is necessary to qualify these provisions with reference to sec. 24 (3), which bears that the provisions of the Sheriff Courts (Scotland) Act, 1907, relating to removings shall in the case of any holding to which this section (24) applies have effect subject to the provisions thereof and it has been held in the Sheriff Court[3] that notice to quit must in all cases be served against 15th May or

---

[1] *Hunter* v. *Barron's Trustees*, 1886, 13 R. 883.
[2] *Duguid* v. *Muirhead*, 1926 S.C. 1078.
[3] Reported in *Stirrat & Anor.* v. *Whyte*, 1968 S.L.T. 157.

11th November because Whitsunday and Martinmas are so defined in the 1907 Act.

There is a statutory form for notice to quit given by the landlord, but not for that given by the tenant. This form must be used by the landlord, while a simple letter by the tenant is sufficient so long as it unequivocally bears that it is a notice of the tenant's intention to quit the farm at the terms provided for quitting.

The statutory provisions for notice to quit do not apply to a notice given under a power reserved in the lease for the landlord to resume land for building, planting, feuing, or other purposes not being agricultural purposes nor to subjects let for less than a year unless the lease takes effect as a lease from year to year under sec. 2. Sec. 24 (6)

In a year to year tenancy a notice to quit part of a holding is not invalid on the ground that it relates only to part if it is given for the purpose of adjusting the boundaries or for the erection of farm cottages, allotments, tree planting, making of reservoirs and certain other purposes. Sec. 32

Where the landlord gives notice to quit part of a holding held on a yearly tenancy, for any purpose specified in sec. 32 the tenant may accept such notice as notice to terminate the tenancy of the entire holding. The tenant may choose to give a counter-notice under sec. 25 (1). See *Hamilton* v. *Lorimer*, 1959 S.L.C.R. 7. Sec. 33

(18) Provisions as to Security of Tenure: Restriction on Operation of Notices to Quit

As a general rule, when a notice to quit has been served on him the tenant may take steps to have the notice referred for the consent of the Land Court. This is done by the tenant, within one month of the giving of the notice to quit, serving on the landlord a counter-notice in writing, requiring that sub-sec. (1) of sec. 25 shall apply to the notice to quit. If this is done the notice to quit is not effective unless it is consented to by the Land Court. A mere statement of objection to the notice or a refusal to go out will not be sufficient. The provision is peremptory and must be strictly followed. Sec. 25 (1)

Several important exceptions detailed in sec. 25 (2) of the 1949 Act as amended remain to this general rule. See the amended text of this section on page 138. Sec. 25 (2)

The exception or exceptions founded on must be stated as the reason(s) for the notice. Any question arising under sec. 25 (2) falls to be determined by arbitration. A tenant in receipt of a notice to quit may within one month from the date of the service of the notice to quit require such a question to be determined by arbitration. If two concurrent notices to quit are given for different reasons and the tenant requires arbitration in regard to the reason stated in one of the notices it has been held in England that the fact that the operation of that notice is suspended does not prevent the other notice from becoming effective—*French* v. *Elliott* [1960] 1 W.L.R. 40. This case involves a number of other points some of which may have a bearing in Scotland.

Sec. 27 (5) Where a notice to quit is given to a sub-tenant by a tenant who has himself been given a notice to quit by his landlord, the sub-tenant cannot invoke the provisions of sec. 25 (1). He is, however, entitled to be a party to any Land Court proceedings if the tenant has served a counter-notice on his landlord.

(19) PROVISIONS AS TO CONSENTS FOR PURPOSES OF SEC. 25

Sec. 26 Section 26 has been amended considerably by the 1958 Act. The Land Court must now consent to the operation of a notice to quit if they are satisfied as to at least one of the following:
   (*a*) that the landlord's purpose is in the interests of good husbandry; or
   (*b*) sound estate management; or
   (*c*) is desirable in the interests of agricultural research, education or the like or for the purposes of the Small-holdings Acts or in relation to the holdings mentioned in sec. 64 of the Agriculture (Scotland) Act, 1948 or allotments; or
   (*d*) that greater hardship would be caused by withholding consent than by giving it; or
   (*e*) that the landlord's purpose is to make a use of the land which is not agricultural and which does not fall within sec. 25 (2) (*c*).

The landlord must specify the purpose for which he seeks to terminate the tenancy. It must be noted that, for the purpose of paragraph (*a*), it is good husbandry in relation to the particular land which the landlord seeks to repossess which alone must be looked to. The proviso to sec. 26 (1) as amended confers upon the Land Court a discretion to withhold consent even where they are satisfied as to one of the above matters 'if in all the circumstances it appears to them that a fair and reasonable landlord would not insist on possession'. See notes to sec. 26. The Land Court has power to impose conditions to secure that the land is used for the Sec. 26 (5) purpose for which it allowed it to be repossessed and to impose penalties of an amount not greater than two years' rent for breach Sec. 30 of such conditions.

The procedure by which application for consent under sec. 25 is made by a landlord and by which a tenant refers questions arising under the section to arbitration is regulated by sec. 27. This Sec. 27 section also deals with the procedure to be followed by a tenant who has received a notice to quit, or who has served a counter-notice on his landlord under sec. 25 (1), in relation to a sub-tenant.

In the case of a 'near relative' successor to a deceased tenant, sec. 25 (1) applies to a notice to quit subject to the exceptions stated in the 1968 Act, sec. 18 (2). See above under 'Succession to Holdings', page 17.

(20) APPLICATION FOR CERTIFICATE OF BAD HUSBANDRY

Formerly such applications were made to the Secretary of State Sec. 28 with a right of appeal by either party to the Land Court. Sec. 28 as amended by the 1958 Act provides that application must be made direct to the Land Court.

(21) NOTICE TO QUIT PART OF HOLDING

In the case of year-to-year tenancies (including tenancies Sec. 32 renewed by tacit relocation) a notice to quit part of an agricultural holding is not to be invalid on that ground, if it is given expressly for any of the purposes detailed in sec. 32. The notice itself is one to which the provisions of sec. 25 (1) apply and it is open to the tenant to serve a counter notice.[1]

(22) PROVISIONS REGARDING COMPENSATION FOR DISTURBANCE

In general, compensation is payable by a landlord to his tenant Sec. 35 where (a) the tenant quits possession after the landlord gives due notice to quit or (b) the tenant gives counter-notice under sec. 33

[1] *Hamilton v. Lorimer*, 1959, S.L.C.R. 7.

of the Act that he accepts a notice duly given to quit part of the holding as notice to quit the entire holding. This is called compensation for disturbance, and amounts to a minimum of one year's net rent (i.e., gross rent less public burdens, but not minister's stipend) or a maximum of two years' net rents. In some cases the landlord is not bound to pay such compensation. These cases are detailed in the proviso to sec. 35 (1). Of these it is useful to refer to one in particular since it is necessary also to examine part of the Agriculture (Scotland) Act, 1948. In this example, the provisions of sec. 25 (1) of the 1949 Act are excluded by the operation of sec. 25 (2) (*d*):—

Where the Land Court, on an application made to them not more than nine months before the giving of the notice to quit, was satisfied in relation to the holding that the tenant was not fulfilling his responsibilities to farm in accordance with the rules of good husbandry, and certified that the Court was so satisfied, and that fact is stated in the notice. The Rules of Good Husbandry are to be found in the Agriculture (Scotland) Act, 1948, at the Sixth Schedule, as follows—

'(1) For the purposes of this Act, the occupier of an agricultural unit shall be deemed to fulfil his responsibilities to farm it in accordance with the rules of good husbandry in so far as the extent to which and the manner in which the unit is being farmed (as respects both the kind of operations carried out and the way in which they are carried out) are such that, having regard to the character and situation of the unit, the standard of management thereof by the owner and other relevant circumstances, the occupier is maintaining a reasonable standard of efficient production, as respects both the kind of produce and the quality and quantity thereof, while keeping the unit in a condition to enable such a standard to be maintained in the future.

(2) In determining whether the manner in which a unit is being farmed is such as aforesaid regard shall be had, but without prejudice to the generality of the provisions of the last foregoing paragraph, to the following:—

(*a*) the maintenance of permanent grassland (whether meadow or pasture) properly mown or grazed and in a good state of cultivation and fertility;
(*b*) the handling or cropping of the arable land, including the treatment of temporary grass, so as to maintain it clean and in a good state of cultivation and fertility;
(*c*) where the system of farming practised requires the keeping of livestock, the proper stocking of the holding;
(*d*) the maintenance of an efficient standard of management of livestock;
(*e*) as regard hill sheep farming in particular:—
  (i) the maintenance of a sheep stock of a suitable breed and type in regular ages (so far as is reasonably possible) and the keeping and management thereof in accordance with the recognised practices of hill sheep farming;
  (ii) the use of lug, horn or other stock marks for the purpose of determining ownership of stock sheep;
  (iii) the regular selection and retention of the best female stock for breeding;

INTRODUCTION

    (iv) the regular selection and use of tups possessing the qualities most suitable and desirable for the flock;
    (v) the extent to which regular muirburn is made;
(f) The extent to which the necessary steps are being taken:—
    (i) to secure and maintain the freedom of crops and livestock from disease and from infestation by insects and other pests;
    (ii) to exercise systematic control of vermin and of bracken, whins, broom and injurious weeds;
    (iii) to protect and preserve crops harvested or in course of being harvested;
    (iv) to carry out necessary work of maintenance and repair of the fixed and other equipment.'

Compensation will be payable unless the notice to quit states one or more of the reasons detailed under sec. 25 (2) for terminating the tenancy and, of course, the reasons stated must be true in fact. In an English case it was held sufficient that the notice referred to reasons as in paragraphs of the Act.[1]

It will be observed that the words 'at the date of the notice' are frequently used in the Act. This generally refers to the conditions existing at or immediately prior to the notice.[2] Anything done after serving the notice could not be assigned as a reason for serving it. If a tenant is to comply with a preliminary demand he must do so before the notice to quit has been served. See *Price* v. *Romilly* [1960] 3 All E.R. 429.

Further conditions must be fulfilled in order to recover compensation for disturbance in excess of one year's rent.

    (a) if the tenant claims loss in respect of the sale of any goods, implements, fixtures, produce, or stock, he must have given the landlord one month's notice of the sale and also a reasonable opportunity of making a valuation thereof before the sale;
    (b) loss or expense directly attributable to the quitting of the holding up to the amount of the claim must be proved.

The additional payments introduced by sec. 9 of the Agriculture (Miscellaneous Provisions) Act, 1968 are payable only where the tenant is entitled to compensation for disturbance, and the exclusion of disturbance payment will exclude the additional payment also. The additional payments are however for the reorganisation of the tenant's affairs and are not, like disturbance payments, in respect of loss or expense incurred.

[1] *In re Digby* v. *Penny,* 1932, 2 K.B. 491.
[2] *Smith* v. *Callander,* 1901, 3 F. (H.L.) 28, and *Taylor* v. *Steel Maitland,* 1913 S.C. 562.

(23) THE AMOUNT OF COMPENSATION FOR DISTURBANCE

Sec. 35 (2)

The compensation normally payable for disturbance is one year's net rent without proof of loss or expense, i.e., one year's gross rent after deducting (1) the amount payable by the landlord in respect of the holding by way of any public burdens the charging of which on the landlord would entitle him to relief in respect of tax under Rule 4 of No. V of Schedule A, to the Income Tax Act, 1918; and (2) the amount recovered from the landlord in pursuance of sub-sec.1 of sec. 47 of the Local Government (Scotland) Act, 1929. Stipend, or standard charge is not a proper deduction.

A maximum of two years' net rent can be claimed where the loss and expense directly attributable to the quitting of the holding which the tenant unavoidably incurs upon or in connection with the sale or removal of his household goods, implements of husbandry, fixtures, farm produce or farm stock exceed one year's net rent. Any tenant contemplating a claim based on such loss would be well advised to have his stock valued independently prior to sale.

Where more than one year's rent is claimed, and it is proved that the loss and expense exceed one year's rent, there does not appear to be room for making any deduction; the arbiter will merely have to fix the compensation at the actual loss and expense.

Where the tenant has lawfully sub-let the whole or part of the holding, and has had to pay compensation for disturbance to the sub-tenant in consequence of a notice to quit given by the landlord, the tenant may, although not himself in occupation, have a claim against the landlord.

Sec. 35 (3)

(24) SUM FOR REORGANISATION OF TENANT'S AFFAIRS

1968 Act, sec. 9

The Agriculture (Miscellaneous Provisions) Act, 1968, sec. 9 provides that where compensation for disturbance becomes payable by the landlord to the tenant of an agricultural holding there shall also be payable by the landlord to the tenant 'a sum to assist in the reorganisation of the tenant's affairs'. The amount of this sum is to be four times the annual rent of the holding, or in the case of part of a holding the appropriate proportion, at the rate at which rent was payable immediately before the termination of the tenancy of the holding or part of the holding.

INTRODUCTION

The question of what is the appropriate proportion is, failing agreement, to be determined by arbitration under the provisions of the 1949 Act or by the Land Court on a joint application.

The additional payment is not due unless the notice to quit was served after 1st November, 1967, or in the case of resumption of part of the holding the resumption took place after 3rd July, 1968.

It will be noted that the additional payment is not due unless compensation for disturbance is payable. Where disturbance compensation is payable the additional sum is due unless the notice to quit contains a statement that it is given on one of the grounds stated in sec. 11 of the Act and either the tenant does not serve counter-notice or the Land Court consents to the operation of the notice to quit on one of these grounds. _{Sec. 11}

The grounds stated in section 11 of the Act which if stated in the notice to quit may exclude payment of the additional sum are:

(a) That the purpose for which the landlord proposes to terminate the tenancy is desirable on any of the grounds mentioned in paragraphs (a) to (c) of sec. 26 (1) of the 1949 Act. These are that it is in the interest of good husbandry, or sound estate management or for the purpose of agricultural research, education or the like or for small holdings. The reference to allotments in sec. 26 (1) (c) is however to be treated as excluded.

(b) That the landlord will suffer hardship.

(c) Where the notice to quit is served on a 'near relative' successor one of the grounds specified in section 18 (2) of the Act. See 'Succession to Holdings' above, (page 17).

Sec. 11 (2) of the Act provides that the exemptions afforded by sec. 11 (1) are not to apply in two circumstances even although the Land Court has consented to the operation of the notice to quit. These are:

(a) Where the Court's reasons include that they are satisfied under sec. 26 (1) (e) of the 1949 Act (use of land for certain non-agricultural purposes); or

(b) Where the Court's reasons are or include that they are satisfied under sec. 26 (1) (b) of the 1949 Act (sound management of estate) or sec. 18 (a) or (c) of the 1968 Act (tenant's lack of training or experience or possession of other agricultural land) but would have been satisfied also under sec. 26 (1) (e) of the 1949 Act (use of land for non-agricultural purposes) if that matter had been specified in the application for consent.

Where the Court would have been satisfied as mentioned in paragraph (b) they must include a statement to that effect in their decision.

Payment of the additional sum is excluded where sec. 25 (1) of the 1949 Act does not apply by virtue of sec. 29 (4) of the Agriculture Act, 1967, which relates to notices to quit served by the Secretary of State or a Rural Development Board for the purpose of amalagamation or reshaping of an agricultural unit. (Sec. 29 (4) applies only where the tenant has signed a lease acknowledging that the tenancy is subject to the provisions of the section).

The additional sum payable under section 9 is payable 'notwithstanding any agreement to the contrary'; contracting out is prohibited.

The additional sum is not payable to a deceased tenant's successor who is not a near relative if notice to quit is served under section 6 (3) of the 1958 Act and contains a statement that the tenancy is being terminated for agricultural purposes only. If there is any question as to the purpose for which the tenancy is being terminated this must be referred to the Land Court and not to arbitration. Sec. 73 of the 1949 Act applies to any such reference.

Schedule 2 of the Act contains transitional provisions relating to notices to quit served between 1st November 1967, and 3rd July, 1968.

The provisions of sec. 11 of the 1968 Act can obviously have important consequences for both landlords and tenants. A landlord will have to consider carefully the effect of any notice to quit which he may serve, as omission of the reason for its service may result in liability to pay a large sum to the tenant. A tenant who receives a notice to quit will require to consider its financial effects, depending on whether or not reasons are stated, and weigh up the advisability of contesting the notice or of accepting it and obtaining payment of the additional compensation.

(25) COMPENSATION FOR EARLY RESUMPTION

1968 Act, sec. 15

The 1968 Act also introduced (sec. 15 (3)) provision for additional compensation to be paid to a tenant where the landlord resumes part of the holding in terms of the lease. The tenant was previously entitled only to a reduction of rent in terms of sec. 34 of the 1949 Act and to compensation for disturbance and improvements in terms of sec. 60 of the 1949 Act. The additional payment due when resumption occurs after 3rd July, 1968, is to be equal to the value of the additional benefit which the tenant would have received if the land had been resumed twelve months after the end of the year of tenancy current two months before the date of the resumption. The current year of a tenancy for this purpose is the year from the term, corresponding to the term of ish, which occurs during the twelve months ending two months before the date of resumption. Thus a tenant with a Martinmas waygoing who has ground resumed at 5th December, 1969, will be entitled to claim for the benefit he would have received if he had been allowed to remain until Martinmas 1970. (The date two months back is 5th October, 1969, and the current year is thus Martinmas 1968 to Martinmas 1969). If the resumption were at 30th January, 1970, the current year would be Martinmas 1969 to Martinmas 1970 and the tenant could claim for the benefit he would have received if he had been allowed to remain until Martinmas 1971.

(26) COMPENSATION FOR IMPROVEMENTS
Extent of the Right to Compensation for Improvements

Where the tenant has made one or more of the improvements comprised in the First, Second or Third Schedules to the Act he may be entitled at the determination of the tenancy on his quitting the holding, to obtain from the landlord, as compensation, such sum as fairly represents the value of these improvements to *an* incoming tenant (not, be it noted, *the* incoming tenant). Compensation is due though there may be no incoming tenant[1]. The 1949 Act divides improvements into 'old improvements' and 'new improvements'. An 'old improvement' is one which: (*a*) is within Schedule II to the Act being 'a 1923 Act improvement', or (*b*) is within Schedule III to the Act being 'a 1931 Act improvement'. 'A 1923 Act improvement' is one begun before 31st July, 1931, and 'a 1931 Act improvement' is one begun before 1st November, 1948, and on or after 31st July, 1931. 'New improvements' are those within Schedule I to the Act and begun on or after 1st November, 1948.

Secs. 36 to 47

[1] *Faviell* v. *Gaskoin*, 1852, 21 L.J. Ex. 85. This case deals with compensation according to the custom of the country, not under the Acts.

Sec. 37 (1)   Where, however, the tenant undertook to execute the improvements in question, he is to receive compensation only where the lease was entered into on or after 1st January, 1921.

Sec. 96   The right to compensation for improvements may depend on the time at which they were made or begun, and, in certain cases, it may be necessary to fall back upon the repealed Acts, beginning with the original Act of 1883. There are cases in which improvements embraced in the Schedules have been made too early to enable compensation to be claimed. For example, if the improvement of laying down temporary pasture was made or begun before 1st January, 1901, no compensation can be claimed for it, because that improvement was scheduled for the first time under the Act of 1900 (sec. 7).

Sec. 45   The fact that an improvement was made during a previous tenancy does not, *ipso facto,* bar the right to claim, so long as the improvement is unexhausted. But see the under-noted case.[1]

Sec. 46   Compensation may be claimed even for improvements made by a previous tenant where, with the landlord's consent in writing and in pursuance of an agreement made before 1st November, 1948, the tenant has paid compensation therefor. Questions have been raised as to whether a tenant who sub-lets part of his farm to a potato-merchant or another farmer is entitled to claim compensation for manurial improvements effected by the sub-tenant. It is sometimes argued that the improvements must, to comply strictly with the terms of the repealed Acts, be made by 'the tenant' himself and by no one else. It is thought, however, that compensation should be awarded if an improvement has, in fact, been made. The principal tenant may have bound his sub-tenant to execute the improvements.

It has been suggested that, before compensation can be claimed, the tenant must have effected an improvement on the general condition of the holding. Reference may be made to the arguments in the three cases undernoted.[2] Though much can be said for this view on equitable grounds, it does not appear to be consistent with the scheme of the Act. Certain things are specified as 'improvements'. If these are effected, compensation is due up to the value to an incoming tenant, notwithstanding that, from some other operation or some neglect of the tenant the holding

---

[1] *Findlay* v. *Munro,* 1917 S.C. 419.
[2] *M'Quater* v. *Fergusson,* 1911 S.C. 640; *Earl of Galloway* v. *M'Clelland,* 1915 S.C. 1062; *Mackenzie* v. *M'Gillivray,* 1921 S.C. 722.

has otherwise suffered deterioration. This may be concluded from the fact that the statute directs that in ascertaining the amount of compensation certain things shall be taken into account in assessing the value of old improvements, viz., 'benefit' ^{Sec. 44} allowed by the landlord, and the manurial value of crops sold off the farm within the last two years of the tenancy. If it had been the intention that the general condition of the holding was to be taken into account, the Act could have so provided. Instead, it directs that only two things shall be taken into account, and it seems clear that nothing else requires to be considered. This does not mean that the landlord has no remedy, because he may make claims for deterioration or dilapidations. Moreover, in certain cases, notably where contributions have been made under Part I of the Agriculture Act, 1937, or otherwise out of moneys provided by Parliament there may be a reduction in the amount of, or exclusion of the right to, compensation for improvements.

Conditions are attached to the tenant's rights to freedom of ^{Sec. 12} cropping and sale of produce. The question whether a tenant is exercising, or has exercised, his rights: (*a*) to dispose of the produce of the holding other than manure produced thereon, or (*b*) to practise any system of cropping of the arable land on the holding in such a manner as to injure or deteriorate his holding or to be likely to injure or deteriorate his holding is determined by arbitration. The certificate of the arbiter as to his determination is to be conclusive proof of the facts therein stated. Where the tenant abides by his contract or by custom (in the absence of contract, for custom cannot override express contract) regarding the method of cropping and disposal of produce, the rules of common law will apply. Where, however, and in so far as the tenant exercises his statutory rights to freedom of cropping or sale of produce, and makes any of the improvements referred to, the arbiter may have to disallow compensation for these improvements in so far as they afford no more than suitable and adequate provision for protecting the holding from injury or deterioration. This subject is dealt with in further detail elsewhere in this Introduction.

(27) Improvements for which the Act allows Compensation

These improvements are comprised in the First, Second and

Third Schedules and as regards market gardens in the Fourth Schedule.

The scheduled improvements in the First, Second and Third Schedules are classified under Part I, II and III in each case.

Sec. 56 provides for compensation for the continuous adoption of a special standard of farming.

(28) PERMANENT IMPROVEMENTS (PART I OF EACH OF SCHEDULES I, II AND III)

Compensation for this class of improvement can be obtained only where the landlord has given consent in writing (which can be informal—sec. 85) to the execution of the improvement before it was begun. Such consent, however, does not imply that in all cases compensation would be payable on the basis of value to an incoming tenant, because the consent may be given 'unconditionally, or upon such terms as to compensation or otherwise as may be agreed upon between the landlord and the tenant'.

Secs. 39 & 50

Sometimes, under a lease, the tenant is expressly *permitted* to execute some of the improvements embraced in this part of the Schedules. If the permissive clause is sufficiently specific it would probably be held equal to the 'consent in writing' which is required by this section. In such case, the tenant, with permission, is free to do as he wishes, and, if he executes the improvements, he does so voluntarily, and under the permission and not under contract. Where the tenant is bound to execute the improvement under any lease or agreement, there is no statutory right to compensation. For conditional consent, see *Turnbull* v. *Millar*, 1942 S.C. 521.

Sec. 44
Sec. 49

In ascertaining the amount of compensation payable in respect of an old or a new improvement, the arbiter must take into consideration 'benefit' given or allowed by the landlord. This may not apply in the case of substituted compensation, which, however, is now confined to market garden improvements, and in other cases to agreements entered into before 1st January, 1921.

Sec. 42

Part I of Schedules First, Second and Third

The landlord's consent is required for 'alteration' of buildings in the Second and Third Schedules, but not for 'repair'. No such consent is required in the First Schedule.

The schedules are alike applicable to market gardens and farms, but in the case of farms the 'erection, alteration or enlargement of buildings', requires consent under Schedules II and III.

It is convenient to have the respective improvements under this heading classified under separate schedules dependent on whether they were begun (as in the First Schedule) on or after 1st November, 1948 (as in the Second Schedule) before 31st July, 1931, or (as in the Third Schedule) on or after 31st July, 1931, and before 1st November, 1948. A comparison of the respective Parts I of the three Schedules clearly indicates a trend in favour of increased freedom for the tenant. The provision regarding the eradication of bracken or other obstacles to cultivation (No. 18) in the Second Schedule was in 1949 transferred in an altered form to Part III where it appears as Item 24 in the Third Schedule and as Item 30 in the First Schedule. The reference to reservoirs is deleted in the First and Third Schedules, but the expression 'Works for the application of water power or for the supply of water' in the First Schedule, Part II, Item 12, and in the Third Schedule, Part II, Item 13, would probably cover reservoirs.

Part I (in which the landlord's consent is required) includes in the First Schedule and since the passing of the 1948 Act, the following improvements.

(1) Laying down of permanent pasture.
(2) Making of water meadows or works of irrigation.
(3) Making of gardens.
(4) Planting of orchards or fruit bushes.
(5) Warping or weiring of land.
(6) Making of embankments and sluices against floods.
(7) Making or planting osier beds.
(8) Haulage or other work done by the tenant in aid of the carrying out of any improvement made by the landlord for which the tenant is liable to pay increased rent.

It cannot be said that the provision of compensation for these improvements has been taken advantage of to any great extent until recently, because most landlords, who wished to have permanent improvements effected, have preferred to make them themselves rather than consent to them being made by tenants, and few tenants were willing to incur the expense of making such improvements without any prospect of receiving compensation therefor till the end of the tenancy.

(29) Drainage, &c (Part II)

The second class of improvements is covered by the single

word 'drainage' in the Second Schedule. The Third Schedule transfers to this part the following, among other, improvements, viz:—Formation of silos, making of roads and bridges, water courses, permanent fences, reclamation of waste land, and provision of sheep dipping accommodation. It also includes the provision of electrical equipment other than moveable fittings and appliances and the repairing and renewal of embankments and sluices against floods. The First Schedule introduces improvements relating to *inter alia* the erection of hay or sheaf sheds, &c., the provision of fixed threshing mills, barn machinery and fixed dairying plant, the improvement of permanent pasture, sewage disposal and necessary repairs to fixed equipment other than repairs which the tenant is under an obligation to carry out. Compensation can be claimed for the improvements in this part of the schedule only where the statutory conditions are complied with. The tenant must have given written notice of his intention to execute the improvement, and of the manner in which he intended to do so, not more than three nor less than two months before beginning operations, in the case of 1923 Act improvements, and not more than six nor less than three months in the case of 1931 Act improvements. In the case of new improvements the tenant must give to the landlord not less than three months' notice in writing. Upon such notice being given, the landlord and the tenant may agree 'on the terms as to compensation or otherwise' on which the improvement is to be executed. The landlord may, however, after service of the notice, undertake to execute the improvement himself in the case of old improvements, and, unless the notice by the tenant was previously withdrawn, charge the tenant, as rent, a sum not exceeding a certain percentage *per annum* on the outlay.

In the case of new improvements specified in Part II of the First Schedule if, within one month of receiving notice from the tenant of his intention to carry out the improvement, the landlord gives notice to the tenant that he objects to the making of the improvement or to the manner in which it is proposed to carry it out, the tenant may notify the landlord and make application to the Land Court for approval of the improvement. The Land Court may (*a*) approve conditionally or unconditionally the improvement (and make terms as to compensation) or (*b*) withhold approval. Thereafter, within one month of receiving notice of a decision of the Land Court approving the improvement the landlord may undertake to carry out the improvement himself. If the landlord does not give notice of his intention to do so, or having given such notice the Land Court finds that he has failed to execute it, the tenant may carry out the work himself and be entitled to compensation therefor.

With reference to improvements under this part, the Act allows parties, by lease or otherwise, to agree to dispense with notice. Mere acquiescence by the landlord after verbal intimation by the tenant does not infer such an agreement. It is open to doubt whether such an agreement can be proved otherwise than by writing.[1]

As regards drainage, the most usual arrangement, in practice, has been for the landlord to supply the drain tiles, and for the tenant to do the work, including carting. Whether such an arrangement would bar the tenant's right to claim compensation is a question of some difficulty, especially where the tenancy is terminated within a few years after the improvement is executed. It is thought that it would not in view of the terms of sec. 64.

(30) Temporary Improvements (Part III)

The third class of improvements, and undoubtedly the class of which the greatest advantage has been taken is that embraced in the third part of the schedules. No consent of, or intimation to, the landlord is required before making improvements of this class. The First Schedule has its provisions as to buildings in Part II, under Items 18 and 23.

It is sometimes found that leases require the tenant to give the landlord or his factor particulars of the manures and feeding-stuffs he proposes to use, along with analysis of the manures, as a condition of receiving compensation. It may also be laid down that compensation shall be limited to the average quantities used during the three years preceding the termination of the tenancy. In general, such conditions may be ignored; they cannot be Sec. 64 enforced, unless they fall within the proviso to sec. 64 (1) which, along with sub-sec. (2). protects the landlord and tenant from limitations, even by agreement, of the right to claim. The right to compensation for improvements is statutory. An usuccessful

[1] *Barbour* v. *M'Douall*, 1914 S.C. 844.

attempt was made to refuse to pay compensation in respect that a tenant had failed to implement an obligation in the lease for a month's notice of intention to claim.[1]

Sec. 42

Nor can agreed compensation (according to a scale in a lease or otherwise) other than for market garden improvements in the Fourth Schedule, be 'substituted' under leases entered into after 1st January, 1921.

First Schedule Part III

The most important of the improvements in this class are—

(1) Liming of land.

(2) Application to land of purchased manure (including artificial manure).

(3) Consumption on the holding of corn (whether produced on the holding or not) or of cake, or other feeding-stuff not produced on the holding by (*a*) horses, cattle, sheep or pigs; or (*b*) poultry folded on the land as part of a system of farming practised on the holding.

(4) Laying down temporary pasture with clover, grass, lucerne, sainfoin, or other seeds, sown more than two years prior to the termination of the tenancy, in so far as the value of the temporary pasture on the holding at the time of quitting exceeds the value of the temporary pasture on the holding at the commencement of the tenancy for which the tenant did not pay compensation.

(5) And, in the case of old improvements only, repairs to buildings, being buildings necessary for the proper cultivation of the holding, other than repairs which the tenant is bound under an obligation to execute. (In this case attention must be paid to the proviso with regard to notice).

(6) In the case of new improvements and the case of old improvements begun on or after 31st July, 1931, the eradication of bracken, whins, or gorse, growing on a farm at the commencement of a tenancy, and, in the case of arable land, the removal of tree roots, boulders, stones or other like obstacles to cultivation.

It may be useful to make some general remarks with regard to several of these improvements.

---

[1] *Cathcart* v. *Chalmers*, 1911 S.C. 292; 1911 S.C. (H.L.) 38.

INTRODUCTION

(31) ARTIFICIAL MANURES AND FEEDING STUFFS

Most outgoing tenants have a claim for the application of artificial or other purchased manures and lime, and the consumption of feeding stuffs not produced on the holding and corn produced and consumed on the holding.

In order to recover compensation for such improvements it is, of course, necessary for the tenant to prove that he has made them, and that a residual value is left unexhausted. The tenant should, therefore, carefully preserve his invoices and receipts for purchased artificial manures and feeding-stuffs, especially those used in the later years of his tenancy. In the case of lime, it may be desirable to produce vouchers for all lime applied during the last seven to ten years; in the case of manures and feeding-stuffs, vouchers for the last two or three years are generally sufficient. (This takes no account of cases where claims are put forward for continuous high farming, which is dealt with later.) A claim having been duly made, it is usual for the tenant to submit his vouchers to the landlord (or his agent or factor) who may make an offer for settlement, and, if this be not accepted, the claim must be referred to arbitration.

Each claim should be judged on its merits. A similar expenditure on each of two farms would not necessarily result in an improvement of the same value on both. Climatic conditions, the method of cropping, exhaustive or otherwise, and the question whether the produce is consumed on or removed from the land, have each an important bearing on the amount of residual value of such 'improvements'. Tables prepared by a Standing Committee appointed by the Secretary of State for Scotland, and brought up to date annually, are available as a guide.

It may be noticed that, in the case of artificial manures, compensation is only allowed where these are 'purchased', while in the case of feeding-stuffs (except corn produced on the holding) all that is required is that they be such as were not produced on the holding (they might be produced on another holding of the claimant). In the case of corn produced and consumed on the holding there is often a difficulty regarding proof. The most satisfactory evidence is the production of a regularly kept barn book or record, but, where that is not available, arbiters are sometimes satisfied otherwise.

Compensation for purchased manures and feeding-stuffs is

generally a proportion of the manurial value, so much per ton based on the unit values of the different constituents at the time when the claim arises in accordance with the Standing Committee's tables.

In dealing with claims for 'manurial improvements', it is frequently of importance to ascertain to what crops the manures were applied, and, as already mentioned, whether the crops have been consumed on or removed from the land. For example, if a certain quantity of manure were applied to an acre of turnips, and the same quantity to an acre of potatoes sold off, most arbiters would agree that the remaining improvement would be greater in the case of the turnips.

It is also necessary, as a general rule, to ascertain what proportion of the feeding-stuffs was in the dung taken over by the incoming tenant at valuation because no compensation is due for such feeding-stuffs where they are included in dung not applied to the land.[1] If a different course were followed, the awaygoing tenant would, in effect, be paid twice over.

A question has been raised whether compensation may be claimed for feeding-stuffs in dung which the tenant was bound to leave steelbow. Compensation has been awarded in such a case but the decision was not tested in Court. In *Davidson* v. *Hunter, supra*, the decision was based largely on the ground that manure taken over at a valuation was moveable, had not been, and might never be, applied to the land. In a question of succession, dunghills or manure prepared for being spread upon the land may be heritable from manifest purpose or intention.[2] In *Reid's Executors* v. *Reid*[3] it was held that the dung on the farm was heritable, in a question between the heir and the executor of the tenant. 'Steelbow' dung left at a waygoing may be merely the equivalent of what was handed over steelbow at entry, and (assuming compensation to be payable) the question of 'benefit' may therefore arise.[4] It may, however, be argued that a steelbow obligation is confined to the dung made from the produce of the farm. For further consideration of Steelbow dung, see the undernoted reference.[5]

[1] *Davidson* v. *Hunter,* 1888 Sh.Ct.Rep. 33; *Brown* v. *Mitchell,* 1910 S.C. 369.
[2] Bell's Prin., 1475.
[3] *Reid's Executors* v. *Reid,* 1890, 17 R. 519.
[4] *Earl of Galloway* v. *M'Clelland,* 1915 S.C. 1062.
[5] *John Chapman* v. *Department of Agriculture* (1938), 26 S.L.C.R. 47.

Questions may arise as to when bracken is 'eradicated'. If the bracken is reduced to such a condition that cultivation of arable or the grazing of pastoral land is restored, it is at least arguable that compensation should be allowed. The amount awarded would vary with the extent of the improvement effected. It is provided that compensation shall only be payable when the bracken or whins were growing at the commencement of the tenancy, this seeming to imply that it is one of the 'rules of good husbandry' for a tenant to deal himself with any new development.

(32) TEMPORARY PASTURE

This improvement consists in laying down temporary pasture with clover, grass, lucerne, sainfoin, or other seeds, sown more than two years prior to the termination of the tenancy, in so far as the value of the temporary pasture on the holding at the time of quitting exceeds the value of the temporary pasture on its holding at the commencement of the tenancy, for which the tenant claiming did not pay compensation. Probably the best definition of 'permanent pasture' is pasture which the tenant cannot break up under the terms of his tenancy, and all other grass may be described as temporary. The tenant's right to freedom of cropping does not include the breaking up of permanent pasture. The mixtures used for sowing down the two species of pasture vary greatly and experts often differ as to what constitutes a 'permanent' grass mixture and what a 'temporary', but, whatever its constituent grasses, the test above referred to is most useful in practice. Grass still remains 'temporary' even though it remains down for many years, provided only it can be broken up by the tenant. In England it has been held that the question of what is temporary pasture has been left unsettled by the Acts, and that it is a practical question for an arbiter.

While the improvement is apparently the excess in value, the question of 'benefit' requires to be taken into account, excluding, however, from 'benefit' the value of the temporary pasture at entry, which has already been debited in the comparison of values. Lord Johnston's views concerning the basis of value may be referred to.[1]

---

[1] *Earl of Galloway* v. *M'Clelland*, 1915 S.C. 1062.

Compensation is not due for the mere leaving of temporary pasture unless the tenant claiming laid it down, otherwise a tenant might claim for temporary pasture laid down not by him, but by his predecessor. On certain soils, well-laid down temporary pasture may improve in value over a course of years. There remains, however, the question of when the improvement was made or begun. It has been held that leaving down pasture which might have been broken up was not 'laying down pasture', to entitle the tenant to compensation.[1] In the same case, an opinion was expressed by Lord Salvesen that, in comparing the condition of a holding at entry and outgo, the entry is to be taken as at the beginning of the last lease. If this view is sound compensation would not be payable for temporary pasture laid down under a previous lease. See, however, sec. 45 which provides that compensation may be allowed at the end of a lease for improvements effected during a previous lease in name of the same tenant.

The Act of 1931 provided that compensation should be payable notwithstanding that the pasture was laid down in contravention of the provisions of the lease or any agreement regarding cropping. The position is unaltered by the 1949 Act. In ascertaining the amount of compensation, however, the arbiter is directed to take into consideration any injury to the holding due to the contravention, but these provisions affect only improvements made after the passing of the 1931 Act. An arbiter has the duty not merely to fix a figure more or less arbitrarily at which the pasture itself may be valued but also, if there is breach of contract, to compare the condition in which the farm was left with the condition in which it would have been if the contract had been fulfilled.

Compensation in the case of 1931 Act improvements and new improvements is based on the increase in the value of the temporary pasture on the holding 'at the commencement of the tenancy for which the tenant did not pay compensation'. The amendment made a material change on the pre-1931 position.[2] It is difficult to say whether, in making the comparison, there should be taken into consideration temporary pasture less than two years old at the commencement of the tenancy. There is a good deal to be said for confining the clause throughout to temporary pasture as described

---

[1] *Findlay* v. *Munro*, 1917 S.C. 419.

[2] *Earl of Galloway* v. *M'Clelland*, 1915 S.C. 1062.

therein. In this view, the clause might be paraphrased thus—

> Laying down temporary pasture with seeds sown more than two years prior to the termination of the tenancy, in so far as the value of the temporary pasture (so laid down) on the holding, at the time of quitting exceeds the value of the temporary pasture on the holding at the commencement of the tenancy *which had been sown more than two years prior to such commencement,* and for which the tenant did not pay compensation.

Here like is being compared with like. Upon the alternative view the value of the whole temporary pasture at entry, so far as not paid for, would fall to be set against the temporary pasture sown more than two years before quitting with a result probably unfortunate for the tenant. He is not entitled to claim, under the Act, for pasture laid down within the last two years of his tenancy, whereas the value of the pasture laid down, say, within the two years immediately preceding his entry, if he paid no compensation or value therefor, would form a deduction. This seems to be contrary to equity. Suppose that, at a claimant's entry, there had been on the farm 50 acres of temporary pasture, and that he entered to that without payment, and that at quitting there were 100 acres of temporary pasture of the same kind, and sown not less than two years before quitting, the arbiter according to the scheme of the Act would allow compensation, not on the basis of the excess of 50 additional acres, but on the basis of the difference in the value of the 100 acres over the value of the 50 acres. (If the tenant had paid for the 50 acres, then he would get compensation for the whole 100.) The value of the 100 acres might be less or more than the value of the 50 acres. The quality of the temporary pasture at quitting has to be compared with the quality of that which was taken over at entry free of compensation.

It is thought that compensation may be awarded for temporary pasture left on the farm, where the tenant had right to break up the pasture, and that even though permanent grasses were used in the process of laying down.

As to the expression 'the commencement of the tenancy', see *Findlay* v. *Munro*.[1]

---

[1] *Findlay* v. *Munro,* 1917 S.C. 419.

Repairs to Buildings, being Buildings Necessary for the Proper Cultivation of the Holding, other than Repairs which the Tenant is Himself under an Obligation to Execute (*Old Improvements*).

As a tenant is bound, even apart from express contract, to keep the farm buildings in a state of tenantable repair, ordinary tear and wear and natural decay excepted, the right to make this 'improvement' would appear to be useful mainly in connection with repairs rendered necessary by causes for which the tenant is not responsible, such, e.g., as extraordinary hurricane, or snowstorm, or accidental fire. It should not be overlooked that what is here referred to is 'repair' not reconstruction. As to the difference between the obligation on a tenant at quitting, and the obligation of the landlord to an incoming tenant, see the undernoted cases.[1] In this connection reference may be made to sec. 23 which provides that sums recovered under insurance policies must be expended on restoration. These observations although framed with the provisions affecting old improvements in mind may be found helpful in considering new improvements, *mutatis mutandis*. Similar provisions affecting these latter are found in Items 18 and 23 of the First Schedule.

Sec. 56

(33) HIGH FARMING

This heading is not used in the Act, but is short for 'the continuous adoption of a special standard or system of farming' for which compensation may be claimed subject to certain conditions. Compensation is awarded to the tenant who proves that the value of the holding to an incoming tenant has been increased during the tenancy by the continuous adoption of a standard of farming, or a system of farming, which has been more beneficial to the holding than the standard or system required by the lease. This goes further than a mere allowance for cumulative fertility.

The conditions precedent to the recovery of such compensation are that (1) there must have been a record of the condition of the holding as provided for under sec. 17—compensation not being payable for anything done prior to the date of the record; (2) the tenant must, before the termination of the tenancy, give the landlord one month's notice of intention to claim. In assessing

[1] *Davidson* v. *Logan,* 1908 S.C. 350; *Johnstone* v. *Hughan,* 1894, 21 R. 777: *Macnab* v. *Willison,* Court of Session, First Division, 5th July, 1955.

the compensation due allowance is to be made for any compensation agreed or awarded in respect of any improvement which has caused or contributed to the benefit. The tenant, therefore, would not be paid twice over for the same thing.

(34) ASCERTAINING THE VALUE OF IMPROVEMENTS

In ascertaining the amount of compensation payable for any scheduled improvement, arbiters are directed to take into account— Sec. 44

(a) In the case of old improvements any benefit which the landlord has given or allowed to the tenant in consideration of his carrying out the improvement. It should be observed that, although it is not essential that the benefit was expressly stated to be given in consideration of the tenant's carrying out the improvements, it is nevertheless necessary to prove that it was so given. The main effect of the words 'whether expressly stated in the lease to be so given or allowed or not' is to allow oral evidence, and they contemplate the possibility of 'benefit' being given during the currency, and not merely at the inception of the lease. Oral evidence may be led to prove that a benefit was given and the arbiter is entitled to decide on the facts adduced.[1] In the case of new improvements, however, the benefit to be taken into account must be 'under an agreement in writing'. 'Benefit' need not be in money. In *Findlay* v. *Munro*,[2] opinions were expressed that the right to take two successive white crops off land which had lain three years in grass was not a 'benefit'. Sec. 49

Sec. 44 (5)

(b) Manuring is omitted as a matter to be taken into account in ascertaining compensation for new improvements. In respect of old improvements, however, manuring is defined in the Act as the application of lime or purchased artificial or other purchased manures, or the consumption of feeding-stuffs not produced on the holding, and corn produced on the holding. The arbiter, in respect of old improvements, should take into account the value of the manure required, by the lease or by custom, to be returned to the holding, in respect of any crops grown on and sold off or removed from the holding during the last two years of the tenancy, or other less time for which the tenancy has endured, which would have been produced by the consumption on the holding of the

[1] *Earl of Galloway* v. *M'Clelland*, 1915 S.C. 1062.
[2] *Findlay* v. *Munro*, 1917 S.C. 419.

crops so sold off or removed. In the general case, and apart from contract, nothing falls to be taken into account in respect of the away-going crop where, as is usual, the tenant has the right to sell or remove that crop, or where it is taken over by the incoming tenant at a valuation. It has not been the custom for the away-going tenant to return to the holding manure in respect of that crop, although, in many cases, the straw of the crop has, by contract, to be left steelbow. It should be noted, however, that this provision is not confined to the crops of the last two years; it extends to any crops sold off or removed from the holding in the last two years, whether grown in these years or not.

Claims are frequently made which include feeding-stuffs purchased and consumed up to the termination of the tenancy, notwithstanding the fact that some part of the manure resulting from such consumption is included in the dung taken over at a valuation by the incoming tenant. As, however, there is no improvement to the holding from the consumption of the feeding-stuffs till the resulting dung has been incorporated with the soil, compensation is not payable in respect of the dung made from feeding-stuffs not produced on the holding (nor from home-grown corn) in so far as that dung, instead of being applied to the land by the away-gong tenant, is taken over by the proprietor or incoming tenant at a valuation. Nor is this inequitable, because if the away-going tenant gets the value of that dung on a valuation, he is not entitled also to recover it as compensation from the landlord, which would result in payment twice over.[1] It is true that, in some cases, the arbiters are instructed to value the dung as if it contained no product of the feeding-stuffs for the consumption of which compensation could be claimed. This is not correct. It is the duty of arbiters to value the dung, with all that it contains, just as they find it. They have no right to proceed on the assumption, which is or may be contrary to fact, that the dung contained nothing resulting from the consumption of the feeding-stuffs in question. Should they act in that way, the incoming tenant might get the dung at less than its intrinsic value. There would then be no obligation on the landlord (on whom rests the obligation to meet the claim for compensation for improvements) to pay compensation for 'improvement' resulting from the

---

[1] *Davidson* v. *Hunter*, 1888 Sh.Ct.Rep. 33; *Brown* v. *Mitchell*, 1910 S.C. 369.

consumption of such feeding-stuffs, in so far as the same was part of the dung heap for which the incoming tenant should have paid full value.

The arbiter has also to take into account, for new improve- Sec. 49
ments, any grant out of moneys provided by Parliament which has been or will be made to the tenant in respect of the improvement. Grants, for example, made in pursuance of the Hill Farming Act, 1946, will clearly have to be taken into account.

(35) 'SUBSTITUTED' COMPENSATION FOR IMPROVEMENTS

Sec. 64 declares that a landlord or tenant shall be entitled to compensation in accordance with the statutory provisions notwithstanding any agreement to the contrary. The section has an important proviso which excepts a written agreement in connection with permanent pasture under sec. 9. It is still competent to have the compensation for the improvements valued on the basis of an Sec. 42
agreed scale. This continues in force as regards market garden improvements in the Fourth Schedule, but as regards improvements under Part III of the Second Schedule, whether on market gardens or farms, it has effect only under agreements entered into on or before 1st January, 1921.

With regard to improvements under Part III of the Second Schedule, executed under agreements entered into prior to 1st January, 1921, substituted compensation is still allowed where it is fair and reasonable, having regard to the circumstances at the time the agreements were entered into.

In the case, however, of permanent improvements and drainage, the substituted compensation may, it is thought, be any sum of money which is not merely nominal or illusory.

In all cases, it is thought that the substituted compensation must be money—something that can be 'awarded' at the termination of the tenancy. Throughout the Act, 'compensation' is to be 'payable'.

A difficult question is the ascertainment of what is fair and reasonable compensation, having regard to the circumstances at the time of making the agreement. That is a matter for the arbiter to decide subject to appeal to the Court on a question of law.[1]

[1] *Bell* v. *Graham,* 1908 S.C. 1060. See also Seventh Schedule to the 1949 Act.

46   AGRICULTURAL HOLDINGS (SCOTLAND) ACTS

The Court held that a stipulation, in a lease, that compensation payable at quitting was to be subject to a deduction of the amount paid by the landlord to the previous tenant was not void.[1]

(36) IMPROVEMENTS MADE UNDER REPEALED STATUTES

Sec. 96

Unless otherwise expressly provided by the Act, the compensation in respect of an improvement made or begun before 1st January, 1909, or made upon a holding held under a lease, not being a lease from year to year, current on the 1st January, 1884, shall be such (if any) as could have been claimed had the Agricultural Holdings (Scotland) Acts, 1923-1948, and the present Acts not been passed, the procedure for ascertaining and recovering the same to be that provided by the 1949 Act.

The effect is to retain in force several of the repealed Acts so far as necessary for regulating (1) what improvement compensation may be claimed, and (2) the conditions under which such compensation may be payable. It will be observed that the procedure for the ascertainment and recovery of compensation is not as prescribed in the repealed statutes, but as provided under the 1949 Act. Apparently, however, the procedure for giving notice remains unaffected.

(37) COMPENSATION FOR DAMAGE BY GAME, i.e., DEER, PHEASANTS, PARTRIDGES, GROUSE AND BLACK GAME

Sec. 15

At common law an agricultural tenant was not entitled to compensation for game damage unless he proved that the damage resulted from an increase in the stock of game. On taking the farm, it was assumed that he agreed to suffer, without compensation, such damage as would naturally result from the stock of game as it was when he entered. Lord Fullerton said—'The true ground of damage seems to be not that the game is abundant, but that its abundance has been materially increased since the date of the lease'.[2] This is not now the law, for since the passing of the Act of 1908, compensation can be claimed where the damage exceeds 1s. per acre of the area over which the damage extends.

[1] *Buchanan* v. *Taylor,* 1916 S.C. 129; see also *Young* v. *Oswald,* 1949 S.C. 412.
[2] *Drysdale* v. *Jameson,* 1832, 11 S. 147 at p. 150.

If the damage does not exceed that figure, no compensation is due. If it does exceed that figure, the full amount of the damage may be claimed without any deduction.[1] It is no defence that the game came from the property of a neighbouring proprietor, even during close time. Practically the only case in which the landlord is not liable is where the agricultural tenant has permission in writing to kill the game. Where the right to kill and take the game is vested in some person other than the landlord, the landlord is entitled to be indemnified by the other person against claims for compensation on this ground. Even should the tenant enter into an agreement with his landlord not to claim compensation, he can repudiate the agreement, contracting out not being permitted under the statute. The tenant is entitled, however, to agree, after the damage has been caused, as to the amount of the compensation, and failing such agreement, the amount is to be fixed by arbitration.

No compensation can be claimed under the statute unless notice in writing is given to the landlord as soon as may be after the damage was first observed by the tenant, and a reasonable opportunity is given to the landlord to inspect the damage—

(a) in the case of damage to growing crop, before the crop is begun to be reaped, raised, or consumed; and

(b) in the case of damage to a crop reaped or raised, before it is begun to be removed from the land.

Notice in writing of the claim, together with the particulars thereof, must also be given to the landlord within one month after the expiration of the calendar year, or such other period of twelve months as by agreement, between the landlord and tenant, may be substituted therefor, in respect of which the claim is made.

The Agriculture (Scotland) Act, 1948, sec. 43 (1) gives the occupier of an agricultural holding the right to kill deer on land other than moorland or unenclosed land. Sec. 52 of that Act states that nothing in the part of the Act which includes sec. 43 shall preclude the occupier of an agricultural holding from recovering any compensation for damage by game which he would have been entitled to recover if that Act had not been passed.

Where the tenant of a holding consisting entirely of arable land and permanent pasture who did not have permission in writing to kill deer claimed compensation from a landlord for damage to

[1] *Roddan* v. *McCowan*, 1890, 17 R. 1056.

crops by deer it was held that he was entitled to compensation.[1] The right conferred by sec. 43 of the 1948 Act is in addition to and is not to derogate from the tenant's right to compensation for damage done. It was pointed out that sec. 52 of the 1949 Act preserved the tenant's right to compensation for damage by deer conferred by sec. 11 of the Agricultural Holdings (Scotland) Act, 1923 of which sec. 15 of the 1949 Act is practically a re-enactment. As the 1949 Act is a consolidating Act it was held that it did not alter the tenant's previous rights. Sec. 99 (10) of the 1949 Act expressly provides that compensation rights conferred by the Act are in lieu of rights to compensation conferred by any former enactment relating to Agricultural Holdings.

(38) NOTICES OF INTENTION TO CLAIM COMPENSATION AND LODGING OF PARTICULARS

Generally such notices must be in writing, but there are some exceptions. Again, generally, the notices must be in the hands of the person to whom they are addressed within the prescribed period, no matter when despatched, and the intervention of a Sunday makes no difference.[2]

Sec. 56

Sec. 59

1. In order to recover compensation for improvements, no notice of intention to claim is necessary, except in the case of compensation for improvement effected by following a special standard or system of farming. In such case notice of intention to claim must be given in writing to the landlord not less than one month before the determination of the tenancy.

2. *Landlord's claim for Compensation for Deterioration of Holding.*—Notice in writing of intention to claim under the Act must be given not later than three months before the termination of the tenancy, but this is not necessary where the claim is made under the lease.

Sec. 15

3. *Compensation for Damage by Game.*—Notice must be given as soon as may be after the damage and the landlord given an opportunity of valuation. Further, notice in writing of the

---

[1] *Lady Auckland* v. *Dowie,* 1965 S.L.T. 76.
[2] *Scott* v. *Scott,* 1927 S.L.T. (Sh.Ct.) 6.

claim, together with particulars, must be given not later than one month after the expiration of the calendar year (i.e. 31st December) or such other period of twelve months as by agreement may be substituted therefor, in respect of which the claim is made. Notice given before the end of the appropriate period of twelve months is competent[1].

(39) CONTRACTING OUT

Sec. 64 provides that notwithstanding any agreement to the contrary, a tenant or landlord shall be entitled to compensation in accordance with the provisions of the Act. This, however, does not apply to improvements which, by contract, the tenant agreed to execute under any lease dated on or prior to 31st December, 1920. Compensation cannot be claimed for such improvements. It would be equivalent to contracting out if conditions were imposed on the tenant's right to compensation, such, for example, as requiring him to give notice within a period not prescribed by the Act;[2] not, however, conditions under which the tenant got certain benefits in consideration of which he undertook to effect the improvement. It is not competent to contract out, with respect to 'agricultural holdings' which fall within the definition under sec. 1. _{Sec. 37 (1)} _{Sec. 1}

Attention is drawn to the variation of the terms of a lease provided for, in reference to permanent pasture, under sec. 9 read in conjunction with sec. 63 (1). Compensation is not payable for an old improvement specified in Part III of the Second and Third Schedules respectively, nor for a new improvement specified in Part III of the First Schedule.

Further, it is competent to contract out of the provisions relating to fixtures and buildings and the bequest of lease.[3] *Obiter dicta* suggest that sec. 20 can not readily be applied except to a direct bequest of a lease to a specific person.[3] If the tenant is bound by his lease to execute all repairs to buildings, he cannot claim compensation for doing so, notwithstanding that secs. 37 and 48 allow compensation for contracted-out improvements under leases entered into after 1st January, 1921. _{Sec. 14 Sec. 20 as amended}

---

[1] *Earl of Morton's Trs.* v. *Macdougall*, 1944 S.C. 10.
[2] *Cathcart* v. *Chalmers*, 1911 S.C. 292; 1911 S.C. (H.L.) 38.
[3] *Kennedy* v. *Johnstone*, 1956 S.C. 39.

It is not competent to contract out of (1) compensation for disturbance; or (2) compensation for damage by game; (3) the additional payments due under Part II of the 1968 Act; (4) the provisions for notice to quit in order to terminate a tenancy; (5) the provisions of sec. 3 for continuation of tenancies by tacit relocation or (6) the provisions of sec. 22 as to payment for implements and other items sold on quitting the holding.

[Sec. 15, Sec. 24 (1)]

It is not expressly provided that it is incompetent to contract out of the security of tenure provisions but the general scheme of the Act suggests this. It is nevertheless possible to grant a let in such a way that it does not continue indefinitely and various methods have been devised for this purpose. These may be summarised as follows:—

1. *Agreement not a 'lease'*

If the agreement does not satisfy the definition of 'lease' in the 1949 Act sec. 93 the subjects let will not be an agricultural holding. See, e.g., *Stirrat & Anor. v. Whyte* 1968 *S.L.T.* 157 where the let was indefinite and to terminate on sale.

2. *Company*

The let may be granted to a limited company in which the landlord holds a share or shares carrying voting rights which enable him to terminate the company's tenancy.

3. *Partnership*

A let to a firm normally terminates when the firm is dissolved. As the firm is in Scots Law a separate person from the partners the let could be to a firm in which the landlord is a partner. For the position in England see the undernoted case.[1] From the landlord's point of view apart from the obvious danger of incurring liability there is the possibility that the Partnership Agreement would be treated as a sham if the parties did not in fact carry on business together. Unless the landlord wishes to be able to terminate the lease at any time it is preferable to form a Limited Partnership (under the Limited Partnerships Act 1907) in which the tenant is the general partner and the landlord the sleeping partner. The agreement will be of a stipulated endurance to end when it is desired to terminate the lease.

[1] *Harrison-Broadley* v. *Smith* [1964] 1 All E.R. 867.

4. Joint Tenancy

In *Smith* v. *Grayton Estates Ltd.*[1] it was held that tacit relocation depends on implied consent and that one of two joint tenants could not insist on continuing when the other wished to give up. It may therefore be possible to avoid creating security of tenure by the landlord's granting a short tenancy jointly to the real tenant and a nominee of the landlord who will be in a position to terminate the tenancy.

5. Sub-tenancy

A notice to quit which terminates a tenancy will terminate also any sub-tenancy.[2] If the holding is let to a nominee of the landlord and he sub-lets to the real tenant the latter is not in a position to prevent termination of the tenancy (and with it his own sub-tenancy).

6. Letter of Removal

Attempts have been made to exclude security of tenure granting a missive of let in exchange for a letter of removal by the tenant. It has, however, been held in England that notice of removal cannot validly be given before the commencement of the tenancy and it was observed that such a notice 'would have the effect of defeating the purpose of the Agricultural Holdings Act'.[3]

(40) CHARGING THE ESTATE WITH COMPENSATION

The Act contains provisions authorising a proprietor to charge his estate with sums paid for compensation for old or new improvements, and also for disturbance. Application for such charge must be made to the Secretary of State. The charge ranks *after* all prior charges and burdens heritably secured on the holding or estate. It is thought that compensation paid by agreement, where there is no arbitration, cannot be charged on the estate, because the necessary certificate must be granted by an arbiter. Sec. 82

Sec. 70

(41) SPECIAL PROVISIONS AS TO MARKET GARDENS

The improvements for which compensation may be claimed by tenants of market gardens are practically the same as those Secs. 65, 66 and 67

---

[1] *Smith* v. *Grayton Estates Ltd.*, 1960 S.C. 349.
[2] *Sherwood* v. *Moody*, [1952] 1 All E.R. 389.
[3] *Lower* v. *Sorrell*, [1963] 1 Q.B. 959 per Ormerod L. J. at page 968.

which may be claimed by ordinary agricultural tenants, with this difference, that in their case certain additional improvements which require neither the consent of nor notice to the landlord, are provided for by the Fourth Schedule, viz.—

1. Planting of standard or other fruit trees permanently set out.
2. Planting of fruit bushes permanently set out.
3. Planting of strawberry plants.
4. Planting of asparagus, rhubarb, and other vegetable crops which continue productive for two or more years.
5. Erection, alteration or enlargement of buildings for the purpose of the trade or business of a market gardener.

Sec. 65 (1) (c)   The tenant of a market garden is entitled to remove all fruit trees and fruit bushes not permanently set out, subject to this condition, that if he fails to remove them before the termination of his tenancy, they remain the property of the landlord without compensation. In short, practically the whole provisions of the statute with reference to compensation for improvements, for disturbance, for game damage, fixtures and buildings, erected or acquired by the tenant, and notices to terminate tenancy, apply with the same force to market gardens as they apply to ordinary agricultural holdings with the additions referred to above. The tenant of a market garden is in a more favourable position than the ordinary agricultural tenant, as he may erect or enlarge buildings without the landlord's consent, and claim compensation therefor at quitting, on the basis of the value to an incoming tenant. Agricultural tenants whose leases fall within the First Schedule are now also in a more favourable position than other agricultural tenants.

The provisions of the Acts relating to substituted compensation continue to apply to market garden improvements—not to other improvements, except under agreements entered into prior to 1st January, 1921.

Sec. 66 as amended   It was found that the compensation which might be claimed by the tenant of market garden land was so onerous as to hinder the letting of land for market garden purposes. With a view to meeting that difficulty new provisions were introduced by the Act of 1920. These were incorporated in the 1949 Act and amended by the 1958 Act. A tenant of agricultural land who wishes to make any of the improvements which may be executed by tenants of market gardens, where the landlord refuses, or within a reasonable time fails to agree in writing, that the land shall be treated as a market garden, may apply to the Land Court who may direct that the land shall, in effect, be treated as market garden land, with right in the tenant to excute all market garden improvements thereon, or some only of these improvements. Such a direction may be given, subject to such conditions, for the protection of the landlord as the Land Court thinks fit to impose. In all cases where a direction is given, the following provision, based on what is known as the Evesham custom, must apply:—

> Where notice to quit is given by the tenant he is not to be entitled to the compensation specified in the direction, unless he produces a substantial and otherwise suitable successor who is willing to accept the tenancy of the holding on the same terms and conditions, and pay to him, the outgoing tenant, all compensation payable under the statute, or under the contract of tenancy, and the landlord fails to accept that tenant within three months after the offer is made to him. [Sec. 66 (2)]

In the event of the landlord accepting the offer, the incoming tenant must pay to him all sums payable by the outgoing tenant, on the termination of the tenancy, for rent, or breach of contract, or otherwise in respect of the holding, and any amount so paid may be deducted by the incoming tenant from any compensation payable by him to the outgoing tenant.

If the direction relates only to part of a holding the direction may, on the application of the landlord, be made subject to the condition that the tenant shall consent to the division of the holding into two parts (one such part being the part to which the direction relates) to be held at rents agreed between landlord and tenant, or failing agreement, determined by arbitration, but otherwise on the same terms and conditions as the original holding, so far as applicable. The new tenancy created by the acceptance by the landlord of a tenant on the terms and conditions of the existing tenancy, is not to be treated as a new tenancy for the purposes of the provisions of the Act relating to demands for arbitration as to rent. [Sub-sec. (4)]

The provisions of the Small Landholders (Scotland) Acts, 1886 to 1931, with regard to the Land Court, are made applicable to references under the Act subject to necessary modifications. [Sec. 73]

ARBITRATION UNDER THE ACT

Claims and Questions which are, or may be, referred to Arbitration, and conditions applicable thereto.

An arbiter may be appointed, even in cases where there is no relevant claim, the question of relevancy being left for him to deal with, subject it may be, to an appeal to the Court by stated case or by an action of suspension and interdict.[1] Practically all questions (except as to payment of rent) and even including the construction of the lease are referred to arbitration (secs. 68 and 74). A question may be raised as to whether the wide terms of sec. 74 affect the landlord's remedy of interdict. (Sec. 12 (2)). Sub-sec. (3) provides that an arbiter shall determine whether or not the tenant has exercised his rights in such a way as to deteriorate the holding. The ascertainment of the damages payable to the landlord is also a matter for arbitration. Interdict is a remedy competent only in the Courts and where invoked for the purposes of sec. 12 (2) it is probable that the Sheriff would pronounce *interim interdict* where a *prima facie* case has been made, and allow a sist to enable an arbiter to decide the real question. The alternative would be to have the arbitration before initiating any Sheriff Court proceedings. It is not a 'question or difference' between the landlord and and tenant, but an administrative act exercisable by the Courts to preserve the landlord from loss.

It is desirable that arbiters have a general knowledge of the conditions, subject to which claims and questions may be competently dealt with by arbitration. In all cases these claims and questions must be between landlord and tenant. Other questions, e.g., between an outgoing and incoming tenant are not dealt with and fall to be settled by application to the Courts or by arbitration at Common Law.[2] Certain questions, e.g., with regard to the carrying out of improvements included in the First Schedule,

---

[1] *Glasgow, Yoker, and Clydebank Railway Co.* v. *Lidgerwood*, 1895, 23 R. 195; *Bennet* v. *Bennet*, 1903, 5 F. 376; also *Hamilton Ogilvy* v. *Elliot*, 1904, 7 F. 1115.

[2] *Cameron* v. *Nicol*, 1930 S.C. 1.

Part II, are referred to the decision of the Land Court (sec. 52 as amended.)

In one case the away-going tenant is required to offer hay, straw, manure, &c., to the landlord or incoming tenant at their market value, but it is thought that, even in that case, the right to enforce the obligation would lie solely with the landlord against the away-going tenant.

In virtue of sec. 78, parties may agree to refer their differences to the Land Court instead of to an arbiter. The Act further provides in sec. 87 for cases where the Secretary of State is the landlord in which circumstances the assistance of the Land Court is invoked. In cases which go to the Land Court, attention must be paid to the Rules of the Land Court. It is thought that the normal statutory provisions as to arbitration procedure do not apply.

Notes on Assessment of Compensation.

(1) No compensation is payable to a tenant for anything done under an order under sec. 9 which enables an arbiter to vary the terms of a lease prohibiting the breaking up of permanent pasture (sec. 63). Where permanent pasture has been ploughed up under an order the value of the tenant's pasture (i.e., what was laid down by him or paid for at his entry) shall be taken not to exceed the average value per acre of the whole of the tenant's pasture at the termination of the tenancy.

(2) A tenant is not entitled to compensation in respect of an old improvement (Second and Third Schedules, Part III) if it was made to restore fertility in respect of the sale of produce off the farm.

We now refer shortly to the different classes of claims and questions and relative conditions:—

CLAIMS BY THE TENANT

In all cases notice of intention to claim must be given within Sec. 68 (2) two months from the termination of the tenancy. It is not necessary to quantify or specify the claim in detail; it is sufficient to refer to the section of the Act under which the claim is made and to state its nature in general terms.

A full statement of case and details of claim must be supplied to an arbiter within 28 days of his appointment. (Sixth Schedule Para. 5 as amended by the Agriculture (Miscellaneous Provisions) Act, 1963, sec. 20).

I. *Scheduled Improvements.*

These are now divided into 'Old Improvements,' i.e., those made before 1st November, 1948 (secs. 36-46) and 'New Improvements', i.e., those made on or after that date (secs. 47-55). 'Old Improvements' are further divided into '1923 Act Improvements,' i.e., those begun before 31st July, 1931, and '1931 Act Improvements,' i.e., those made on or after 31st July, 1931, and before 1st November, 1948 (sec. 36). Reference should be made to the First, Second and Third Schedules which each detail the improvements under three separate headings.

Schedule II (*Old Improvements under the* 1923 *Act, i.e., those begun before* 31*st July,* 1931).

Conditions:

(1) A tenant under a lease dated before 1st January, 1921, is not entitled to compensation for an improvement which he was bound by his lease to carry out (sec. 37).

(2) Compensation is not payable unless before carrying out an improvement under Part I of the Schedule the tenant obtained the landlord's consent in writing. (Agreed compensation may be substituted for compensation under the Act) (sec. 39).

(3) Compensation is not payable for an improvement under Part II unless the tenant, not more than three nor less than two months before he began to carry out the improvement, gave the landlord notice in writing of his intention and either: (a) the parties agreed as to compensation; or (b) the landlord failed to carry out the improvement himself (sec. 40).

(4) In the case of repairs to buildings the tenant must have given notice in writing of intention to carry out the repairs and the landlord failed to carry them out himself within a reasonable time (sec. 41).[1]

Schedule III, Old Improvements under the 1931 Act, i.e., those begun on or after 31st July, 1931, and before 1st November, 1948.

Conditions:

(1) The tenant is not entitled to compensation for improvements which he was bound by a lease entered into before 1st January, 1921, to carry out (sec. 37).

(2) Compensation is not payable unless before carrying out an improvement in Part I of the Schedule the tenant obtained the landlord's consent in writing. Agreed compensation may be substituted for compensation under the Act (sec. 39).

(3) In the case of an improvement under Part II of the schedule the tenant must have given not more than six and not less than three months' notice in writing of his intention and either: (a) the parties have agreed to the terms as to compensation; or (b) the landlord failed to carry out the improvement; or (c) the appropriate authority for the time (now the Land Court) determined that the improvement should be carried out (sec. 40).

(4) In the case of repairs to buildings the tenant must have given notice in writing to the landlord of his intention and the landlord failed to carry out the repairs himself within a reasonable time (sec. 41).

Schedule I, New Improvements, i.e., those carried out after 1st November, 1948.

Conditions:

(1) The tenant is not entitled to compensation for improvements which he was bound to carry out by a lease entered into before 1st January, 1921 (sec. 48).

[1] See paragraph 29 of Second and Third Schedules.

58   AGRICULTURAL HOLDINGS (SCOTLAND) ACTS

    (2) Improvements under Part I. Landlord must consent in writing (sec. 50).

    (3) Improvements under Part II. The parties may dispense with notice and provide also as to the terms of compensation. Otherwise the tenant must have given written notice of his intention not less than three months before carrying out the improvement and either: (a) the parties have agreed to the terms of compensation (sec. 51); or (b) the landlord has himself failed to carry out the improvement after it has been approved, on the tenant's application (the landlord having lodged notice of objection) by the Land Court (sec. 52).

II. *Compensation for Continuous Good Farming* (*sec.* 56).

    Conditions:

    (1) There must be a record of the fixed equipment on and the cultivation of the holding, and only so far as the improvement is made after the date of that record is compensation payable.

    (2) Notice of intention to claim, one month before termination of tenancy.

III. *Compensation for Disturbance* (sec. 35).

    Conditions:

    (1) Notice to quit by the landlord in ordinary form, or in form required to exclude additional payments under 1968 Act[1] followed by tenant quitting (sec. 35).   or
Notice to quit part of the holding followed by a counter notice by the tenant that he is to treat the notice as a notice to quit the whole (sec. 33).

    (2) Landlord to be given one month's notice of sale of implements, stock, &c., and a reasonable opportunity of making a valuation of goods, implements, fixtures, produce, and stock (sec. 35), if claim exceeds one year's rent.

[1] See the exceptions in the proviso to sec. 35 (1) and forms of Notice to Quit in Appendix.

ARBITRATION UNDER THE ACT

(3) Notice of intention to claim within two months after termination of tenancy (sec. 68 (2)).

(4) Delivery to arbiter within 28 days of his appointment of a statement of case and all necessary particulars (Sixth Schedule Para. 5 as amended by the Agriculture (Miscellaneous Provisions) Act, 1963 sec. 20).

IV. *Additional Payments under Sec. 9 of 1968 Act.*

Conditions:

(1) Entitlement to Compensation for Disturbance. See III above.

(2) Notice to Quit served after 1st November, 1967.

(3) Payment not excluded under Sec. 11 of Act (See (24) 'Sum for reorganisation of tenant's affairs' above, page 26).

(4) Notice of intention to claim within two months after termination of tenancy (Sec. 68 (2)).

V. *Compensation for Early Resumption* (1968 *Act Sec.* 15)

Conditions:

(1) Resumption after 3rd July 1968 on less than 12 months' notice.

(2) Notice of intention to claim within two months after termination of tenancy (Sec. 68 (2)).

VI. *Claims (by the Tenant) in respect of any matter arising out of the Tenancy (sec. 74).*

Condition:

Notice of intention to claim must be given within two months after termination of tenancy. In the case of claims arising during the currency of the tenancy it is clear that particulars must be given within 28 days of the arbiter's appointment (Sixth Schedule Para. 5 as amended by the Agriculture (Miscellaneous Provisions) Act, 1963, sec. 20).

VII. *Fixtures, Machinery or Buildings* (sec. 14).

These are not, properly speaking, 'improvements' for which compensation may be claimed but where the landlord agrees to take them over, their value falls to be fixed by arbitration under the Act. It may be noted that the basis is the value to an incoming tenant, being the same basis as applies to permanent improvements to which the landlord consented in writing.

Conditions:

(1) A month's notice in writing of intention to remove the fixtures.

(2) All rent must be paid and other obligations under the lease performed.

(3) Unless the landlord elects to take them over, they may be removed (subject to making good damage) before or within six months after the termination of tenancy.

VIII. *Damage by Game* (sec. 15).

Damage by game can form the subject of a claim during, as well as at the termination of, a tenancy.

Conditions:

(1) Notice of the damage must be given in writing as soon as may be after its discovery, and a reasonable opportunity given to the proprietor to inspect the damage.

(2) The claim must be made within a month after the end of year in which it occurs.

IX. *Variation of Rent* (sec. 7).

Conditions:

(1) Notice in writing demanding arbitration as to rent payable as from the next ensuing day on which the tenancy could have been terminated by notice to quit given at the date of demanding the reference.

(2) Demand not effective if alteration in rent would take effect earlier than 5 years from: (a) the commencement of the tenancy; (b) the date of a previous alteration in

rent; or (c) the date of a direction in an arbitration as to rent under the section. *Note:* There are certain exceptions (sec. 7 (3)).

Where a tenant quits part of a holding under sec. 32 or the landlord resumes possession under the provisions of a lease the reduction of rent is determined by arbitration.

Conditions:

(1) The reduction is proportionate to the part given up and in respect of the depreciation of the remainder.

(2) Where there is resumption under a lease any benefit or relief allowed to the tenant is taken into account.

Where a landlord has carried out certain improvements specified in sec. 8, the rent may be increased by arbitration.

Condition:

Landlord to serve notice in writing on the tenant within six months from the completion of the improvement.

X. *Adjustment of terms of Leases.*

(a) Where land is let without the consent of the Secretary of State for less than one year the lease will be treated as one from year to year, if the circumstances are such that if the tenant were a tenant from year to year the tenancy would come under the Act. Any question falls to be decided by arbitration (sec. 2). (b) Where liability for the maintenance of any fixed equipment is transferred from the tenant to the landlord any claim by the tenant in respect of the landlord's previous failure is also settled by arbitration (sec. 5 (5)). (c) An arbiter may be appointed to settle the terms of a lease (sec. 4).

Conditions:

(1) Either: (1) there is no lease; or (2) a lease has been entered into after 1st November, 1948, or the tenant is sitting under tacit relocation and the lease contains no provision for one or more of the matters contained in the Fifth Schedule or provisions inconsistent with sec. 5.

(2) Either party must give to the other six months' notice requesting a lease to be entered into (sec. 4).

(3) Every new lease is held to have incorporated in it an undertaking by the landlord to put the fixed equipment in order (based on a record to be made in every case) the tenant's liability being to maintain it in as good a state as it is put in by the landlord. Any question arising is referred to arbitration (sec. 5).

XI. *Market Garden 'Direction'* (sec. 66).

The Land Court may direct that a holding, or part thereof, be treated as a market garden.

Conditions:

(1) Intimation in writing by tenant of desire to carry out market gardening improvements.

(2) Tenant may not claim compensation if he is removed on account of bankruptcy unless he provides a suitable tenant within one month of the notice to quit being served.

XII. *Payment for implements, fixtures, produce or stock agreed to be purchased by landlord on termination of tenancy.*

Where payment is not made within one month after the tenant has quitted the holding or within one month after the issue of the award the outgoing tenant may sell or remove the goods and claim compensation equal to any loss or expense unavoidably incurred thereby, including the expense of preparing his claim (sec. 22).

XIII. *Restriction on Operation of Notices to Quit* (sec. 25).

The provision that a tenant on receipt of a notice to quit may serve a counter-notice requiring sec. 25 (1) to apply, in which case the notice is subject to the consent of the Land Court, does not apply in certain circumstances, e.g., where the Court has already consented to the notice to quit.

ARBITRATION UNDER THE ACT

Claims by Landlord

I. *Compensation for Deterioration of Particular Parts of Holding* (sec. 57).

II. *Compensation for General Deterioration of Holding* (sec. 58).

   Conditions:

   (1) Tenant quitting holding.

   (2) Notice in writing of intention to claim not less than three months before termination of tenancy. If the claim is under a lease (sec. 57 (3)) it is arguable that no such notice is required, but the safest course is to give notice. See sec. 59 (1).

   (3) Record of fixed equipment and cultivation to have been made if claim is under sub-sec. 1 of sec. 57 or under sec. 58 when the lease was entered into after 31st July, 1931, or the claim is under a lease entered into after 1st November, 1948. Sec. 59 (2) (b).

   (4) Case with particulars to be lodged with arbiter within 28 days of his appointment (Sixth Schedule Para. 5 as amended by the Agriculture (Miscellaneous Provisions) Act, 1963 sec. 20).

III. *Any other Matter arising between the Landlord and the Tenant arising out of the Tenancy which is referred to arbitration by the lease or under the Act.*

   Conditions:

   (1) Notice of intention to claim must be given within two months after termination of tenancy where the claims arise at the away-going (sec. 68 (2)).

   (2) Case with particulars to be lodged with arbiter within 28 days of his appointment (Sixth Schedule Para. 5 as amended by the Agriculture (Miscellaneous Provisions) Act, 1963, sec. 20).

IV. *Questions as to Rent* (sec. 7).

Conditions:

(1) As under Claims by Tenant.

(2) Under sec. 8, landlord may demand arbitration by notice in writing as to rent to be paid in respect of improvements carried out by him as specified in that section.

Condition:

Landlord must serve notice on tenant within six months of executing the improvements.

Particulars of Claims

In general, claims arising out of the termination of the lease can be dealt with by arbitration under the Act only if notice of intention to claim in general terms (not full 'particulars') has been given by the claimant before the expiration of two months from the termination of the tenancy. If such notice is not given, the claims lapse. Notice does not require to be given after if given before termination of the tenancy.

Within 28 days of the appointment of an arbiter, however, each party must deliver to him 'a statement of that party's case with all necessary particulars' (Sixth Schedule, Para. 5 as amended by the Agriculture (Miscellaneous Provisions) Act, 1963, sec. 20). This would appear to indicate that something more than was formerly understood by particulars is necessary. The words 'statement of case' seem to involve what amounts to a narrative of facts and averments corresponding to the condescendence in a case in Court. It is noted that no amendment or addition to the case can be made without the arbiter's consent.

Relevancy and Competency of Claims

An explanatory statement under this heading may be desirable for the information of arbiters, especially as most of them are practical farmers and not lawyers.

ARBITRATION UNDER THE ACT

An arbiter may be interdicted from proceeding where the claim is not relevantly founded[1], but the Secretary of State cannot be interdicted from appointing an arbiter—see sec. 21 (1) (a) of the Crown Proceedings Act, 1947, as applied to Scotland by sec. 43 (a) of that Act. Where, however, it is thought that the Secretary of State is about to exceed his powers it is open to either party to seek an order in the Courts declaratory of his rights. 'When such an order is asked it should be formulated with precision.'[2]

A claim or question to be competent must be one which falls to be dealt with by arbitration under the Act, and, accordingly, it must arise in connection with the tenancy of a holding as defined in the Act; further, it must be between the landlord and tenant of such a holding. This excludes questions and claims between away-going and incoming tenants, even for compensation of the kinds embraced under the Act. See sec. 11, 1949 Act.[3] If an incoming tenant were taken bound to settle the away-going tenant's claims against his landlord for compensation under the Act, the outgoing tenant can refuse to deal with the incoming tenant, and insist on dealing with the landlord against whom the claims fall to be made in terms of the Act. Where a tenant is bound to hand over his waygoing crops to the landlord or incoming tenant he, as the debtor in an alternative obligation, is entitled to elect with whom he will deal.

All claims between landlord and tenant in connection with the holding, whether arising out of the termination of the tenancy, or during its currency (except a difference as to liability for rent) are now referred to arbitration in terms of sec. 74.[4]

Questions relating to the valuation of sheep stocks, dung, fallow, and general away-going valuations are excluded from the arbitration procedure of the Act, under sec. 75 (4).

The subject of relevancy frequently provides questions of difficulty. In litigation, the pursuer must state a relevant case, i.e., his averments of fact, as stated, must be such as, if established, would justify a decision in his favour. Assuming a relevant case for the

---

[1] *Roger* v. *Hutcheson*, 1921 S.C. 787; 1922 S.C. (H.L.) 140.

[2] *MacCormick* v. *Lord Advocate*, 1953 S.C. 396 per Lord President Cooper at p. 409.

[3] *Sinclair* v. *Clynes' Tr.*, 1887, 15 R. 185; *Hamilton Ogilvy* v. *Elliot*, 1904, 7 F. 1115; *Donaldson's Hospital* v. *Esslemont*, 1925 S.C. 199; *Cowdray* v. *Ferries*, 1919 S.C. (H.L.) 27; *Hoth* v. *Cowan*, 1926 S.C. 58.

[4] *Brodie* v. *Ker*, 1952 S.C. 216. See also 'Claims and Questions which are referred to Arbitration,' *supra*, p. 47.

pursuer, the defender would have decree given against him unless he states a relevant defence, i.e., a defence which, if established, would effectively answer the pursuer's claim.

Though arbitrations under the Act are not litigations, and need not be conducted with the same rigid adherence to form and procedure, nevertheless, it is the duty of the arbiter to ascertain if there be a relevant claim, for, as soon as he is satisfied to the contrary, he ought to rule it out, and thus save parties the trouble and expense of needless proof and other procedure. As an example, suppose a tenant were to claim for improvement by sowing wild white clover (to which many farmers attach importance), the arbiter would have no difficulty in seeing at once that such an improvement is not specifically included as an improvement in the Schedule to the Act. He might, however, consider whether, consistently with the terms of the claim as stated, it could be taken into consideration under the provision for compensation for temporary pasture. If—but only if—the claim were well based on that provision, he would be entitled to hold it relevant; otherwise not—in the latter case for the reason that there would be no statutory authority for it.[1]

Arbiters are entitled to satisfy themselves that they have jurisdiction to proceed.[2] If an arbiter is inclined to decide that he has no jurisdiction, it is most important that he make a proposed finding to that effect in order that the point may be tested in Court.

APPOINTMENT OF ARBITER

The parties are entitled to agree to the appointment of any person as arbiter. Agricultural questions are usually submitted to a practical man, but legal questions, such as the interpretation of a lease, are sometimes submitted to counsel or solicitors. Many arbitrations involve both types of questions when the usual course is to have a practical arbiter advised by a lawyer as clerk.

Where the parties—landlord and tenant—are unable to agree with reference to claims and questions under the Act, or under the lease, there is provision for arbitration before an arbiter agreed on by them or appointed by the Secretary of State on application

---

[1] *Brodie-Innes* v. *Brown*, 1917, 1 S.L.T. 49.
[2] *Christopher Brown Ltd.* v. *Genossenschaft Oesterreichischer Waldbesitzer, etc.*, 1953, 2 All E.R. 1039.

by either of them. Claimants should adopt the safeguard provided by sec. 68 (4) of applying for the appointment of an arbiter.[1] The leading provisions are contained in secs. 68, 74 and 75 and the Sixth Schedule. These provisions, however, operate under other sections which deal with particular claims and questions. Where an application is made to the Secretary of State to appoint an arbiter, the appointment must be made from a panel of arbiters drawn up by the Lord President of the Court of Session in consultation with the Secretary of State in accordance with sec. 76. It was held that in making such appointment the Department was acting in an administrative capacity and that its selection of an arbiter from the Panel could not be challenged.[2] Once the Secretary of State has made the appointment his functions end and he has no further duties. The appointment of an arbiter by the Secretary of State does not imply either that the arbiter has jurisdiction to deal with all the possible disputes between the parties or that any claims made are valid.

In sec. 77 of the Act special provisions are made in cases where the Department of Agriculture are landlords in order that they shall not require to act in their own case. Accordingly, any matter in which they require to act will be dealt with by an arbiter appointed by the Land Court.

Appointments must be in writing. The document of appointment should be dated and forwarded to the arbiter. An agreement to appoint a named arbiter is not an appointment.[1] An appointment determined by drawing one of several names by lot is not objectionable.[3] In any case the arbiter must observe the rules laid down by the Act.

The Department has been more disposed to appoint than to decline to do so. It has been their custom to appoint on all *bona fide* applications, and where there was *ex facie* a claim or question between landlord and tenant in connection with the holding. In this attitude they received encouragement from a dictum of Lord President Dunedin.[4] Where the dispute is whether a man is in fact a tenant it is thought that the determination of such a question would be for the Courts.[5] A landlord or tenant who was anxious

[1] *Chalmers Property Investment Co., Ltd.* v. *MacColl*, 1951 S.C. 24.
[2] *Ramsay* v. *M'Laren and ors.*, 1936 S.L.T. 35.
[3] *Smith* v. *Liverpool and London and Globe Insurance Co.*, 14 R. 931.
[4] *Christison's Trs.* v. *Callender-Brodie*, 1906, 8 F. 928.
[5] *Brodie* v. *Ker*, 1952 S.C. 216 at p. 224.

not to have an arbiter appointed to deal with such a question might require to invoke the Crown Proceedings Act, 1947, sec. 21. It would appear that the Secretary of State ought to satisfy himself whether the case falls under the exception (sec. 75 (4)). It should be noted that sub-sec. (4) does not apply where the stock belongs to the landlord.

An arbiter appointed by the parties is in all respects—except two—in the same position, and must act in the same way, as an arbiter appointed by the Secretary of State—that is to say, he must act conform to the Sixth Schedule to the Act. The exceptions are that the fee of an arbiter appointed by the Secretary of State must be fixed by him and that an arbiter appointed by the Secretary of State is a Tribunal within the meaning of the Tribunals and Inquiries Act, 1958, and can be required to give reasons for his award.

If the arbiter dies, or is incapable of acting,[1] or for seven days after notice requiring him to act fails to do so, a new arbiter may be appointed to take his place, just as if no arbiter had been appointed (Sixth Schedule para. 2).

In an arbitration under the Agricultural Holdings (Scotland) Acts, 1923 and 1931, it was held that the arbitration still subsisted notwithstanding the death of one of the parties.[2]

Every appointment (whether by the parties or by the Secretary of State), notice, revocation, and consent must be in writing, and an appointment is irrevocable except of consent.

The better practice is for an arbiter to accept office in writing, but this is not strictly necessary.[3] Once he has accepted office he is not entitled to resign, and, unless parties agree to revoke his appointment, he must apply to the Court for leave to resign on cause shown, e.g., ill-health, intention to go abroad for a long time, or emerging interest.[4]

The Court of Session alone can compel an arbiter to proceed with his task.[5] The proper course is by an action in the Outer House.[6] On the other hand, interdict is competent if the arbiter

---

[1] From any cause. See *Dundee Corporation* v. *Guthrie*, 1969, S.L.T. 93.
[2] *Robert Coutts* v. *William Smith's Trs. and Hugh Kellas*, 1940 (not reported).
[3] Formal acceptance of office not necessary. *Sheriff* v. *Christie and anor.*, 1953 Sh. Ct. Rep. 88.
[4] *Edinburgh and Glasgow Railway Co.* v. *Miller*, 1853, 15 D. 603; Erskine's Inst. IV, 3, 20.
[5] *Forbes* v. *Underwood*, 1886, 13 R. 465.
[6] *Watson* v. *Robertson*, 1895, 22 R. 362; see also *Sinclair* v. *Fraser*, 11 R. 1139.

Disqualification and Removal of Arbiter

An arbiter may be removed by the Sheriff where he 'misconducts' himself by refusing to act on the opinion of the Court.[1] Misconduct here is evidently misconduct in the arbitration, but in extreme circumstances the Court would remove an arbiter on personal grounds, such as conviction of grave crime.

An arbiter is not disqualified by having an interest in the subject in dispute if, where that interest existed previous to his appointment, the fact is known to both parties and neither takes objection.[2] If his interest supervened at a subsequent date, he would be disqualified in the absence of the parties agreeing to his continuance in office.[3] In any event the interest must be substantial.

Acting as adviser to one of the parties in a different matter is no disqualification.[4] It is, however, not desirable for one of the parties' advisers to act as arbiter if he has already applied his mind to the particular dispute in issue. He must be in a position to approach questions submitted with an open mind. It would probably be held that an arbiter whose award had been reduced could not act again on the ground that he could not free his mind of his earlier experience. An arbiter has been interdicted from acting when he had previously made a record of the holding and improperly expressed therein an opinion on the subject matter of the dispute on which he was subsequently called on to arbitrate. On the other hand the mere making of a record is not in itself a disqualification.[5]

An objection to the competency of an arbiter acting, or to his conduct in the arbitration, or to the competency or propriety of the proceedings must be timeously stated, otherwise it will be rejected.[6]

---

[1] *Mitchell-Gill* v. *Buchan*, 1921 S.C. 390.
[2] *Sellar* v. *Highland Railway Co.*, 1919 S.C. (H.L.) 19.
[3] *M'Kenzie* v. *Clark*, 1828, 7 S. 215.
[4] *Addie & Sons* v. *Henderson and anor.*, 1879, 7 R. 79.
[5] *Sheriff* v. *Christie and anor.*, 1953 Sh. Ct. Rep. 88.
[6] *Johnston* v. *Cheape*, 17th Dec., 1818, F.C.; *Drew* v. *Drew*, 1851, 14 D. 212.

Minute of Submission

There is no provision in the statute for the execution of a formal minute of submission. Where the arbiter is appointed by the Secretary of State, the form of appointment constitutes the only submission that is necessary. That form always follows on and embodies the claim or question at issue as set forth in the application for the appointment of the arbiter.

Where the arbiter is appointed by the parties, it is, of course, necessary to have a joint minute of appointment, because the Schedule provides that the appointment of an arbiter must be in writing. The form may be similar to that issued by the Department, but no particular form is required. It is, however, necessary to state clearly what claims and questions are submitted to arbitration, and it should always be stated that the arbitration is one under the Act. The arbiter must, of course, confine himself to the questions and claims stated, unless the parties agree to his going further. Obviously, without such agreement, an arbiter appointed to deal only with a claim for compensation for, say, improvements cannot deal with a claim for compensation for disturbance.

A submission should bear a 10s. stamp.

(For Form of Appointment of Arbiter by the Secretary of State, see Appendix).

Procedure in Arbitration

The arbiter has a very wide discretion as to his procedure. He must decide what he considers are the subjects and questions submitted to him, but in the last resort it lies with the Court to decide matters of jurisdiction.[1] An arbiter may be interdicted if he proposes to deal with matters not referred to him, or if there be no question to try. Interdict would probably not be granted against findings unless the arbiter made it clear that, despite representations, he was to incorporate the findings in his award. The jurisdiction of the arbiter may be extended by the actings of parties when he has entered on the business of the submission, but writing is always preferable. As a rule, he should follow the order

---

[1] *Adams* v. *Gt. North of Scotland Railway Co.*, 1890, 18 R. (H.L.) 1. See *Christopher Brown Ltd.* v. *Genossenschaft, etc.*, 1953, 2 All E.R. 1039.

adopted in the Courts—not that the Courts will insist on anything like rigid adherence to such procedure. The arbiter should not overlook the fact that the arbitrations are statutory, and should be careful to adhere to the code of procedure laid down in the Sixth Schedule. Over and above everything, he should be careful to be strictly impartial, to show that he is impartial and to do nothing which might be calculated to prejudice either party or to defeat the ends of justice.

(1) *Appointment of Clerk*

In terms of paragraph 18 of the Sixth Schedule an arbiter may not include a sum in respect of remuneration or expenses of a clerk in the arbitration expenses unless the clerk was appointed (a) after submission to him of the claims and answers,[1] and (b) with either the consent of the parties or the sanction of the Sheriff. The only decision on the question indicates that the appointment of a clerk will be approved if there are questions of a legal character which would properly be referred to an assessor. If an application has to be made to the Sheriff, the statement of the case should be lodged and intimation of the application should be served on the parties to give them an opportunity to oppose it. The application should narrate the appointment, the lodging of the claims, and the reason, such as the complexity of the case or the legal questions involved, which necessitates the services of a clerk being retained.

(2) *Receiving Claims and Objections*

With the statement of case and particulars of claim before him, the arbiter will be in a position to decide on further procedure. Sometimes a period for adjustment of pleadings is allowed, but this may not be necessary. Para. 5 of the Sixth Schedule provides that each of the parties to an arbitration must, within 28 days from the arbiter's appointment, deliver to him 'a statement of that party's case with all necessary particulars'. No amendment or addition to this is allowed after the period of 28 days has elapsed without the arbiter's consent and parties are confined to the

---

[1] There is no provision for answers in the Sixth Schedule, paragraph 5.

matters stated in the claim.[1] The arbiter may ask for production of documents (leases, records, &c.), and may consider that answers to the claim are necessary. Formerly he could issue an order to that effect but there is no specific authority for this procedure under the present Act. If answers are allowed it should be of consent. It is obviously desirable in many cases to have the parties' respective averments of the fact incorporated into the document like a Court Record but the co-operation of both parties is required for this, as there is no statutory warrant for it.

(3) *Hearing and Inspection*

The arbiter's next step will probably be to fix a hearing of the parties, and an inspection of the farm. In certain circumstances he is not even bound to hear parties,[2] and an inspection may be dispensed with.[3]

In many cases the first hearing may, with advantage, be at the farm, and, unless the parties are sharply in conflict on important facts and questions of law, a very informal hearing may suffice. After an informal hearing and inspection (where necessary), the arbiter may, in very simple cases, be sufficiently informed to enable him to issue his award,[4] There is, however, a danger in allowing proceedings to be too informal and if there are questions of fact in dispute it is better to proceed with a proof with opportunity for examination and cross-examination. The matters which fall within an arbiter's jurisdiction are now more important than ever before. In rent arbitrations a proof is desirable.[5]

If notice be issued for a hearing and inspection, the arbiter should not *then* take any evidence except with the consent of both parties, because one party might say he had had no intimation that evidence was to be taken, and that, therefore, he had not prepared for it. The inspection is, of course, for the purpose of enabling the arbiter to see the condition of the farm, its buildings, fences, and drains, &c., so far as necessary to enable him to deal

---

[1] Matters not in the statement are inadmissible even if raised in evidence by the other party: *Stewart* v. *Brims*, 1969 S.L.T. (Sh. Ct.) 2.

[2] *M'Nair's Trs.* v. *Roxburgh*, 1855, 17 D. 445; *M'Kenzie* v. *Hill*, 40 J. 499; *M'Gregor* v. *Stevenson*, 9 D. 1056 (repair of houses and fences).

[3] *Ledingham* v. *Elphinstone*, 1859, 22 D. 245; *Logan* v. *Leadbetter*, 1887, 15 R. 115.

[4] *Davidson* v. *Logan*, 1908 S.C. 350; *Nivison* v. *Howat*, 1883, 11 R. 182.

[5] *Secretary of State for Scotland* v. *Young*, 1960 S.L.C.R. 31.

with the claims. He should have the fields, and parts of buildings, fences, or drains, in respect of which claims are made, pointed out to him. The Court has held that the mere fact that the tenant met the arbiters upon the ground and gave them information outwith the presence of the landlord was not sufficient reason for reducing an award,[1] but that is a practice which should be avoided. So far as the arbiter can see anything with his own eyes, he is entitled to refuse to listen to evidence of or concerning it, but this step should rarely be taken. He cannot, however, refuse to receive evidence of previous condition (which he could not himself be aware of) if a comparison of past with present conditions be relevant.

As a rule, the arbiter should not proceed with the arbitration or even with the inspection in the presence of one only of the parties.[2] If, however, due notice has been given, a party cannot complain if the arbiter proceeds in his absence.[3]

(4) *Evidence*

While there are cases in which no proof is required or should be allowed—and the arbiter is generally entitled to decide whether, and what, proof should be taken—there are cases in which it would be wrong to refuse to hear evidence.[4] Wherever relevant facts, not within his knowledge, are in question, proof should be allowed, and he may at any time before making his award permit or require the production of further evidence.[5] In fact, the arbiter should never fail to follow such procedure as is essential to justice.[6] In one case the Court set aside an award on account of the arbiter taking into account evidence by one side only where he had decided no proof was necessary.[7] In the Sheriff Court an arbiter was held justified, on the evidence of one witness and after an inspection of the farm, in holding that the tenant was justified

---

[1] *Campbell* v. *M'Holm*, 1863, 2 M. 271.

[2] *In re O'Connor and Whitelaw's Arbitration*, 1919, 88 L.J. K.B. 1242; *Earl of Dunmore* v. *M'Inturner*, 1835, 13 S. 356.

[3] *Drew* v. *Drew*, 14 D. 559.

[4] *e.g.* rent arbitrations; *Secretary of State for Scotland* v. *Young*, 1960 S.L.C.R. 31.

[5] *Bignall* v. *Gale*, 1841, 10 L.J. C.P. 169.

[6] *Ledingham* v. *Elphinstone*, 1859, 22 D. 245; *Mowbray* v. *Dickson*, 1848, 10 D. 1102; *Macdonald* v *Macdonald*, 1843, 6 D. 186; *Holmes Oil Co., Ltd.* v. *Pumpherston Oil Co., Ltd.*, 1891, 18 R. (H.L.) 52.

[7] *Mitchell* v. *Cable*, 1848, 10 D. 1297.

in refusing to accept the landlord's offer to withdraw the notice to quit.¹ No evidence should be taken except at a place, date, and hour, of which both parties have had reasonable notice. If the arbiter has ruled out any question or claim as irrelevant, he should not hear evidence on it. He may, however, defer his decision on relevancy until after hearing evidence, and, where he finds anything relevant, he must allow proof on it, provided he has not all the necessary facts within his own knowledge as an expert. If proof be allowed, the parties must also be heard on it, should they so desire. Questions of relevancy may come up at the proof if there is no preliminary hearing on the question and may either be disposed of or 'proof before answer' may be allowed.

The arbiter may rely on his personal knowledge of facts relevant to the question at issue. He is not bound to hear expert evidence on any matter on which he is able to judge without such evidence,² but the better practice is to hear the evidence in case the arbiter's *prima facie* view is mistaken.

He may take the opinion or advice of a man of skill, such as an architect, builder, or engineer.³ He is not, however, bound to act on such opinion or advice. He cannot bind himself in advance to adopt the report of any expert, for that would practically amount to substituting an expert for himself as arbiter. If he has appointed a layman as his clerk, the arbiter is entitled to call in the services of a law agent for his guidance on questions of law, and for framing a stated case, award, or other document.⁴ He is not bound by the views either of his clerk or of such agent. With the consent of parties, he can submit a question of law to counsel, but parties would not necessarily be bound by counsel's opinion.

The arbiter is not bound to have evidence taken down in shorthand, but where both parties request him to do so, should have this done.

The parties to the arbitration, and all persons claiming through them respectively, must, 'subject to any legal objection, submit to be examined on oath or affirmation in relation to the matter in dispute' (Sixth Schedule para. 6).

---

¹ *Buchanan* v. *Smith*, 1929 Sh. Ct. Rep. 299.

² *Fletcher* v. *Robertson*, 1918, 1 S.L.T. 68; *Macdonald* v. *Macdonald*, 1843, 6 D. 186; *North British Railway Co.* v. *Wilson*, 1911 S.C. 730; *Paterson* v. *Glasgow Corporation*, 1901, 3 F. (H.L.) 34.

³ *Caledonian Railway Co.* v. *Lockhart*, 1860, 3 Macq. 808.

⁴ *Burnet* v. *Henry*, 1866, 5 M. 96.

'Parties claiming through them respectively' may possibly apply to cases where the incoming tenants take over the obligations of the landlord, or where assignees come in place of the landlord or the tenant.

'Subject to any legal objection' would appear to refer to confidentiality, mental competency, or the age of the witness. It is thought that, just as in ordinary judicial proceedings, agent and client are not bound to disclose communications between them in relation to the subject-matter of the arbitration.

'On oath or affirmation.'—The parties and persons claiming through them 'must' submit to examination on oath or affirmation, and it would appear that the arbiter should insist on their doing so.

As regards witnesses, there is no absolute 'must' under the Schedule, which reads: 'The arbiter shall have power to administer oaths and take the affirmation of parties and witnesses appearing, and witnesses shall, *if the arbiter thinks fit,* be examined on oath or affirmation.'

In both cases the party may choose between oath and affirmation.

The usual form of oath is as follows:—

'I swear by Almighty God that I shall tell the truth, the whole truth, and nothing but the truth.'

and the corresponding form of affirmation—

'I solemnly, sincerely, and truly affirm and declare that I shall true answer make to all questions as shall be asked of me touching the matters in difference in the arbitration.'

The parties and witnesses must also produce all samples, books, deeds, papers, accounts, writings, and documents within their possession or power respectively, which may be required or called for, or do all other things which, during the proceedings, the arbiter may require. This is all subject to any legal objection (already referred to) and to relevancy to the question at issue.

When a witness objects to produce documents on the ground of confidentiality, the party who has obtained an order from the arbiter for the recovery of documents may apply, by summary petition, to the Sheriff to compel production. The Sheriff, in dealing with the application, will determine the question of confidentiality. An application may also be made to the Sheriff to order a witness to produce documents for the recovery of which diligence had been granted.[1]

It is thought that a formal specification of documents called for is not necessary, though it may be desirable to have one, especially where the production of numerous documents is required. In any case, an order by the arbiter for production must be obtained before applying to the Sheriff.

If application for warrant to cite witnesses and havers (i.e. person holding documents or other productions) be made to the Sheriff Court, it should be by initial writ; if to the Court of Session, by petition; and in the latter case, where it is desired to obtain the evidence of witnesses in England, application should be made for the appointment of a commission for that purpose. The Court will not grant warrant to cite a witness resident out of Scotland to attend before an arbiter in Scotland.[2] Reference may be made to the undernoted cases where witnesses reside in other sheriffdoms of Scotland.[3] Where the application is by either party, it should be made with the concurrence of the arbiter.[4]

Parties are entitled to be present along with their agents and counsel during the whole proceedings, notwithstanding that they may intend to give evidence themselves. Unless the parties otherwise agree, the arbiter should follow the usual judicial course of excluding all witnesses (other than the parties or their agents) not examined while the evidence of other witnesses is being taken. A request is frequently made to allow expert witnesses to remain during the proof. This may shorten proceedings, but if there is opposition to the request it is better for the experts to withdraw. A witness who gives false evidence in an arbitration is guilty of perjury, and may be punished (False Oaths (Scotland) Act' 1933, sec. 8, which repealed sec. 16 (4) of the 1923 Act).

STATED CASE

The arbiter is, generally speaking, final on questions of fact, but on questions of law, or mixed law and fact, he may, either on his own initiative or at the request of either party, and must, if so ordered by the Sheriff on the application of either party, state a

---

[1] *Blaikie* v. *Aberdeen Railway Co.*, 14 D. 590.
[2] *Highland Railway Co.* v. *Mitchell*, 1868, 6 M. 896.
[3] *Caird*, 1865, 3 M. 851; *Harvey* v. *Gibsons*, 1826, 4 S. 809.
[4] *Crichton* v. *North British Railway Co.*, 1888, 15 R. 784.

ARBITRATION UNDER THE ACT

case for the opinion of the Court (Para. 19 Sixth Schedule). In England it has been decided that a stated case cannot be made after the final award is issued.[1]

It is sometimes difficult to determine whether a question of law is involved. The interpretation of statute is a matter of law, but interpretation of a lease or agreement may involve questions of fact and law. It is a question of law whether facts found are sufficient to justify the arbiter in arriving at a particular decision, e.g., whether a tenancy has been terminated.[2] In general, the Court will not interfere, even though the arbiter has erred, where a pure question of fact alone is involved. Where it is clear that the arbiter had facts before him on which he might competently arrive at his decision the Court will not interfere.[3]

If the arbiter is of opinion that no question of law has arisen, he may decline to state a case, but either party may apply to the Sheriff to have him ordered to do so. It has been held that it is not the function of the Court to entertain the kind of representations appropriate before an arbiter.[4]

The duty is on the arbiter to state the case, which is generally drafted by his clerk, and should be submitted to the parties for revisal.

A case should be in three parts, viz.:—
1. A narrative containing a clear statement of the relevant facts found proved or admitted.
2. The contentions of parties.
3. The question or questions of law.

No. 1 should be confined to such facts as are relevant to the question of law, and no superfluous narrative should be given. If either party requires a particular fact to be stated, and the fact has been proved or admitted, the arbiter should state it unless he is clearly satisfied that it is not relevant. If the arbiter should decline to state a material and relevant fact, which has been proved or admitted, he may be ordered to do so by the Sheriff. It is important to note that to state the facts does not require an

---

[1] *Tabernacle Permanent Building Society* v. *Knight,* 1892 A.C. 298. See also *Johnston* v. *Glasgow Corporation,* 1912 S.C. 300, and *Steele* v. *M'Intosh Bros.* (1879) 7 R. 192.

[2] *Euman* v. *Dalziel & Co.,* 1912 S.C. 966.

[3] *Nelson* v. *Allan Brothers & Co. (U.K.) Ltd.,* 1913 S.C. 1003; *Henderson* v. *Corporation of Glasgow,* 1900, 2 F. 1127.

[4] *Chalmers Property Investment Company Ltd.* v. *Bowman,* 1953 S.L.T. (Sh. Ct.) 38.

elaborate recital of the evidence. The bare facts based on the evidence led, and on which the questions of law depend, should be stated by the arbiter as he finds them,[1] but the arbiter should state enough to make it plain how the question of law arises.[2]

No. 2 should be the contentions of the parties as nearly as may be in their own words—or altered by the arbiter merely for the purpose of making them free from ambiguity.

No. 3 should generally be in the precise terms in which the question is put forward by the parties, though there may be variation or adjustment by the arbiter for the purpose of making the point at issue clear.

If the case, as stated, is ambiguous, or if it be not in terms to enable the question of law to be fairly dealt with, the Sheriff may remit it back to the arbiter for amendment.[3] In England a case was remitted back to enable one party to instruct counsel where the other party had done so.[4] It may be open to doubt whether the Courts in Scotland would follow the last case.

It would be misconduct on the part of the arbiter if, when requested, he refused to state a case or to delay his award till an application could be made to the Court, provided that his findings upon a question of fact did not render immaterial the question of law.[5]

When the case had been adjusted and signed, it falls to be lodged in Court by the arbiter's clerk, along with a fee of fifteen shillings, and thereafter the arbiter need take no further steps regarding it till either party informs him of the Court's decision.

The Sheriff's decision is final, unless an appeal is taken by either party to the Inner House of the Court of Session, within twenty-one days, as provided by the Rules of Court, V. 268.

Where a case was stated and answered, it was held that a further case was incompetent regarding questions suggested by the decision on the first case arising out of other claims acquiesced in at the stating of the earlier case.[6]

---

[1] *North and South-Western Junction Railway Co.* v. *Brentford Union,* 1888, 13 A.C. 592.
[2] *Windsor Rural District Council* v. *Otterway & Try,* 1954, 3 All E.R. 721.
[3] *Lendrum* v. *Ayr Steam Shipping Co.,* 1914 S.C. (H.L.) 91: *Palmer & Co.* v. *Hosken & Co.,* 1898, 1 Q.B. 131.
[4] *Whatley* v. *Morland,* 3 L.J. Ex. 58.
[5] *Buerger & Co.* v. *Barnett,* 1920, 35 T.L.R. 260.
[6] *Earl of Galloway* v. *M'Clelland,* 1917 Sh. Ct. Rep. 351.

NOTE OF PROPOSED FINDINGS

It is usual for the arbiter to issue a note of his proposed findings, though he is not bound to do so.[2] It is generally prudent to issue such a note, and, if desired, to hear parties or their agents thereon, or on the representations with reference thereto,[3] although the arbiter is not bound to do so.[4]

A note of proposed findings is generally expedient where questions of law are involved, because it is desirable that parties should know how the arbiter proposes to answer such questions, in order that they may have an opportunity of applying for a stated case for the opinion of the Court. The note of proposed findings, which may be in the form of a draft award, should be stated in clear language—preferably in numbered paragraphs arranged in logical order, and ought to cover all the questions included in the arbitration. It is sometimes convenient to include the arbiter's proposals regarding expenses. It is usual to allow a week to a fortnight for making representations.

The Tribunals & Inquiries Act, 1958, applies to agricultural arbitrations. The effect of sec. 12 of the Act is that an arbiter, appointed by the Secretary of State, is bound to furnish a statement of the reasons for his decision, if requested to do so by a party to the arbitration. Since it is the established practice of arbiters to issue a note along with their proposed findings or awards and since there is access to the Courts on questions of law the provisions of this Act did not greatly innovate on previous arbitration practice.

There may however be questions as to the adequacy of reasons given. It has been held that reasons where given are to be treated as part of the award and must be adequate and substantial and that if they do not comply with the statutory provisions there is an

---

[1] *Mitchell-Gill* v. *Buchan*, 1921, S.C. 390.
[2] *M'Callum* v. *Robertson*, 1826, 2 W. & S. 344.
[3] *M'Callum* v. *Robertson, supra; Baxter* v. *Macarthur*, 1836, 14 S. 549.
[4] *Morison* v. *Thomson's Trs.*, 1880, 8 R. 147.

error of law on the face of the award.[1] On the other hand it has been pointed out that 'it must have been realised that certain tribunals could do no more than state that the case was not made or that the figure was a certain sum;[2] and in a case regarding an award which found that a tenant had failed to remedy a breach of his lease it was held that there was no need to detail the precise extent of the failure to remedy the breach or reasons for minor findings of fact.[3]

Expenses

The expenses of and incidental to the arbitration and award are in the discretion of the arbiter, who is entitled to direct to, and by, whom, and in what manner, the expenses or any part thereof shall be paid.

The arbiter must deal with the expenses in his award, otherwise it may be set aside.[4] It is sufficient, however, if he do so in the usual way by setting forth the proportions of expenses payable to or by the respective parties.

In the general case, the expenses commence with the appointment of the arbiter, and do not include anything done prior to that, except (1) the procedure for making the appointment; and (2) the framing and lodging of the claim. It is unlawful to include in the expenses of a statutory arbitration any expenses to a clerk unless such appointment was made after submission of the claim and answers to the arbiter and with either the consent of the parties or the sanction of the Sheriff. The expense of preparing a claim for compensation for disturbance is expressly referred to in sec. 35 (2). It is thought that expenses do not extend to the preceding statutory procedure such as giving notice of intention to claim (where notice is required), nor to the making of the claim, because these are essential conditions of, and precedent to, a claim, and not expenses of, or incidental to, the arbitration. They may be said to correspond to extrajudicial expenses.

The arbiter has a general discretion in dealing with expenses and

---

[1] *Re Poyser and Mills Arbitration* [1964] 2 Q.B. 467: *See also Earl of Iveagh* v. *Minister of Housing and Local Government* [1961] 3 All E.R. 98.

[2] *Parker L. J. in ex parte Woodhouse* (Times 18th June 1960).

[3] *Horswell* v. *Alliance Assurance Co.* [1967] 205 E.G. 319.

[4] *Richardson* v. *Worsley*, 5 Ex. 613.

he is not bound to follow the procedure of the Courts regarding them.[1] His discretion must be properly exercised. Where he departs from the settled practice of the Courts, that of awarding expenses to the successful party, such an exercise of his discretion must be judicial and not capricious.[2] He need not, however, state in terms the reason for his departure from the usual practice.[3] The House of Lords held that it was not incompetent for the arbiter to find the successful party liable in expenses where he claimed £740 for dilapidations and was only awarded £71.[4]

The arbiter is directed to take into consideration, in dealing with expenses, the reasonableness or unreasonableness of the claim of either party, either in respect of amount or otherwise and any unreasonable demand for, or refusal to supply, particulars, and generally all the circumstances of the case, and he may disallow the expenses of any witness whom he considers unnecessarily called, and any expenses he may consider to have been unnecessarily incurred. This direction is a salutory one, and is obviously intended to put a check on the too common practice of making extravagant claims. Unfortunately, few arbiters appear to act on it effectively. The direction should, of course, be applied in a commonsense way. Parties cannot be expected to estimate their claims with exactitude, and, in some cases, they may be greatly in excess of the amount ultimately awarded. But it is thought that the arbiter is entitled to consider whether there was a reasonable ground for stating the claim as lodged. In a recent case it was observed that the arbiter in considering his award as to expenses must direct his mind to the most important element in the case.[5]

The expenses are subject to taxation by the Auditor of the Sheriff Court, on the application of either party, and the taxation is subject to review by the Sheriff.

It is thought than an arbiter should not award expenses on the basis of agent and client.[6]

The remuneration of an arbiter appointed by the Secretary of State is to be fixed by him, and, if so fixed, it is thought that there

[1] *Breslin* v. *Barr & Thornton*, 1923 S.C. 90.
[2] *Lewis* v. *Haverfordwest R. D. C.* [1953] 2 All E.R. 1599; see also *Smeaton Hanscomb & Co.* v. *Sassoon I. Setty Son & Co.* (No. 2) [1953] 2 All E.R. 1588, and *Heaven & Kesterton Ltd.* v. *Etablissements Francois Albiac et Cie* [1956] 2 Loyds Rep. 316.
[3] *Perry* v. *Stopher* [1959] 1 All E.R. 713.
[4] *Gray* v. *Ashburton* [1917] A.C. 26.
[5] *Smeaton Hanscomb & Co. Ltd.* v. *Sassoon I. Setty Son & Co., supra.*
[6] *Griffiths* v. *Morris* [1895] 1 Q.B. 866.

can be no taxation of the amount by the Auditor of the Court of Session or Sheriff Court. Otherwise, the arbiter's fees, as part of the expenses in the arbitration, are subject to taxation by the Auditor of the Sheriff Court.

The fees, &c., of the clerk are held to be earned, though the arbiter died before issuing an award, and the parties cannot deny the clerk's right to his fee by delaying to take up the award.[1]

The expenses of a stated case fall to be dealt with by the Court, not by the arbiter.[2] These expenses do not include the preparation of the stated case and work incidental thereto, which work is part of the expenses incidental to the arbitration, and ought to be covered by the arbiter's award of expenses, apart from the expenses proper of the stated case. Generally, he should deal with the expenses of preparation of the case consistently with the Court's decision regarding expenses.

In one case, where, contrary to general practice, a conclusion for expenses was sought against the arbiter, he offered not to appear if that conclusion was withdrawn. On this being refused he lodged defences confined entirely to the question of expenses, and was successful.[3] An English case decided that if the arbiter takes part in proceedings to set aside his award, costs can be given against him.[4]

The Award

(Sixth Schedule, 8-15).

The arbiter is required to make and sign his award within two months after his appointment (whether by the parties or by the Secretary of State), or within such longer period as (either before or after the expiry of the period of two months) the parties may agree on, or the Secretary of State may direct. (Sixth Schedule, Para. 8).

The arbiter should be careful to cease acting after the period has expired and until an extension is granted.

A question may arise in the case where the arbiter has issued an interim award which has been acted on, and the arbitration lapses

---

[1] *Jackson* v. *Galloway*, 1867, 5 S.L.R. 130.
[2] *M'Quater* v. *Fergusson*, 1911 S.C. 640.
[3] *Roger* v. *Hutcheson and ors.*, 1921, 58 S.L.R. 546; 1922 S.C. (H.L.) 140.
[4] *Lendon* v. *Keen*, 85 L.J. K.B. 1237; [1916] 1 K.B. 994.

ARBITRATION UNDER THE ACT

in consequence of the award not being made and signed within the statutory period. It is thought that the interim award would stand in that case as in the case where the arbitration lapses by the death of the arbiter after an interim award has been made.

A distinction has been drawn between interim and partial or part awards. It is thought that the provision for interim awards was mainly intended to allow the whole sum payable in respect of any particular head or item of claim to be awarded, and not merely so much to account thereof. If this course be followed, the interim award would be practically final as regards the heads or items embraced in it and regarding which no difficulty need arise in the event of the arbitration lapsing from any cause and a new arbiter being appointed. In order to have this effect, it is well to state in the interim award that though otherwise it is intended to be final as regards the items with which it deals—it is issued under reservation of other items and all questions of expenses.[1] An interim award, which would show clearly that it is interim, may be sustained even though the final award be reduced or invalid.[2]

The award must be in the form prescribed by the Secretary of State (see form in Appendix), varied to meet the circumstances. An award in a form materially different from the statutory form could probably be reduced. Awards must be either holograph or tested.

Alternative awards, e.g., awarding a certain sum in a particular event and a different sum in another event, are not competent under the statute.[3]

The arbiter is now bound (Sixth Schedule, Para. 11 (a)) to state separately in his award, the amounts awarded in respect of the several claims referred to him. This does not mean that he must give a detailed statement, e.g., if there be long and detailed claims for manures and feeding-stuffs, it is thought that he need only give the amount awarded under (1) purchased artificial manures, and (2) feeding-stuffs. There are usually separate items for temporary pasture, for disturbance, manures, feeding-stuffs, &c., and each of these should be regarded as a separate claim, and the same course should be followed regarding the other claims respectively. In short, there should be an item in the award for each claim according to its class and character, and generally following the relative

[1] *Taylor* v. *Neilson*, 1822, 1 S. 253.
[2] *M'Kessock* v. *Drew*, 1822, 2 S. 13.
[3] *Glendinning* v. *Board of Agriculture*, 1917, 54 S.L.R. 234.

language of the Schedules. In an English case an award was set aside for not stating the amounts awarded for each item of claim.[1]

The arbiter must also, if required by either party, specify the amounts awarded in respect of any particular improvement, or any particular matter the subject of the award, and the award must fix a day, not later than one month after its delivery, for the payment of the sum or sums awarded as compensation, expenses, or otherwise. (Paras. 11 (b) and 13).

An arbiter has power at common law to alter or cancel his award, even after it has been signed, so long as it has not been issued. It is thought that this power can be exercised by an arbiter under the Act, so long as the period for making the award has not expired. He is expressly given power to correct any clerical mistake or error arising from an accidental slip or omission. This is supplementary to the common law power already referred to, and is evidently intended to operate even after the award has been issued, but this may not be quite free from doubt.[2] When the award is issued the arbiter is *functus officio*, and if his award does not embrace the matters really in issue, he cannot of his own motion treat it as no award and issue another.[3] In England it was held that an arbiter could supplement his award after issue by a letter disposing of expenses with which he had omitted to deal. A report and valuation of a house was not probative and was held not to be an effective award and could not be sued on.[4] In an arbitration under the Agricultural Holdings Act, however, an award was not probative—i.e., signed before witnesses—but it was held that the claimant could sue on it on proof that the document was in fact the award of the arbiter and oversman.[5]

An award or agreement under the Act as to compensation, expenses, or otherwise, if any sum payable thereunder is not paid within a month after it becomes due, may (if it contains a consent to registration) be recorded in the Books of Council and Session or in the Sheriff Court Books, and is enforceable in like manner as a recorded decree-arbitral (sec. 69).

The award does not now require to be stamped.

[1] *Hamilton* v. *Peeke*, 1926 L.J., C.C.R. 54.
[2] *Sutherland & Co.* v. *Hannevig Brothers* [1921] 1 K.B. 336.
[3] *Re Stringer & Riley Brothers' Arbitration* [1901] 1 Q.B. 105.
[4] *M'Laren* v. *Aikman*, 1939 S.L.T. 267.
[5] *Cameron* v. *Mackay*, 1938 Sh. Ct. Rep. 276.

ARBITRATIONS UNDER THE ACT

REDUCTION OF AWARD

Although the Act and Schedule do not refer to the reduction of awards they may, whether by an arbiter appointed by the Secretary of State or not, be reduced on various grounds.

There is no appeal in the ordinary sense, against an award under the Act, but there are grounds on which an award may be set aside or reduced.

It may be challenged, because (1) it is contrary to the Act of Regulations, 1695; (2) it is bad at common law; (3) it does not provide details required in terms of the Sheep Stocks Valuation (Scotland) Act, 1937, or the Hill Farming Act, 1946.

Under (1) the grounds are corruption,[1] bribery, and falsehood. The Court gives a very wide interpretation to the word corruption, and has allowed a reduction where there was a palpable failure of the arbiter in his duty, or such misconduct as would lead to manifest injustice.[2]

As to bribery, it has been held improper for an arbiter to accept hospitality from one of the parties; and, if the hospitality be given with the object, or have the effect, of inducing the arbiter to act unfairly, the Court will set aside the award; but the mere dining or lunching with one of the parties and the witnesses, in the absence of the other party, will not, of itself, necessarily invalidate the award.[3] Treating the arbiter, where there was no evidence of corrupt intention to influence him, was not enough to reduce the award.[4]

As to falsehood, it is thought that dishonesty or deceit on the arbiter's part, which affected the award, would afford ground of reduction.

The chief grounds on which reduction of an award may be granted at common law are—

(1) Where the award is bad on the face of it, ambiguous, uncertain, or improperly executed according to what formalities are appropriate.

(2) Misconduct or improper conduct of the arbiter, e.g., examining a witness on one side in the absence of the other party or

---

[1] *Mitchell* v. *Cable*, 1848, 10 D. 1297; *Adams* v. *Great North of Scotland Railway Co.*, 1890, 18 R. (H.L.) 1.
[2] *Miller* v. *Millar*, 1855, 17 D. 689; *Cameron* v. *Menzies*, 1868, 6 M. 279.
[3] *Re Hopper* [1867] 2 Q.B. 367; *Moseley* v. *Simpson*, 1873 L.R. 16 Eq. 226; *Crossley* v. *Clay*, 5 C.B. 581; *Harvey* v. *Skelton*, 1844, 13 L.J. Ch. 466.
[4] *Hopper, supra.*

his agent,[1] in which case, however, the award was not reduced because the pursuer had been successful on the point at issue. Failure to issue the award within the time allowed under Para. 8 of the Sixth Schedule is misconduct and a late award was set aside.[2]

(3) Where there has been a mistake which the arbiter admits.[3] An error of judgment is not a ground for reduction.

(4) When the arbiter has acted *ultra vires*—has exceeded his powers by, e.g., including matters not referred to him.[4]

(5) Where the award does not exhaust the reference.[5]

(6) An arbiter cannot make an order *ad factum praestandum*.[6]

In the Sheriff Courts two attempts to upset awards on the ground of misconduct which have been unsuccessful may be mentioned. It was averred in one that the arbiter appointed to fix a new rent had not made an adequate inspection of the subjects.[7] In the other the principal averments were that the arbiter had not called for claims or evidence or had a hearing, but the Sheriff held that the submission was sufficiently specific to inform the arbiter what questions he had to decide and that otherwise his procedure was not open to challenge.[8]

While it is true that an arbiter has a wide latitude as to what procedure he follows and as to what proof he may allow, he will generally be well advised to call for detailed claims and to hesitate before he refuses any not unreasonable request by a party for proof or hearing. He may be competent to judge for himself in many cases, but if the parties be allowed opportunity of stating their own case, they cannot be heard to make any effective complaint at a later stage.

The following dictum of Lord President Clyde is useful as a summary of one aspect of the law: 'If it could be proved that in arriving at his award an arbiter had invented the facts to suit some

---

[1] *Black* v. *Williams & Co.* (*Wishaw*), *Ltd.*, 1924 S.C. (H.L.) 22.
[2] *Halliday* v. *Semple* 1960 S.L.T. (Sh. Ct.) 11.
[3] But see *Welch* v. *Jackson*, 1864, 3 M. 303.
[4] *Carruthers* v. *Hall*, 1830, 9 S. 66; *Steele* v. *Steele*, 1809, 15 F.C. 345.
[5] *Thallon* v. *Wemyss*, 1855, 18 D. 80; *Lawrence* v. *Bristol and Somerset Railway Co.*, 1867, 16 L.T. 326; *Napier* v. *Wood*, 1844, 7 D. 166; *Pollich* v. *Heatley*, 1910 S.C. 469; *Miller & Son* v. *Oliver & Boyd*, 1906, 8 F. 390; *Donald* v. *Shiell and anor*, 1937 S.L.T. 70.
[6] *Chalmers Property Investment Company Ltd.* v. *Bowman*, 1953 S.L.T. (Sh. Ct.) 38.
[7] *Sim* v. *M'Connell and anr.*, 1936 Sh. Ct. Rep. 324.
[8] *Dundas* v. *Hogg and anor.*, 1936 Sh. Ct. Rep. 329.

view of his own, or had fashioned the law to suit his own ideas, then however innocent might be the eccentricity which had seduced him into such a travesty of judicial conduct, his behaviour would naturally imply that justice had not been done; he would be guilty of . . . misconduct, and his award would be reduced.' Moreover, the arbiter must observe certain understood rules of conduct to ensure that injustice is not done. He must act judiciously and not capriciously.[1]

A general submission is limited by the claims of parties and cannot be held unexhausted, simply because a possible claim has not been dealt with. Where one of the parties reserved a competent claim, and the other party did not object, a plea that the award (which contained a reservation of the claim) did not exhaust the submission was repelled.[2]

Where it is possible, from the award itself, to explain it to the effect that part is unobjectionable, that part may be upheld, and only what is objectionable reduced.[3]

An award may be cut down where it is obtained by fraud by one of the parties. Irregularities in conducting the arbitration may be waived by the parties' consent, after the irregularities have been discovered;[4] and parties may, either expressly or by their actings, so extend the limits of a submission as to bar the plea of *ultra vires*.[5]

A refusal to make a reference to oath, the admission of incompetent evidence, oral or documentary, or the rejection of evidence which a Court of law would have admitted, are not always sufficient grounds for reducing an award.[6]

A refusal to state a case is not good ground for reduction because the proper remedy is an application to the Sheriff to order a stated case.

In England it has been held that an award can be reduced (1) where there has been wrongful admission of evidence on a material point;[7] (2) where there has been refusal to admit material evidence;[8] where the arbitrator took evidence in the absence of

[1] *Lewis* v. *Haverfordwest R. D. C.* [1953] 2 All E.R. 1599.
[2] *Perrens & Harrison* v. *Borron*, 1869, 6 S.L.R. 581.
[3] *Kyd* v. *Paterson*, 1810 F.C.; *Napier* v. *Wood*, 1844, 7 D. 166.
[4] *Moseley* v. *Simpson*, 1871 L.R. 16 Eq. 226.
[5] *North British Railway Co.* v. *Barr*, 1855, 18 D. 102.
[6] *Alston* v. *Chappell*, 1839, 2 D. 248; *Mowbray* v. *Dickson*, 1848, 10 D. 1102; *Grant* v. *Girdwood*, 23rd June, 1820, F.C.
[7] *Walford Baker & Co.* v. *Macfie & Sons*, 48 L.J. K.B. 2221.
[8] *Williams* v. *Wallis and Core*, 1914, 2 K.B. 478.

one party and did not deal with a claim submitted to him;[1] (4) where the arbiter heard the evidence of each party in the absence of the other;[2] where the arbitrator inspected the holding in the absence of the parties but accompanied by an architect from whom it was assumed he received evidence.[3]

Where it was contended that an arbiter had shown bias, Lord President Clyde held that the Court ought not to decide a matter concerning the *bona fides* of an arbiter without giving him an opportunity to be heard in the process.[4]

[1] *Re O'Connor and Whitelaw's Arbitration*, 1919, 88 L.J. K.B. 1242.
[2] *Ramsden & Co.* v. *Jacobs*, 91 L.J. K.B. 432.
[3] *Ellis* v. *Lewin* (1963) 107 S.J. 851.
[4] *Bellshill & Mossend Co-operative Society Ltd.* v. *Dalziel Co-operative Society Ltd.*, 1959 S.L.T. 150.

ARBITRATIONS AND VALUATIONS (OUTSIDE THE ACT) GENERALLY BETWEEN AWAY-GOING AND INCOMING TENANTS.

The usual away-going valuations do not fall to be dealt with under the Act but at Common Law. Accordingly it is competent for parties to have these valuations (except for sheep stocks, &c., taken over by the proprietor of an entailed estate under the Entail (Scotland) Act, 1914, 4 & 5 Geo. V. c. 43) made by one arbiter or two arbiters and an oversman as they may decide. In either case the parties must accept the decision of the arbiters or oversman, whether it be in accordance with law or not.[1] There is no power to have a stated case on a question of law, except under the Sheep Stocks Valuation (Scotland) Act, 1937, or under the Hill Farming Act, 1946.

All claims arising out of the tenancy (either during currency or at the termination of the lease) at the instance of landlord or tenant against the other must be dealt with by arbitration under the Act (secs. 68 and 74).

There has long been recognised a distinction between 'valuation' and 'arbitration'. Where 'arbiters' are merely appointed as skilled men to put a value on such things as dung, crops, fallow, or sheep stock, which they can inspect, though called arbiters they are really valuers.[2] The Court will not require them to follow strictly the rules of arbitration, or reduce an award or valuation merely on the ground of irregularity of form or procedure.[3] At the same time, it is generally desirable, where there is a formal minute of submission, to have a formal award. (See forms in Appendix.) As a rule, it is safe to act in accordance with long-established practice in these matters.

It is frequently the case that, in connection with such valuations, questions arise which might appear to be appropriate for settlement in a proper and formal arbitration, but where these questions

---

[1] *Holmes Oil Co., Ltd.* v. *Pumpherston Oil Co., Ltd.,* 1891, 18 R. (H.L.) 52.
[2] *Nivison* v. *Howat,* 1883, 11 R. 182; *Logan* v. *Leadbetter,* 1887 15 R. 115; *Hopper* [1867] 2 Q.B. 367; *M'Gregor* v. *Stevenson,* 1847, 9 D. 1056; *Davidson* v. *Logan,* 1908 S.C. 350.
[3] *Gibson* v. *Fotheringham,* 1914, S.C. 987; *Cameron* v. *Nicol,* 1930 S.C. 1.

are incidental to the valuations and must be dealt with in order to enable the valuations to be made, it is generally in the power of the arbiter (or valuer) to decide them, and even to hear evidence to enable him to do so. Such incidental arbitration questions are expressly excluded from the arbitration provisions of the Act by sec. 75 (4). That sub-section not only excludes the valuations of sheep stocks, dung, fallow, straw, crops, fences, and other specific things the property of an outgoing tenant agreed to be taken over at the termination of a tenancy by the proprietor or incoming tenant, but also any questions which it may be necessary to determine in order to ascertain the sum to be paid in pursuance of such agreement, and that whether such valuations and questions are referred to arbitration under the lease or not. This exclusion has the effect of throwing these questions back on the common law.

In illustration—an arbiter (or valuer) appointed to value sheep stock has been held to have jurisdiction for the purposes of the valuation to determine what is the *bona fide* sheep stock.[1] So also, where there is a reference to arbiters to value dung made after the sowing of the last green crop, in the event of dispute on the point, the arbiters or oversman may hear evidence as to when the dung was made, and base the valuation on the facts as they find them. In short, it is assumed that where arbiters are appointed to value a thing, they are entitled to decide such incidental questions as may be necessary in order to enable them to make the valuation. This is consistent with the principles laid down by the Court.[2]

A submission should specify clearly how much dung is to be taken over. All dung up to one year old should generally be taken over. Arbiters should state in their award (or in an explanatory note) if they are including the value of unexhausted feeding stuffs in their valuation of the dung or if that value is excluded. Different arbiters may have to settle the two questions and the tenant should not receive payment on two counts.

THE SUBJECTS OF COMMON LAW ARBITRATION OR VALUATION

It was formerly common practice for the landlord to assign to his incoming tenant all his rights and claims against the away-going tenant, in respect of failure to maintain and leave buildings, fences, drains, &c., in tenantable repair, and in respect of failure to

[1] *Fletcher* v. *Robertson,* 1919, 1 S.L.T. 260.
[2] *Bell* v. *Graham,* 1908 S.C. 1060.

fulfil the terms of the lease regarding cultivation. Sec. 11 of the Act provides that an agreement between a landlord and an incoming tenant made after 1st November, 1948, shall be null and void if it provides that the latter shall pay to the outgoing tenant compensation under the Act. The landlord and incoming tenant may, however, agree in writing that the incoming tenant shall pay up to a maximum specified sum in respect of compensation for improvements under Part III of the First Schedule, i.e., for the unexhausted value of manure and feeding stuffs and similar claims. There appears to be nothing, however, to prevent the outgoing and incoming tenant making any arrangement they wish with the landlord's consent either tacit or otherwise. It has been customary to refer these matters to two arbiters and an oversman, in a submission between away-going tenant and incoming tenant. Sec. 74 refers comprehensively enough to such matters, but nevertheless they are outside these sections which only apply to questions and claims between landlord and tenant.[1]

An incoming tenant is not under obligation to take over anything at valuation unless, as is usual, he is expressly taken bound to do so, with the exception of (1) grass seeds sown with the last white crop of the away-going tenant,[2] and the cost of harrowing and rolling them. If the tenant fails to sow good seed or to hain the young grass he will be liable; (2) land left in bare fallow for which the away-going tenant is entitled to receive payment;[3] (3) dung. Where a tenant on entry received, without payment, fallow ground prepared with a certain quantity of manure, he was bound to leave the same state of things at away-going, and was held entitled to payment for the surplus only.[4] An away-going tenant is only entitled to the value of such dung as is lawfully withheld from application to the land, and generally only for what is made after the sowing of the last green crop. But there may be variation according to the known custom of the district. In an old case a tenant was bound to consume the whole fodder, except hay and straw of the last crop, and the practice had been to reserve part of the dung to be laid on the wheat crop land in the autumn instead of applying it all in the preceding spring. The outgoing

[1] *Roger* v. *Hutcheson*, 1922 S.C. (H.L.) 140.
[2] *Simson's Trs.* v. *Carnegie*, 1870, 8 M. 811; *M'Intyre* v. *Anderson*, 1872, 10 S.L.R. 59.
[3] *Purves* v. *Rutherford*, 1822, 2 S. 59; *Simson's Trs., supra;* Bell's Prin., 1263.
[4] *Brown* v. *College of St. Andrews*, 1851, 13 D. 1355.

tenant was held entitled to compensation for (a) the reserved dung left on the farm unused at the outgoing Whitsunday, being no more than he had been in use to reserve; and (b) for straw, being no more than would have been required to fodder his bestial till separation of crop.[1]

All these rules, which are largely based on equity, may be altered by contract

If the entry be a Whitsunday one, the incoming tenant is sometimes bound to take over, at a valuation, first and second years' grass, fallow (rent and labour of ploughings), dung, straw (sometimes also grain) of the away-going crop, and occasionally subdivision fences or threshing-mill. He may also have to pay for the potato or turnip crop based on the cost of seed, labour, and manure applied at the date of quitting. In some cases the straw of the away-going crop is steelbow, and must be left without payment, while, in other cases, the incoming tenant must take the whole away-going crop. Difficulties have been experienced with the introduction of combine harvesters which leave the straw on the ground. The following suggestions for dealing with the valuation of such straw may be considered:—

1. Any provision in the lease as to the delivery of steelbow straw must be observed.
2. If the lease makes no such provision well established custom must be observed.
3. Consequently steelbow straw will usually fall to be delivered as it comes from the mill. In the case of combine harvesting the outgoing tenant may be liable for loss or damage resulting from (a) shortage caused by the straw being cut high; (b) deficiency of or damage to young grass; (c) deterioration of straw as fodder; (d) expense of gathering straw off the ground.
4. When combine harvesting is carried out, the tenant should notify his landlord in due time before the away-going; and the incoming tenant should also be notified by the landlord in due course in order that suitable arrangements can be made as to the disposal of the straw.

If the entry be Martinmas, the incoming tenant has generally less to take over. In most cases he has to take at valuation

[1] *Berry* v. *Allen*, 1827, 5 S. 212; aff. 3 W. & S. 417.

(1) dung made after the sowing of the last green crop; (2) green crop, so far as remaining in the land; (3) threshing mill, &c. Here, again, there is variation, according to contract or the custom of the district.

With regard to sheep stocks, there is no obligation, apart from contract, to take these over at valuation, but such an obligation is often imposed.

Silage valuations depend on the nature and quantity of the nutrients it contains and its suitability in other respects as a food for livestock. Tables have been prepared showing the approximate food values of various types of silage which are grouped as high, medium and low protein types. High protein is obtained from leafy young grass, young clover and lucerne cut before flowering. The medium group includes grass cut at the early stage, lucerne and clover in flower and the common mashlum mixtures; low protein is produced from grass cut at the hay stage and crops such as beet tops. The high quality is comparable to good cattle cakes, the medium to good hay, and the low to turnips or straw or poor hay. 'Fodder' has different meanings in different localities and may include hay or oats. In South-West Scotland fodder often includes the oat crop and meadow hay but not rye grass hay.

The property in implements, stock, crop, &c., does not now pass to the buyer (incoming tenant) until payment of the price which must be made within one month after quitting or within one month of delivery of the award (sec. 22).

Appointment of Arbiters and Oversman

In all these cases, it is usual to have the valuations made by two arbiters mutually chosen, or by an oversman appointed by the arbiters in case of their differing in opinion.

Where there is difficulty concerning the appointment of arbiters or oversman, parties can fall back on the provisions of the Arbitration (Scotland) Act, 1894, 57 & 58 Vict. c. 13, the leading clauses of which are—

'II. Should one of the parties to an agreement to refer to a single arbiter refuse to concur in the nomination of such arbiter, and should no provision have been made for carrying out the reference in that event, or should such provision have failed, an arbiter may be appointed by the Court, on the application of any party to the agreement, and the arbiter so appointed shall have the same powers as if he had been appointed by all the parties.

III. Should one of the parties to an agreement to refer to two arbiters refuse to name an arbiter, in terms of the agreement, and should no provision have been made for carrying out the reference in that event, or should such provision have failed, an arbiter may be appointed by the Court on the application of the other party, and the arbiter so appointed shall have the same powers as if he had only been duly nominated by the party so refusing.

IV. Unless the agreement to refer shall otherwise provide, arbiters shall have power to name an oversman, on whom the reference shall be devolved in the event of their differing in opinion. Should the arbiters fail to agree on the nomination of an oversman, the Court may, on the application of any party to the agreement, appoint an oversman. The decision of said oversman, whether he has been named by the arbiters or appointed by the Court shall be final.'

Parties having agreed upon the appointment of arbiters, or arbiters having been appointed under the above statute, it is usual to enter into a minute of submission. It is important to refer to the lease for its terms. If the lease is not referred to the submission is final and regulates the basis on which the arbitration must proceed. But if the submission is not itself explicit on any question it should be competent for the arbiters to refer to the lease.

The minute should be drawn by the agent for the party who is to receive the money (generally the away-going tenant), and revised by the agent for the other party. This course should be followed even though, as is sometimes the case, there are counter claims by the incoming tenant either on his own behalf or as coming in place of the proprietor.

The submission having been executed, the arbiters should sign a short minute (to be written on the submission) accepting office and appointing an oversman, who should also sign a short minute accepting office. These minutes need not be holograph or tested.[1]

PROCEDURE AND PRINCIPLES OF VALUATION

Where no directions are given, in the minute of submission, or in the lease, as to procedure to be followed, the arbiters and oversman should follow the course generally adopted in the district.

[1] *Kirkcaldy* v. *Dalgairns*, 16th June, 1809, F.C. 318 No. 112.

Where arbiters, as men of skill, were appointed to value awaygoing crops, it was observed that it would be incompetent to take evidence.[1]

As regards the principle of valuation, in valuing manure and fences, the actual value at the time and place in question should be taken and not the cost. In the case of crops, market value is usually the basis of valuation.[2] It may be, however, that a tenant bound by his lease to consume fodder such as turnips on the farm will only get consuming value.[3] In order to arrive at market value, account must be taken of the expense of harvesting, marketing, &c., and to arrive at consuming value the manurial benefit of the crop will have to be kept in view. Generally, where the lease and submission are silent market value should rule, i.e., the market price the produce would have fetched on the farm at the time, less, of course, any expense which the away-going tenant would have had to bear, such as for reaping, harvesting, delivering, &c. Turnips left in terms of a lease, at its expiry, must be paid for at actual market value, and not merely consuming value.[4]

Even if the crop to be valued at consuming value is not usually consumed (e.g., white crop) it should be valued on the lower basis. Wholesale price is not the same as consuming value, wholesale being sale in bulk as contrasted with sale by retail, i.e., the sale of individual articles. Crops consumed on the land benefit the land on account of the manurial value of the residue. Consuming value is the value to an occupant who gets the benefit of the crop without having to incur the expense of lifting, storage, transport and realisation. In the case of turnips fed on the land all expense is saved. Market value includes remuneration to the grower and all expense of cultivation, reaping, &c., down to the date of sale.

VALUATION OF SHEEP STOCKS

In the case of sheep stocks the Court of Session held that, where the tenant was bound to leave the sheep stock on the farm to the proprietor or incoming tenant, according to the valuation of men

---
[1] *Logan* v. *Leadbetter*, 1887, 15 R. 115.
[2] *Erskine's Trs.* v. *Crombie*, 1870, 9 M. 54; see also *Lindsay* v. *Bell*, 1862, 1 M. 39.
[3] *Scott* v. *Ritchie*, 1869, 7 S.L.R. 135.
[4] *Erskine's Trs.* v. *Crombie*, supra; *Scott* v. *Ritchie*, supra.

mutually chosen, with power to name an oversman, it was the duty of the arbiters (or oversman) to value the sheep upon the basis of their value to an occupant of the farm, in view of the arbiter's estimate of the return to be realised by such occupant from them in accordance with the course of prudent management, in lambs, wool, and price when ultimately sold, and not upon the basis either (1) of market value only, or (2) of the cost and loss which would be involved in the restocking of the farm with a like stock, if the present stock were removed. The arbiter was entitled to take into account both current market prices and the special qualities of the sheep, both in themselves and in relation to the ground, which in his opinion would tend either to enhance or to diminish the said return to be realised from them by an occupant of the farm.[1] In a submission to value a sheep stock at 'the actual price and simple market value of the sheep without any deductions or additions of any kind' there was a clause that 'if the arbiter either intentionally or by error in judgment or by a clerical error or for any other cause whatsoever' included in his award anything beyond market value the award should be null and void. The valuation was made on the basis of its value to the incoming tenant without regard to the prices which would have been got in the ring and the award was reduced.[2] In a case where the lease provided that the sheep stock was to be taken over at valuation and the submission instructed the arbiter to determine (1) the number of sheep constituting the fair and regular sheep stock of the farm and (2) the sums payable by the proprietor, it is considered that the proper basis of valuation was 'going concern' and not 'break-up' value.

Sheep stock valuations have been subject to a great deal of criticism. In fact, it has been said that arbiters, by making excessive allowances for hefting and acclimatisation, had been fixing the valuation of hill sheep stocks at wholly artificial figures far above market value. While the Court had decided that something more than ring price was justified when the value of a stock, as such, of regular ages and established on the farm, was to be fixed, some arbiters undoubtedly had made quite excessive allowances. The result had been detrimental to the industry by putting an excessive burden on the landlord or incoming tenant, who had to

---

[1] *Williamson* v. *Stewart*, 1912 S.C. 235.
[2] *M'Intyre* v. *Forbes and Fraser*, 1939 S.L.T. 62.

pay an uneconomic price for the sheep stock they took over at entry.

In an effort to minimise the effect of these practices the Sheep Stocks Valuation (Scotland) Act was passed in 1937.[1] The first provision (sec. 1) applies to valuations under all leases, whether entered into before or after the passing of the Act and the arbiter is to state not only the basis of valuation of each class but also has to show separately any amounts included for acclimatisation or hefting or any other special allowance. If he does not do so, his award may be reduced. Sec. 2, dealing with a stated case on a point of law,[2] has a proviso that in circumstances which justify it an interim award may be made. There is a further provision (sec. 3) that the parties may agree to refer the question of valuation to the Land Court.

The difficulty is to break a practice of the kind above referred to once it has been firmly established. Tenants who have paid high prices naturally wish to be recouped when they, in turn, have to leave. The usual practice is to add a definite sum of, say, £3 or £4 a head to each sheep over and above the price they would be expected to realise in the market. Such figures require to be shown in all awards. An attempt has even been made to add a further sum for hefting as distinct from acclimatisation. It is said that hefting, interpreted as getting the sheep 'used to the run of the hill', involving extra herding during a period of, say, nine months after they first come to the ground, is something separate from, although an essential prerequisite of, acclimatisation, which is a gradual adjustment of the health of the sheep to the climatic conditions of the district. You might, therefore, have a stock hefted to ground which has not had time to become acclimatised. It may be said, too, that a stock can be acclimatised to a certain district where hefting is not necessary at all. What must be admitted is that a flock considered as a unit has a value considerably in excess of the individual sheep which compose it and which, as individuals, they would realise in the auction mart. This principle was clearly recognised in the case of *Williamson* v. *Stewart*[3] where the judges agreed that two meanings can be given to the expression 'market value'. There is the market value which an individual sheep will bring if it is exposed in the auction mart. On the other

[1] *M'Intyre* v. *Forbes and Fraser*, 1939 S.L.T. 62.
[2] See *Pott's Jud. Factor* v. *Glendinning*, 1949 S.C. 200.
[3] *Williamson* v. *Stewart*, 1912 S.C. 235.

hand, if a sheep stock is sold on the ground and as a whole, it has a considerably greater 'market value' in the sense of the price an incoming tenant competing with others will be prepared to pay for the stock. The measure of the additional value should be the expenditure of the previous tenant in building up the stock at a time when it could not earn its full income qualified by conditions existing at the time which will determine what price it is economically possible for an incoming tenant to pay. A flock can only be said to be complete when the stock is not only healthy and acclimatised, but of regular ages and properly graded according to numbers. If a flock has to be built up by the gradual purchase of sheep in the markets, there is no doubt that a considerable period will elapse before it will produce the best results and it is this loss of income which an incoming tenant taking over a complete stock is saved and which he has properly to pay for.

In some districts the practice of stocks being taken over at fixed prices has been adopted. That, however, has the disadvantage that it may have obviously artificial results. Prices are bound to change over a period of years, and may change very substantially and there is a further, perhaps more serious, disadvantage that the system does not encourage a tenant to keep up his stock during the last year of his tenancy. He may get just the same price for a poor as for a good stock if the prices are fixed. Again, this system appears to be quite out of keeping with varying economic conditions and might over a period of years result in very considerable inconsistences and inequalities. The only remedy in the past (prior to the passing of the Hill Farming Act in 1946) was to define the terms on which a stock is to be valued as carefully as possible. It may, for example, be at the price which would be realised in a ring in an auction mart in the county. For this method there are frequently no appropriate sales at a Whitsunday entry with which comparison can be made and, accordingly, it must again be left very largely to the arbiters to fix the prices. An effort may be made to define 'market value' in the lease in such a way as to exclude acclimatisation, but yet allow something for the stock being sold as a whole.

Valuations of sheep stocks usually divide the sheep into the different classes, ewes, cast ewes, and lambs, eild ewes, hoggs, &c., and a certain allowance of so many per score is made for shotts and prices fixed for each class. The counting is a matter of agreement, and the sheep having been marked, a date is fixed (usually the end of August in Whitsunday valuations) for the addition of stragglers which have not been in the first count.

HILL FARMING ACT, 1946

Sections 28 to 31 and the Second Schedule to the Hill Farming Act, 1946, made entirely new provisions with regard to sheep stock valuations. The Act does not apply in questions between two proprietors but only between landlord and tenant and only in the case of leases[1] entered into after the passing of the Act, viz:— 6th November, 1946. In such cases an arbiter is directed to value the sheep stock in accordance with the provisions of the Second Schedule to the Act and instead of being required in his award to 'show the basis' of his award (as provided in the Sheep Stocks Valuation (Scotland) Act of 1937) he must state separately the particulars set out in Part III of the Second Schedule. The schedule should be referred to but the particulars include *inter alia* the three year average price of ewes and of lambs adopted in making the valuation; any amount added or taken away as an adjustment in arriving at basic value, and the grounds on which the adjustment is made; the number of each class of stock and the value placed on each class; and any amount added or taken away in fixing the price of ewe hoggs at a Whitsunday valuation or the price of ewe lambs at a Martinmas valuation and the ground on which the adjustment is made. Section 29 provides that either party may insist that the valuation shall be carried out by the Land Court; under the 1937 Act the application had to be joint.

Section 30 lays on the outgoing tenant a duty to submit to the arbiter (or Land Court) 'within 28 days before the determination of the valuations' a statement of all sales of sheep from the farm during the three preceding years in a Whitsunday valuation or during the current year and two preceding years in a Martinmas valuation. He must also submit such sale notes or other evidence as may be required. The other party is entitled to inspect any such documents. It is not clear from what date the 28 days is to be counted. The valuation might be arranged for a date other than 28th May or 28th November. This is an uncertain date usually left to the arbiters, but presumably the date to take into account is the away-going date provided in the lease or submission.

[1] See *Pott's Jud. Factor* v. *Glendinning*, 1949 S.C. 200.

The Second Schedule lays down elaborate provisions with regard to procedure—Part I dealing with Whitsunday and Part II with Martinmas away-going. In the former case the arbiter has to ascertain the 'three year average price' of ewes and of lambs sold off the farm, basing his figures on the sale prices and allowing for shotts as these are ascertained by him. Where, however, the number of ewes or lambs sold in the three preceding years has been less than half the total number sold the 'three year average' price is determined by the Land Court on the application of the parties, the Court basing its findings on the experience of similar farms in the same district. What happens if one party refuses to make an application is not stated. Presumably the arbiters report to the parties and leave the matter in their hands. No form of application is provided. It is assumed that once it has established the 'three year average price' the Court refers further procedure back to the arbiters—an extremely dilatory provision. Having arrived at the 'three year average price' the arbiters may adjust the figure by 10s. more or less having regard to 'the general condition of the stock and the profit which the purchaser may reasonably expect it to earn.' In the case of leases entered into after the passing of the Agriculture (Miscellaneous Provisions) Act, 1963 the adjustment permitted is not 10/- but 20% of the three year average price. There are corresponding percentages for other adjustments referred to below and these are shown in brackets in each case. The resultant figure is the 'basic ewe value' and the 'basic lamb value.' The arbiter then makes the following further adjustments. He adds 15s. (or 30% of basic ewe value) per head in the case of ewes; adjusts the price of twin lambs as he considers proper (single lambs being at 'basic lamb value'); ewe hoggs are valued at two-thirds of the combined basic value of ewe and lamb, adjusted within a limit of 5s. (or 10%) up or down having regard to quality and condition; tups are taken at the value put on them by the arbiter, allowing for acclimatisation or any other factor he thinks fit. Wool is not taken into account and no specific allowance can be made for acclimatisation (except for tups) or hefting or regular ages, although presumably these factors fall to be taken into account in adjusting the three year average. Eild sheep are valued at the ewe value, subject to adjustment for quality and condition, and shotts at not more than two-thirds of the value of good sheep of the same age and class.

There are certain variations necessary in the case of Martinmas valuations. After fixing the 'basic ewe value' ewes (including gimmers) are taken at that value plus 15s. (or 30%) per head; and ewe lambs at basic ewe value subject to an adjustment of 5s. (or 10%) per head up or down, having regard to quality and condition.[1]

The Duties of Arbiters and Oversman

The oversman has an overriding jurisdiction, but it should not be exercised unless the arbiters have failed to agree. Frequently certain questions are devolved and others are not, and although the practice is not to be commended, it has been decided that an award signed by both arbiters and the oversman could not, on that ground, be reduced.[2]

The arbiters, or where there has been a devolution, the oversman, having accepted office, must proceed and issue an award disposing of the matters submitted to them. They are not entitled capriciously to decline to do so.[3] The only excuses the Court would be likely to accept are ill-health, intention to go abroad for a long time, or emerging interest. The submission may bear that the award is to be issued within a stated time or 'between this ... and the ... day of ... next to come.' Where the date is thus left blank, the submission lasts for a year and a day. If no reference to endurance is made, there is no period except that imposed by the negative prescription. The parties can themselves agree on an extension and in certain circumstances their actings may be held to imply an agreed-on extension. An arbitration in the absence of special provisions in the submission comes to an end by the death of a party[4] (nor merely one of a body of trustees)[5] or of an arbiter or oversman.

The Court of Session alone can compel an arbiter to proceed with his task.[6] The proper procedure is by way of action in the Outer House.[7]

[1] *Mackinnon* v. *Secretary of State for Scotland*, 1949 S.L.C.R. 19.
[2] *Cameron* v. *Nicol*, 1930 S.C. 1. Cf. *Davidson* v. *Logan*, 1908 S.C. 350;.
[3] *Edinburgh and Glasgow Railway Co.* v. *Marshall & Miller*, 1853, 15 D. 603; Erskine's Inst. IV, 3, 20.
[4] Erskine Inst. IV, 3, 29; *Robertson* v. *Cheyne*, 1847, 9 D. 599; *Ewing & Co.* v. *Dewar*, 19 Dec., 1820 F.C. But as to position in Statutory Arbitration see p. 68.
[5] *Alexander's Trs.* v. *Dymock's Trs.*, 1883, 10 R. 1189.
[6] *Forbes, &c.* v. *Underwood*, 1886, 13 R. 465.
[7] *Watson* v. *Robertson*, 1895, 22 R. 362.

Where one arbiter is obstructive, and will neither proceed nor devolve, the Court may order him to devolve;[1] and where there is a difference of opinion between the arbiters, even on a point of procedure only, the oversman is entitled to act, at the request of one of the arbiters, notwithstanding the dissent of the other.[2] An arbiter has power, by implication, to award expenses.[3] He has, however, no power to assess damages unless expressly empowered to do so.[4] He is entitled to require and obtain assistance from men of skill, e.g., engineers or valuers, or to consult a law agent or counsel.[5] He is not affected by the provisions of the Act with regard to the appointment of a clerk.

PROCEDURE IN MAKING VALUATIONS

General

The arbiters and oversman generally go together to make the valuation, but the oversman should not act unless and until the arbiters differ and devolve upon him.[6] In a reference to two arbiters named and an oversman to be named by them, the arbiters appointed the oversman before entering on the reference and the oversman accompanied them at their inspection of the subjects. The oversman died before the arbiters issued their award. Held that the award was valid as the reference did not devolve on the oversman except in the event of the arbiters differing.[7] But as these valuations are not, properly speaking, arbitrations, the Court allows very wide latitude as regards procedure, and will not interfere merely because the oversman settles points in dispute as they arise, and thereafter the arbiters sign a general award.[8]

AWAY-GOING CROP VALUATIONS

Where the submission does not specify the precise procedure to be adopted in making the valuations, the arbiters and oversman

---

[1] *Sinclair* v. *Fraser*, 1884, 11 R. 1139.
[2] *Gibson* v. *Fotheringham*, 1914 S.C. 987
[3] *Ferrier* v. *Alison*, 1843, 5 D. 456.
[4] *Mackay & Son* v. *Leven Police Commissioners*, 1893, 20 R. 1093.
[5] *Caledonian Railway Co.* v. *Lockhart*, 1860, 3 Macq. 808.
[6] *Bryson* v. *Mitchell*, 1823, 2 S. 382; *Gordon* v. *Abernethy*, M. 655.
[7] *Stuart* v. *Smith*, 1925, 41 Sh. Ct. Rep. 185.
[8] *M'Gregor* v. *Stevenson*, 1847, 9 D. 1056; *Nivison* v. *Howat*, 1883, 11 R. 182; *Davidson* v. *Logan*, 1908 S.C. 350.

ARBITRATIONS & VALUATIONS (OUTSIDE THE ACT) 103

can follow any course they think best suited to the case. They will generally proceed in the manner usual in the district. There is a common practice in valuing such things as manure, fences, and grass, but it is different with the corn crop, regarding which methods vary. The arbiters must, of course, follow the method (if any) provided in the submission.

Four methods of procedure at valuation of crops may be explained, viz:—

Before Reaping—Whitsunday Entry

1. *By Entire Valuation.*—In this case the arbiters walk through the fields a few days before the crop is ready for cutting, and decide (1) how many quarters (or hundredweights) of grain each acre under crop will yield; (2) the money value of each quarter (or hundredweight) of grain (and straw, unless it is steelbow); and (3) the sum to be paid to the incoming tenant for harvesting the crop. When the fields under crop have been measured, the sum of the valuations can be calculated and the award completed.

Where combine harvesters are used and the crop has to be ripe before cutting questions may arise regarding loss through 'shake' after the valuer's inspection but before separation of crop. There is a variety of practice, some valuers holding that the risk passes when they leave the field, others taking the view that an equitable adjustment may be made and it may be advisable to deal with this point in submissions.

2. *By Valuation and Fiars Prices.*—In this case the arbiters also walk through the fields as before, and decide the points (1) and (3) as above. They do not fix the money value of the grain, because that is to be calculated according to the fiars prices. Till these prices are struck, generally about February following the reaping, the amount of the valuations cannot be ascertained. In some counties no fiars prices of straw are struck. It is best to guard against such a contingency by inserting a clause in the submission empowering the arbiters to fix the value of the straw in the event of no fiars prices being struck, or the parties may agree to accept the fiars prices of a neighbouring county.

It is usual to issue an interim valuation or order for a payment to account.

The Grant Committee Report on the Sheriff Court (Cmnd.

3248) has recommended that the striking of fiars prices be discontinued and suggested (Para. 313) that they be replaced by a calculation based on either the average Scottish market price or on the Guaranteed Price.

After Reaping—Whitsunday or Martinmas Entry

3. *By Proof from the Stack.*—In this case the arbiters select an average stack of the different kinds of crop—corn, barley, or wheat—and have it threshed out, and it is held that each of the other stacks of the respective kinds of grain will yield the same quantity as that stack. An average stack of the whole stacks from each field is generally taken. The arbiters, having thus arrived at the quantity, next proceed to fix the money value of each quarter of grain (and straw unless it be steelbow); or the fiars prices may be the standard of the price where that is agreed on. With the prevalence of combine harvesters this method is now uncommon.

4. *By Threshing Out.*—In some cases the quantities are ascertained by threshing out the whole crop, the away-going tenant having someone constantly in attendance during the operation, and the price is either fixed by the arbiters or conform to the fiars' prices. This system is not generally considered advantageous.

THE AGRICULTURAL HOLDINGS (SCOTLAND) ACT, 1949 (12, 13 & 14 Geo. 6 Ch. 75) as amended by *inter alia* the AGRICULTURE ACT, 1958 (6 & 7 Eliz. 2 Ch. 71). and the SUCCESSION (SCOTLAND) ACT, 1964 (1964 Ch. 41)

The Amendments made by the 1958 Act are underlined.

PREAMBLE TO 1949 ACT

An Act to consolidate the Agricultural Holdings (Scotland) Act, 1923, Part II of the Small Landholders and Agricultural Holdings (Scotland) Act, 1931, Part I of the Agriculture (Scotland) Act, 1948, and certain of the enactments relating to Agricultural Holdings, save, with respect to rights to compensation, in their application to certain cases determined by past events.

[24th November, 1948]

MEANING OF 'AGRICULTURAL HOLDING'

1.—(1) In this Act the expression 'agricultural holding' means the aggregate of the agricultural[1] land comprised in a lease,[2] not being a lease under which the said land is let to the tenant during his continuance in any office, appointment or employment held under the landlord.

Meaning of 'agricultural holding'

(2) For the purposes of this and the next following section, the expression 'agricultural land' means land used for agriculture[1] which is so used for the purposes of a trade or business[3] and includes any other land which, by virtue of a designation of the Secretary of State under sub-section (1) of section eighty-six of the Agriculture (Scotland) Act, 1948, is agricultural land within the meaning of that Act.

---

[1] **'Agriculture'** as defined in sec. 93 (1) has a very wide meaning. See *Dunn v. Fidoe*, [1950] 2 All E.R. 685, C.A.; *Howkins v. Jardine*, [1951] 1 All E.R. 320 C.A.; *R. v. Agricultural Land Tribunal* (South Eastern Province) *ex. p. Palmer*, 1954 J.P.L. 181. For circumstances in which the Land Court held that a holding under a lease consisted only partly of agricultural subjects and determined the extent of the agricultural holding by excision of non-agricultural subjects see *McGhie & Anr. v. Lang & Anr.*, 1953 S.L.C.R. 22. The extent of the subjects is not material. *Stevens v. Sedgeman*, [1951] 2 K.B. 434.

[2] **'Lease'**. See definition in sec 93 (1) and notes. A farm let for 'a rotation of cropping' subject to termination in event of sale was held not to be an agricultural holding; *Stirrat & Anr. v. Whyte*, 1968 S.L.T. 157.

[3] **'Trade or business'**: not necessarily agricultural. See *Rutherford v. Maurer*, [1962] 1 Q.B. 16 (let for grazing to owner of a riding school).

106 AGRICULTURAL HOLDINGS (SCOTLAND) ACTS

PROVISIONS AS TO LEASES

Restriction on letting agricultural land for less than from year to year

2.—(1) Subject to the provisions of this section, where under a lease[1] entered into on or after the first day of November, nineteen hundred and forty-eight, any land is let to a person for use as agricultural land for a shorter period than from year to year,[2] and the circumstances are such that if he were a tenant from year to year he would in respect of that land be the tenant of an agricultural holding, then, unless the letting was approved by the Secretary of State before the lease was entered into, the lease shall take effect, with the necessary modifications, as if it were a lease of the land from year to year:

Provided that this sub-section shall not have effect in relation to a lease of land entered into (whether or not the lease expressly so provides) in contemplation of the use of the land only for grazing or mowing during some specified period of the year[3] or in relation to a lease of land granted by a person whose interest in the land is that of a tenant under a lease which is for a shorter period than from year to year and which has not by virtue of this section taken effect as a lease from year to year.[4]

(2) Any question arising as to the operation of the foregoing sub-section in relation to any lease shall be determined by arbitration.[5]

---

[1] **'Lease'.** *Hollings* v. *Swindell*, (1950) 155 E.G. 269. Held that an award of an arbitrator amounted to an agreement between the parties for a tenancy from year to year.

[2] **'A shorter period than from year to year'.** It has been held in England that a tenancy of 18 months is greater than from year to year: *Gladstone* v. *Bower*, [1960] 2 Q.B. 384. It is thought that subjects in Scotland let for more than one year but less than two would not be an agricultural holding, cf. *Stirrat & Anr.* v. *Whyte*, 1968 S.L.T. 157. A tenancy for one year was held to be for a shorter period than from year to year: *Bernays* v. *Prosser*, [1963] 2 Q.B. 592.

[3] **'A period of the year'** may be adequately specified without definite dates being laid down as *termini*. *Mackenzie* v. *Laird*, 1959 S.C. 266. Let for 364 days—*Reid* v. *Dawson*, [1954] 3 All E.R. 498. Let for 'six months' periods' was held to be for at least one year and so not within the proviso—*Rutherford* v. *Maurer*, [1962] 1 Q.B. 16.

[4] A lease under this sub-section which has not yet taken affect as a lease from year to year is disregarded in a question of additional payments on compulsory acquisition. See Agriculture (Miscellaneous Provisions) Act, 1968 ss. 12 and 14.

[5] It was held in a question arising out of Sec. 2 of the English Act that the applicability of the Act as distinct from its operation is a matter for the Courts. *Goldsack* v. *Shore*, [1950] 1 K.B. 708 C.A. See also *Houison-Craufurd's Trustees* v. *Davies*, 1951 S.C. 1 and *Brodie* v. *Ker*, 1952 S.C. 216.

PROVISIONS AS TO LEASES

3.[1]—(1) The tenancy of an agricultural holding shall, instead of coming to an end on the termination of the stipulated endurance[2] of any lease,[3] be held to be continued in force by tacit relocation for another year and thereafter from year to year, unless such notice to terminate the tenancy as is mentioned in section *twenty-four* of this Act has been given by either party[4] to the other. *Tacit relocation*

(2) The provisions of the foregoing sub-section shall have effect notwithstanding any agreement or any provision in the lease to the contrary.

[1] **Contracting out.** It is not competent to contract out of this section—*Duguid* v. *Muirhead*, 1926 S.C. 1078. It is designed to prevent contracting out by letting land for such periods as 11 months. See also sub-section (2).

[2] **'Stipulated'.** If no period be 'stipulated,' the period would be implied as for a year and thereafter from year to year. The 'termination' of the stipulated endurance may be at a 'break' in the lease.

[3] **'Lease'.** See definition sec. 93 and notes.

[4] **'Either party'.** See case of notice by one of two joint tenants (*Graham* v. *Stirling*, 1922 S.C. 90; *Montgomerie* v. *Wilson*, 1924 S.L.T. (Sh.Ct.) 48, and *Wight* v. *Earl of Hopetoun*, 1864, 2 M (H.L.) 35). See also *Walker* v. *Hendry and Anor.* 1925 S.C. 855. *Smith* v. *Grayton Estates Ltd.*, 1960 S.C. 349.

4.—(1) Where in respect of the tenancy of an agricultural holding— *Provisions for securing written leases and for the revision of certain leases*

    (*a*) there is not in force a lease in writing[1] embodying the terms of the tenancy, or

    (*b*) there is in force such a lease, being either—

        (i) a lease entered into on or after the first day of November, nineteen hundred and forty-eight, or

        (ii) a lease entered into before that date, the stipulated period of which has expired and which is being continued in force by tacit relocation,[2]

and such lease contains no provision for one or more of the matters specified in the Fifth Schedule to this Act or contains a provision inconsistent with that Schedule or with the next following section, the landlord or the tenant may give notice in writing to his tenant or his landlord requesting him to enter into such a lease containing provision for all of the said matters or a provision not inconsistent with the said Schedule or the said section, as the case may be; and if within the period of six months after the giving of such notice no such lease has been concluded, the terms of the tenancy shall be referred to arbitration.[3]

108 AGRICULTURAL HOLDINGS (SCOTLAND) ACTS

(2) On any such reference the arbiter shall by his award specify the terms of the existing tenancy, and, in so far as those terms make no provision for all the matters specified in the Fifth Schedule to this Act or make provision inconsistent with[4] that Schedule or with the next following section, make such provision for those matters as appears to the arbiter to be reasonable.[5]

(3) On any such reference the arbiter may include in his award any further provisions not inconsistent with the provisions of this Act relating to the tenancy which may be agreed between the landlord and the tenant.

[1] **'Lease in writing'** is not limited to probative leases—*Grieve & Sons* v. *Barr*, 1954 S.C. 414.

[2] The important question which arises under this section is to determine what effect it may have on existing leases. Reference may be made to section 10, which provides that a lease is not to be regarded as brought to an end only by reason of some new term being added to it by the provisions in the Act. Sub-section (6) of section 5 provides that that section does not apply to any lease entered into before 1st November, 1948. For tacit relocation see section 3.

[3] The desirability of having a written lease cannot be too strongly emphasised as it enables both parties at any given time to determine their rights.

[4] **'Inconsistent with'** in this context is thought to involve a positive and contradictory provision.

[5] Sub-section (2) gives arbiters powers to include matters which may have been outwith the consideration of the parties at the time of making a reference to arbitration. For arbiters' powers to vary rent see section 6.

Respective liabilities of landlord and tenant for provision and maintenance of fixed equipment and for payment of insurance premiums.

5.—(1) Where a lease has been entered into for the letting of an agricultural holding, a record of the condition of the fixed equipment on the holding shall be made forthwith,[1] and on being so made shall be deemed to form part of the lease; and the provisions of section *seventeen* of this Act shall apply to the making of such a record and to the cost thereof as they apply to a record made under that section.[2]

(2) There shall be deemed to be incorporated in every lease for the letting of an agricultural holding—

(*a*) an undertaking by the landlord that, at the commencement of the tenancy or as soon as is reasonably possible thereafter,[3] he will put the fixed equipment on the holding into a thorough state of repair, and will provide such buildings and other fixed equipment as will enable an occupier reasonably skilled in husbandry to maintain efficient production as respects both the kind of produce specified in the lease, or (failing such specification) in use to be produced on the holding,[4] and the quality and

PROVISIONS AS TO LEASES 109

quantity thereof, and will during the tenancy effect such replacement or renewal of the buildings or other fixed equipment as may be rendered necessary by natural decay or by fair wear and tear,[5] and

(b) a provision that the liability of the tenant in relation to the maintenance of fixed equipment shall extend only to a liability to maintain the fixed equipment on the holding in as good a state of repair (natural decay and fair wear and tear excepted) as it was in immediately after it was put in repair as aforesaid or, in the case of equipment provided, improved, replaced or renewed during the tenancy, as it was in immediately after it was so provided, improved, replaced or renewed.

(3) Nothing in the last foregoing sub-section shall be deemed to prohibit any agreement made after the lease has been entered into between the landlord and the tenant whereby one of the parties undertakes to execute on behalf of the other party, and wholly at his own expense or wholly or partly at the expense of the other party, any work which the other party is required to execute in order to fulfil his obligations under the lease.[6]

(4) Any provision in a lease requiring the tenant to pay the whole or any part of the premium due under a fire insurance policy over any fixed equipment on the holding shall be null and void.[7]

(5) Any question arising as to the liability of a landlord or of a tenant under this section shall be determined by arbitration.[8]

(6) This section shall not apply to any lease entered into before the first day of November, nineteen hundred and forty-eight.

[1] **'Record'.** There is no provision as to the responsibility for having the record made and no penalty except that the parties lose their rights under sections 56 and 57-58 respectively.

[2] Although it is desirable that a comprehensive record as to the condition of the holding be made, it is only in respect of fixed equipment that it is obligatory.

[3] **'As soon as is reasonably possible thereafter'.** It is impossible to give guidance as to what length of time may be reasonable in any set of circumstances. It will clearly depend on the condition of the fixed equipment at date of entry, on the availability of labour and materials in the district and the prevailing legislation or regulations as to building, &c. There may be other local circumstances which can assist in the determination of what is reasonable and comparison with like cases will obviously be helpful.

[4] These words, **'in use to be produced on the holding'**, are very important and limit the arbiter's discretion. For example, it is not contemplated that

a stock farm should be equipped for dairy purposes, unless the lease is silent on the kind of produce to be produced on the holding. See *Taylor* v. *Burnett's Trustees*, 1966 S.L.C.R. App. 139 where the farm was let as 'an ordinary holding or general purpose farm' and the lease stipulated that the landlords were not to be liable for improvements necessary for any other type of farm. It was held that, although dairying was carried on, the landlords were bound to provide fixed equipment only for the type of farming stipulated in the lease, and not that required for dairying.

[5] This is a peremptory provision: see *Secretary of State for Scotland* v. *Sinclair*, 1960 S.L.C.R. 10.

[6] It is not provided that such an agreement must be in writing.

[7] See section 23 for the application of sums recovered under fire insurance policy in the case of earlier leases where the tenant is liable in payment of part of the premium.

[8] See sec. 7 re rent arbitration.

Provisions supplementary to s. 4 and s. 5.

6.—(1) Where by virtue of section *four* of this Act the liability for the maintenance or repair of any item of fixed equipment is transferred from the tenant to the landlord, the landlord may within the prescribed period[1] beginning with the date on which the transfer takes effect require that there shall be determined by arbitration, and paid by the tenant, the amount of any compensation which would have been payable under section *fifty-seven* of this Act or in accordance with sub-section (3) of that section, in respect of any previous failure by the tenant to discharge the said liability, if the tenant had quitted the holding on the termination of his tenancy at the date on which the transfer takes effect.

(2) Where by virtue of section *four* of this Act the liability for the maintenance or repair of any item of fixed equipment is transferred from the landlord to the tenant, any claim by the tenant in respect of any previous failure by the landlord to discharge the said liability shall, if the tenant within the prescribed period[1] beginning with the date on which the transfer takes effect so requires, be determined by arbitration, and any amount directed by the award to be paid by the landlord shall be paid by him to the tenant.

(3) Where it appears to the arbiter—
   (a) on any reference under section *four* of this Act that, by reason of any provision which he is required by that section to include in his award, or
   (b) on any reference under sub-section (5) of section *five* of this Act that, by reason of any provision included in his award,

it is equitable that the rent of the holding should be varied, he may vary the rent accordingly.[2]

PROVISIONS AS TO LEASES

(4) The award of the arbiter under section *four* or *five* of this Act shall have effect as if the terms and provisions specified and made therein were contained in an agreement in writing entered into by the landlord and the tenant and having effect as from the making of the award or, if the award so provides, from such later date as may be specified therein.[3]

[1] '**Prescribed period**'. Under Regulation 5 of the Agricultural Holdings (Scotland) Regulations, 1950, the time limit was one month from the date on which the transfer took effect.
[2] This confers a wide discretion on the arbiter who should exercise it with great care. It would seem that he should determine the annual value of the changes made to the party benefited by them and vary the rent by a similar amount. The provision of section 2 of the 1958 Act should be applied so far as practicable.
[3] The object of postponing the operation of an award in certain cases to a specified date may be to obviate hardship in these cases. Where no hardship would result and no other reason for postponing is apparent, the arbiter will usually cause the award to have immediate effect.

7.[1]—(1) Subject to the provisions of this section the landlord[2] or the tenant of an agricultural holding may, whether the tenancy was created before or after the commencement of this Act, by notice in writing served on his tenant or his landlord demand a reference to arbitration of the question what rent should be payable in respect of the holding as from the next ensuing day on which the tenancy could have been terminated by notice to quit given at the date of demanding the reference, and the matter shall be referred[3] accordingly.

Variation of rent.

For the purposes of this sub-section the rent properly payable in respect of a holding shall be the rent at which, having regard to the terms of the tenancy[4] (other than those relating to rent), the holding might reasonably be expected to be let in the open market[5] by a willing landlord to a willing tenant, there being disregarded (in addition to the matters referred to in the next following sub-section) any effect on rent of the fact that the tenant who is a party to the arbitration is in occupation of the holding.[6]

(2) On any reference under the last foregoing sub-section the arbiter—

    (*a*) shall not take into account any increase in the rental value of the holding which is due to improvements[7] which have been executed thereon in so far as they were executed wholly or partly at the expense of the tenant (whether or not that expense has been or will be reimbursed by a grant out of moneys provided by Parliament) without any equivalent allowance or benefit made or given by the landlord in consideration of their execution, and have not been executed under an obligation imposed on the tenant by the terms of his lease,[8] or to improvements which have been executed thereon by the landlord in so far as the landlord has received or will receive grants out of moneys provided by Parliament in respect of the execution thereof, or fix the rent at a higher amount than would have been properly payable if these improvements had not been so executed;

(b) shall not take into account the relief in respect of rates to occupiers of agricultural lands and heritages effected by the Local Government (Scotland) Act, 1929,[9] nor the amounts recoverable by occupiers from owners under section forty-seven of that Act, nor any benefit that may accrue to the tenant from the operation of the Agricultural Marketing Act, 1931,[10] and

(c) shall not fix the rent at a lower amount by reason or any dilapidation or deterioration of, or damage to, fixed equipment or land caused or permitted by the tenant.[11]

Subject as aforesaid, the arbiter shall determine what rent should properly be payable in respect of the holding as from the day mentioned in the last foregoing sub-section.

(3) A reference to arbitration under sub-section (1) of this section shall not be demanded in such circumstances that any increase or reduction of rent made in consequence thereof would take effect as from a date earlier than the expiration of five years from the latest in time of the following dates, that is to say—

(a) the commencement of the tenancy,[12] or

(b) the date as from which there took effect[13] a previous increase or reduction of rent (whether made under this section or otherwise), or

(c) the date as from which there took effect a previous

direction under this section that the rent should continue unchanged:

Provided that there shall be disregarded for the purposes of this sub-section—
   (i) an increase or reduction of rent under sub-section (3) of the last foregoing section;
   (ii) an increase of rent under sub-section (1) of the next following section or such an increase as is referred to in sub-section (2) of that section;
   (iii) a reduction of rent under section *thirty-four* of this Act.

(4) The continuous adoption by the tenant of a standard of farming or a system of farming more beneficial to the holding than the standard or system required by the lease or, in so far as no system of farming is so required, than the system of farming normally practised on comparable holdings in the district, shall be deemed, for the purposes of sub-section (2) of this section, to be an improvement executed at his expense.[14]

[1] See Opencast Coal Act, 1958, section 14.

[2] **'Landlord'**. The landlord need not be infeft: *Alexander Black & Sons v. Paterson*, 1968 S.L.T. (Sh.Ct.) 64.

[3] **'The matter shall be referred'**. Views have been expressed that the arbiter must be appointed before the date when the alteration of rent is to take effect: *Graham* v. *Gardner*, 1966 S.L.T. (Land Ct.) 12, following *Sclater* v. *Horton*, [1954] 2 Q.B. 1. But see *Dundee Corporation* v. *Guthrie*, 1969 S.L.T. 93.

[4] **'Having regard to the terms of the tenancy'**. The arbiter or the Land Court must have regard to all the terms even if they are not enforced. *Secretary of State for Scotland* v. *Davidson and others*, 1969 S.L.T. (Land Ct.) 7.

[5] **'Open market'**. The discretion of the arbiter or the Land Court is strictly limited. See *Secretary of State for Scotland* v. *Young*, 1960 S.L.C.R. 31; *Secretary of State for Scotland* v. *Sinclair*, 1962 S.L.C.R. 6.

[6] This paragraph added by the Agriculture Act, 1958, section 2, provides as a criterion for arbiters fixing rents the market rent c.f. *Guthe* v. *Broatch*, 1956 S.C. 132. Proof as to market rent must be adduced. *Secretary of State for Scotland* v. *Young*, 1960 S.L.C.R. 31. *Crown Estate Commissioners* v. *Gunn* 1961 S.L.C.R. App. 173. *Secretary of State for Scotland* v. *Sinclair*, 1962 S.L.C.R. 6.

[7] **'Improvements'**. Even if a sitting tenant is not entitled to compensation for improvements, he is not to have his rent increased in consequence of these improvements. It is thought that this provision is not confined to improvements scheduled under the Act. Improvements made by a former tenant are not excluded: *Kilmarnock Estates Ltd.* v. *Barr*, 1969 S.L.T. (Land Ct.) 10.

[8] It is generally understood that, when a new lease is entered into, all common law claims by the landlord or by the tenant against each other, under the expired lease, are extinguished except such as are maintained in force by statute, or by agreement in the new lease. The same rule applies in the case of a renunciation (*Lyons* v. *Anderson*, 1886, 13 R. 1020; *Walker's Trustees* v. *Manson*, 1886, 13 R. 1198). But where a tenancy has been renewed by tacit relocation the case is different (Lord President Clyde in *Cowe* v. *Millar*, 1923, unreported). Where rent has been fixed in an arbitration under the Act the other provisions of the lease continue to apply: the arbiter has no power to vary them.

Questions may arise as to what is 'equivalent allowance or benefit'. Apparently the improvements referred to are such as the tenant voluntarily effected, and for which, by the contract of tenancy or otherwise, he received no concession or benefit. A tenant's fixtures are not improvements.

[9] This refers to the de-rating of agricultural land formerly effected by the rating of land on one-eighth of its annual rent and now by the assessment of the farm house only as a separate subject. Agricultural land no longer enters the valuation roll. See Valuation and Rating (Scotland) Act, 1956, section 7.

[10] The 1931 Act was repealed by the Agricultural Marketing Act, 1958.

[11] This, it is thought, applies only to deterioration or dilapidation during the currency of the lease last entered into, including renewals thereof by tacit relocation. (See *Findlay* v. *Munro*, 1917 S.C. 419.) Deterioration made or permitted by the tenant appears to be confined to deterioration in breach of contract.

[12] **'Commencement of the tenancy'.** Is this the commencement of the current lease or does it refer to the term at which the tenant first entered to the holding? It is thought that the latter is intended. The phrase 'the date as from which there took effect a previous increase, &c'. appears to support this view.

[13] **'Took effect'.** Assuming that a reduction had been granted on the rent of the year from Martinmas, 1969, to Martinmas, 1970, it is thought that the reduction took effect as at Martinmas, 1969, irrespective of the date at which the rent was payable, and that the five years would be counted from that term and not from Martinmas following.

[14] This refers to section 56.

Increases of rent in respect of certain improvements carried out by landlord.

8.[1]—(1) Where the landlord of an agricultural holding has, whether before or after the commencement of this Act, carried out on the holding an improvement (whether or not one for the carrying out of which compensation is provided under the following provisions of this Act) being either an improvement carried out—

(*a*) at the request of, or in agreement with, the tenant; or

(*b*) in pursuance of an undertaking given by the landlord under sub-section (3) or paragraph (*b*) of sub-section (6) of section three of the Agricultural Holdings (Scotland) Act, 1923,[2] or under sub-section (3) of section *fifty-two* of this Act; or

(*c*) in compliance with a direction given by the Secretary of State under powers conferred on him by or under any enactment; or

(*d*) in accordance with a provision in that behalf included in a [livestock rearing][3] land improvement scheme approved under the Hill Farming Act, 1946,[7] being a provision so included at the instance or with the consent of the tenant;

or works for the supply of water to the holding executed in

pursuance of directions given by the Agricultural Executive Committee under Defence Regulations or of a scheme approved by the Agricultural Executive Committee, then, subject to the provisions of this section, the rent of the holding shall, if the landlord by notice in writing served on the tenant within six months from the completion of the improvement so requires, be increased as from the completion of the improvement or, where the improvement was completed before the first day of November, nineteen hundred and forty-eight, as from that day, by an amount equal to the increase in the rental value of the holding attributable to the carrying out of the improvement:

Provided that where any grant has been made to the landlord in respect of the improvement out of moneys provided by Parliament, the increase in rent provided for by the foregoing provisions of this sub-section shall be reduced proportionately.[4]

(2) No increase of rent shall be made under the foregoing sub-section if before the first day of November, nineteen hundred and forty-eight, the landlord and the tenant agreed on any increase in rent or other benefit to the landlord in respect of the improvement, or if before that day any sum became payable under sub-section (3) of section three of the Agricultural Holdings (Scotland) Act, 1923,[5] or section nine of the Agriculture (Miscellaneous Provisions) Act, 1943,[6] or section nine of the Hill Farming Act, 1946,[7] in respect of the cost of executing it.

(3) Where interest on the cost of works for the supply of water, or rent in respect of such an improvement as is mentioned in paragraph (*d*) of sub-section (1) of this section, became payable under the provisions of section nine of the Agriculture (Miscellaneous Provisions) Act, 1943,[6] or of sub-section (3) of section nine of the Hill Farming Act, 1946,[7] as the case may be, before the first day of November, nineteen hundred and forty-eight, or became payable under the said provisions after that day by virtue of an agreement between the landlord and the tenant entered into before that day, it shall continue to be recoverable notwithstanding that the said provisions are by virtue of the Agriculture (Scotland) Act, 1948, no longer in force.

(4) Any question arising between the landlord and the tenant of the holding under this section shall be determined by arbitration.

(5) In this section the expression 'Agricultural Executive

Committee' means the Agricultural Executive Committee for any area to whom the Secretary of State has delegated any of his powers under Defence Regulations.

[1] '(5) Where the landlord of an agricultural holding has executed thereon works of the nature of fixed equipment which are required to be executed as mentioned in the last foregoing sub-section or has executed similar works at the request of, or in agreement with, the tenant, section eight of the Agricultural Holdings (Scotland) Act (which provides for increases of rent in respect of improvements carried out by the landlord) shall have effect as if the works so executed were such an improvement as is mentioned in sub-section (1) of that section.' Agriculture (Safety, Health and Welfare Provisions) Act, 1956 section 25. Similar provisions are contained in section 38 (1) of the Housing (Scotland) Act, 1964, and section 80 of the Housing (Scotland) Act, 1966, in regard to improvements carried out in compliance with notices or undertakings under these Acts. Provision is made for proportionate reduction of the increase in rent if the tenant has contributed to the cost of the improvement. See also Opencast Coal Act, 1958 section 14 (8) (*a*).

[2] This refers to an undertaking to execute the improvements given after receipt of notice by a tenant that he intended to execute improvements in the First Schedule, Part II, of the Act, and there being no agreement as to compensation if they were carried out by the tenant.

[3] **'Livestock rearing'.** This term was substituted for 'Hill Farming' by the Livestock Rearing Act, 1951, section 1 (2) (b).

[4] This refers to such grants as are made under the Hill Farming Act, 1946.

[5] Where under that section the landlord undertook to execute improvements under Part II of the First Schedule to the 1923 Act he was entitled to charge 5 per cent on his expenditure.

[6] This refers to interest (at a rate to be fixed, in the absence of agreement, by the Treasury) on expenditure by the landlord on executing works for the supply of water.

[7] As amended by Livestock Rearing Act, 1951 section 1 (2) (b).

Variation of terms of tenancy as to permanent pasture.

9.[1]—(1) Where under the lease of an agricultural holding, whether entered into before or after the commencement of this Act, provision is made for the maintenance of specified land, or a specified proportion of the holding, as permanent pasture, the landlord or the tenant may, by notice in writing served upon his tenant or landlord, demand a reference to arbitration under this Act of the question whether it is expedient in order to secure the full and efficient farming of the holding that the amount of land required to be maintained as permanent pasture should be reduced.

(2) On a reference under the foregoing sub-section the arbiter may by his award

(*a*) direct that the lease shall have effect subject to such modifications of the provisions thereof as to the land which is to be maintained as permanent pasture or is to be treated as arable land, and as to cropping as may be specified in the direction; and

(b) if he gives a direction reducing the area of land which under the lease is to be maintained as permanent pasture, order that the lease shall have effect as if it provided that on quitting the holding on the termination of the tenancy the tenant should leave as permanent pasture, or should leave as temporary pasture sown with seeds, mixture of such kind as may be specified in the order, such area of land (in addition to the area of land required by the lease, as modified by the direction, to be maintained as permanent pasture) as may be so specified, so however that the area required to be left as aforesaid shall not exceed the area by which the land required by the lease to be maintained as permanent pasture has been reduced by virtue of the direction.[2]

[1] This section has been amended by the Agriculture Act, 1958, Schedule 1, Part II. In the 1949 Act questions arising under this section were dealt with by the Secretary of State; but now landlord or tenant may demand a reference to arbitration of all such questions.

[2] See section 63 (1) which excludes compensation for restoration of pasture in accordance with an order under Section 9 (2) (b) and which requires averaging of the value of pasture.

10.[1]—The lease of an agricultural holding shall not be deemed to have been brought to an end, and accordingly neither the landlord nor the tenant of the holding shall be entitled to bring proceedings to terminate the lease or, except with the consent of the other party, to treat it as at an end, by reason only that any new term has been added to the lease or that any of the terms of the lease (including the rent payable thereunder) have been varied or revised in pursuance of any of the foregoing provisions of this Act in that behalf.

Leases to continue in force notwithstanding variation of terms, &c.

[1] The effect of this section is not clear. It may be that, if the terms of a lease are varied by arbitration—e.g., with regard to rent, or with regard to the respective liabilities of landlord and tenant for the maintenance of fixed equipment the existing lease is continued and there is no new lease. The provisions of section 5 are not held to be incorporated in the lease. They would be, however, if the lease, as varied, was regarded as a new lease. In view of the terms of section 4 these last mentioned provisions could be incorporated in any lease continued by tacit relocation, but, if they are, an arbiter would require to vary the rent payable in order to take into account any transfer of liability.

Miscellaneous Provisions Affecting the Relationship of Landlord and Tenant

Certain agreements by incoming tenant to pay compensation due to outgoing tenant to be void.

11.—(1) Subject to the provisions of this section, any agreement made after the first day of November, nineteen hundred and forty-eight, by the incoming tenant of an agricultural holding with his landlord whereby the incoming tenant undertakes to pay to an outgoing tenant any compensation payable by the landlord under or in pursuance of this Act or the Agricultural Holdings (Scotland) Acts, 1923 to 1948, in respect of improvements or to refund to the landlord any compensation payable as aforesaid which has been paid by the landlord to an outgoing tenant, shall be null and void.[1]

(2) This section shall not apply to an agreement in writing entered into by the incoming tenant of a holding with his landlord whereby the incoming tenant undertakes to pay to an outgoing tenant, up to such maximum amount as may be specified in the agreement, any compensation payable by the landlord under or in pursuance of this Act or the Agricultural Holdings (Scotland) Acts, 1923 to 1948, in respect of the whole or part of any improvement of the kind specified in Part III of the First Schedule to this Act,[2] or to refund to the landlord, up to such maximum amount as aforesaid, any compensation so payable which has been paid by the landlord to an outgoing tenant.

[1] This section refers only to agreements between landlord and incoming tenant. Probably the incoming and outgoing tenants are free to make such bargain as they think fit with the landlord's consent. If such consent was formal the agreement might be regarded as void under the section.

[2] This refers to temporary improvements such as compensation for feeding tuffs and manures.

Freedom of cropping and disposal of produce.

12.[1]—(1) Subject to the provisions of this section, the tenant of an agricultural holding shall, notwithstanding any custom of the country or the provisions of any lease or of any agreement respecting the disposal of crops or the method of cropping of arable lands,[2] have full right, without incurring any penalty, forfeiture or liability,—[3]

(a) to dispose of the produce of the holding,[4] other than manure produced thereon;

(b) to practise any system of cropping[5] of the arable land on the holding:

Provided that this sub-section shall not have effect unless,

RELATIONSHIP OF LANDLORD AND TENANT

before exercising his rights thereunder or as soon as may be after exercising them, the tenant makes[6] suitable and adequate provision,[7] in the case of an exercise of the right to dispose of produce, to return to the holding the full equivalent manurial value to the holding of all crops sold off or removed from the holding in contravention of the custom, contract or agreement,[8] and, in the case of an exercise of the right to practice any system of cropping, to protect the holding from injury or deterioration.

(2) If the tenant of an agricultural holding exercises his rights under the foregoing sub-section in such a manner as to injure or deteriorate, or to be likely to injure or deteriorate, the holding,[9] the landlord shall have the following remedies, but no other, that is to say—

(a) should the case so require, he shall be entitled to obtain an interdict restraining the exercise of the tenant's rights under that sub-section in that manner;[10]

(b) in any case, on the tenant quitting the holding on the termination of the tenancy the landlord shall be entitled to recover damages[11] for any injury to or deterioration of the holding attributable to the exercise by the tenant of his rights under that sub-section.

(3) For the purposes of any proceedings for an interdict brought under the last foregoing sub-section, the question whether a tenant is exercising, or has exercised, his rights under sub-section (1) of this section in such a manner as to injure or deteriorate his holding, or to be likely to injure or deteriorate his holding, shall be <u>determined by arbitration; and a certificate of the arbiter</u> as to his determination of any such question as aforesaid shall, for the purposes of any proceedings,[12] (including an arbitration) brought under this section, be conclusive proof of the facts stated in the certificate.

(4) Sub-section (1) of this section shall not apply—

(a) in the case of a tenancy from year to year, as respects the year before the tenant quits the holding or any period after he has given or received notice to quit which results in his quitting the holding;[13] or

(b) in any other case, as respects the year before the expiration of the lease.[14]

(5) In this section the expression 'arable land' does not include

land in grass which, by the terms of a lease, is to be retained in the same condition throughout the tenancy.[15]

In this sub-section any reference to the terms of a lease shall, in a case where the Secretary of State has directed under section twelve of the Agriculture (Scotland) Act, 1948, or under section *nine* of this Act, or an arbiter has directed under the said section nine, that the lease shall have effect subject to modifications, be construed as a reference to the terms of the lease as so modified.

[1] **'Cropping, etc.'.** See Introduction. Where there is no written lease, and where the lease is silent as to cropping, the tenant must leave the land in the rotation customary in the district.

[2] **'Arable land'.** See definition in sub-section (5), *infra*. All land may be considered arable which is being or has been cultivated, and has not returned to its natural state. Rough hill pasture is not 'arable'.

[3] **'Any penalty, forfeiture, or liability'.** Any attempt to impose such by lease would be invalid. A stipulation, for example, that a tenant who sold off turnips should bring back, say, twice their manurial value could not be enforced to the full extent. To the extent of one-half, it would be a penalty.

[4] **'Produce of the holding'** is very general, and would cover crops which according to usual contract, and even according to what were wont to be considered the rules of good husbandry, had to be consumed on the holding—e.g., turnips, straw, hay, &c. Reference should be made to section 93, which defines 'produce' as including anything (whether live or dead) produced in the course of agriculture. It would not be right for an arbiter to hold that a tenant should pay compensation for removing such produce off the farm in respect merely that he was thereby infringing the rules of good husbandry. It would, however, be competent and right to hold a tenant liable in compensation, not for removing the crops but, after having removed the crops, for failing to satisfy the requirements of this section by bringing back equivalent manurial value. It is not for taking away, but for failing to bring back, that compensation is due. There is a difference here in principle which should not be overlooked. A tenant has the right to keep all kinds of farming livestock in accordance with normal practice and cannot be prevented by his lease from doing so.

[5] **'System of cropping'.** The tenant is free to grow any crops he may think proper, and that with or without following a regular rotation. An express contract to follow a particular rotation can be ignored except so far as it applies to the last year. It is in doubt whether the tenant may cease cropping or lay down grass on a large scale over practically the whole of a holding.

[6] Where practicable, this provision should be made before departing from contract or custom, although the tenant is entitled to make an infringement provided that, 'as soon as may be', he makes provisions in the nature of restoration. 'As soon as may be' would be interpreted with reference to usual agricultural conditions and practice.

[7] **'Suitable and adequate provision'** will, of course, vary according to circumstances. Sometimes it is attempted, in leases, to specify what would be suitable and adequate provision, but apart from the removal of crops the conditions are so varied that it is scarcely possible to meet them all by specification, and it is generally safer not to attempt to do so. In the case of removal of crops the case is different.

[8] Here we have a standard imposed, and so long as that standard is followed the landlord would not be entitled to find fault with the tenant for failure to protect the holding from injury or deterioration arising from the disposal of crops contrary to custom or contract (*Stark* v. *Edmonstone*, 1826, 5 S. 45).

As to 'full equivalent manurial value'—see Introduction. It is thought that this includes mechanical as well as chemical properties.

'In contravention of custom, lease, or agreement'. In order to secure that full and adequate provision be made, the wise course would be to adhere, in leases, to the usual clauses prohibiting the removal of crops which are commonly consumed on the land, and especially straw and turnips. Where potatoes are grown, it is generally necessary that they be sold for removal off the farm, and that would not be prevented by a prohibitory clause which would merely have the effect of requiring the full manurial value of the potatoes to be applied to the land. Such a provision is not at all unreasonable in view of the fact that the potato crop is very exhaustive of the fertility of the soil.

With reference to 'custom', it may be pointed out that it is not the custom of the particular estate that is referred to, unless in some way that custom had been imported into the contract of tenancy (*Allan* v. *Thomson*, 1829 7 S. 784; *Officer* v. *Nicolson*, 1807 Hume 827; *Anderson* v. *Tod*, 1809 Hume 842). It is the custom of the country or district that is meant. This custom to be effective must be general or, if local, must be known to and relied on by both parties (*Armstrong & Co.* v. *M'Gregor & Co.*, 1875 2 R. 339; *Anderson* v. *McCall*, 1866, 4 M. 765; *Holman* v. *Peruvian Nitrate Company*, 1878, 5 R. 657).

9 The Act assumes the right of the tenant to injure or deteriorate the holding temporarily, because it contemplates the making of 'adequate provision' after the event. Further, it allows 'the return' to the holding of 'equivalent manurial value' of crops sold off. Accordingly, it is thought that interdict would only be granted against gross acts, and especially such as could not readily be made good. See Rankine on Law of Leases, 3rd Edn., p. 418.

10 It would appear that, failing agreement, the amount of the damages could be settled by arbitration under section 74.

11 The measure of damages is the injury to the reversion on the determination of the tenancy, i.e., the diminution of the rent which the landlord can obtain on reletting (*Williams* v. *Lewis*, [1915] 3 K.B. 493).

12 Reference to arbitration was introduced by the Agriculture Act, 1958, Schedule 1, Part II. Formerly questions arising under this sub-section were decided by the Secretary of State.

13 As, in order to terminate all tenancies, including those renewed from year to year on tacit relocation, notice for not less than one nor more than two years is required, there will be very few cases in which the quitting tenant will not either receive or give such notice. He will generally know what is the last year of the tenancy before beginning to lay down the last crop. Do the tenant's powers under this section cease the moment notice to quit is given in the case where the landlord gives longer than one year's notice? It seems unreasonable to suppose that the rights of the tenant under (a) were less than under (b), and the latter extends only to 'the year before the expiration of the lease'. In this view, freedom of cropping and sale of produce may operate even in the major part of the penultimate year where nearly two years' notice to terminate the tenancy has been given. It is difficult to give a meaning to (a) in the context of a notice to quit which must be at least one year. It is suspected that there may have been an inappropriate repetition of the provisions of the English Act. See Muir Watt on Agricultural Holdings, 12th Edn., p. 260.

See also section 13, which prevents sale or removal of hay, straw, roots, &c., grown in the last year. It should be noted, however, that the present section prohibits the sale or removal of any crops in the last year, whether grown in that year or not (*Gale* v. *Bates*, 1864 33 L.J. Ex. 235; *Meggeson* v. *Groves*, [1917] 1 Ch. 158).

14 See preceding note. What is the year before the expiration of the lease? It is thought that, in the case of an arable farm, if a lease provides for entry at Whitsunday as to houses and grass and at separation of crop thereafter

122 AGRICULTURAL HOLDINGS (SCOTLAND) ACTS

as regards land under crop, the expiration 'of the lease' would be the Martinmas in the last year of the tenancy. The doubt arises from the different expressions, *'the year before the tenant quits, &c.,' 'the year before the expiration of the lease,'* but this section is concerned with cropping, and it would be strange indeed if the period of restriction were terminated prior to reaping. *'Expiration of the lease'* may be at a *'break'*. (See *Edell* v. *Dulieu*, 1924 A.C. 38: *Alston's Trustees* v. *Muir*, 1919 2 S.L.T. 8). (See also *'expiration of the stipulated endurance of any lease'* in section 2 (1).)

In respect of (a) and (b) the tenant should be careful to see that, towards the natural termination of his lease, whether at a break or not, he has his cropping arrangements in such order that they would permit of his leaving the land in the rotation prescribed by his lease or, if there be none prescribed, in the rotation customary in the district (*Carron Co.* v. *Donaldson*, 1858 20 D. 681; *Hunter* v. *Miller*, 1862, 24 D. 1011; *Marquis of Tweeddale* v. *Brown*, 1821, 2 Mur. 563; *Meggeson* v. *Groves, supra*). (See also Rules of Good Husbandry.) It is not, however, clear whether a tenant is bound to do this or only to take such steps as he reasonably can after giving or receiving notice to quit. He cannot say when the tenancy will end until such notice is given.

[15] See note 2, *supra*.

Prohibition of removal of manure, &c., after notice to terminate the tenancy.

13.—Where notice to terminate the tenancy of an agricultural holding is given, either by the tenant or by the landlord, the tenant shall not, subject to any agreement to the contrary, at any time after the date of the notice, sell or remove from the holding any manure or compost, or any hay or straw or roots grown in the last year[1] of the tenancy, unless and until he has given the landlord or the incoming tenant a reasonable opportunity[2] of agreeing to purchase on the termination of the tenancy at their fair market value, or at such other value as is provided by the lease, the said manure, compost, hay, straw or roots.[3]

---

[1] As hitherto, this will generally be regulated by the conditions of the lease but, in so far as that is not done, this section will operate, unless there is contracting out. In the case of a Whitsunday and separation of crop entry, the last year would, it is thought, be Martinmas to Martinmas just as in the case of a Martinmas entry, because separation of the last crop must be included.

[2] What is a reasonable opportunity? In *Barbour* v. *M'Douall*, 1914 S.C. 844, it was held, in connection with a claim for compensation for unreasonable disturbance under the 1912 Act, that the tenant had no duty to give any intimation to the landlord before advertising his stock for sale. Nevertheless, it would be prudent for the tenant to write to the landlord and incoming tenant, offering to sell at a fair market value or at the value provided by the contract of tenancy. It will be noted that there is no direct provision here for referring questions as to the value to arbitration, but this would apparently be covered by section 74. The point is not, however, clear, especially where the incoming tenant agrees to take over from the away-going tenant, because the scheme of the act is generally confined to questions and claims between landlord and tenant, and section 74 is expressly so confined. (See also *Roger* v. *Hutcheson*, 1922 S.C. (H.L.) 140.)

[3] 'Roots' includes mangolds, swedes, turnips, and cabbage, but not potatoes. See introduction for procedure and principles of valuation.

14.[1]—(1) Subject to the provisions of this section— Tenant's right to remove fixtures and buildings.
  (a) any engine, machinery, fencing or other fixture[2] affixed to an agricultural holding by the tenant thereof; and
  (b) any building (other than one in respect of which the tenant is entitled to compensation under this Act or otherwise) erected by him on the holding;

not being a fixture affixed or, as the case may be, a building erected, in pursuance of some obligation in that behalf or instead of some fixture or building belonging to the landlord, as the case may be, shall be removable by the tenant at any time during the continuance of the tenancy or before the expiration of six months, or such longer period as may be agreed, from the termination of the tenancy[3] and shall remain his property so long as he may remove it by virtue of this sub-section.

(2) The right conferred by the foregoing sub-section shall not be exercisable in relation to a fixture or building unless the tenant—
  (a) has paid all rent owing by him and has performed or satisfied all his other obligations to the landlord in respect of the holding;[4] and
  (b) has, at least one month before both the exercise of the right and the termination of the tenancy, given to the landlord notice in writing of his intention to remove the fixture or building.[5]

(3) If, before the expiration of the notice aforesaid, the landlord gives to the tenant a counter-notice in writing electing to purchase a fixture or building comprised in the notice, sub-section (1) of this section shall cease to apply to that fixture or building, but the landlord shall be liable to pay to the tenant the fair value thereof to an incoming tenant of the holding.[6]

(4) In the removal of a fixture or building by virtue of sub-section (1) of this section, the tenant shall not do to any other building or other part of the holding any avoidable damage, and immediately after the removal shall make good all damage so done that is occasioned by the removal.

[1] See Introduction, and section 17, as to record of holding. It is competent to contract out of this section (*Premier Dairies* v. *Garlick*, [1920], 2 Ch. 17). See also Opencast Coal Act, 1958, section 27.
[2] **'Fixture'**. The Act gives no definition of 'fixture'. The definition of 'fixed equipment' is now much wider (section 93) as it includes buildings, fences, drains, &c. It must be assumed, therefore, that this provision only applies to things which are of the nature of landlord's fixtures; such as

shelving, grates, &c., which can be removed entire, and are only slightly fixed do not come under this section. But 'fixtures' erected for the purpose of trade, though excepted in the general case fall to be dealt with under the section (*Elwes* v. *Maw*, 2 Smith's Leading Cases 188, 12th edn.). The words 'Engine, machinery, fencing' appear to be qualified by the words 'or other fixture'. Again 'any building' applies, it is thought, only to structures of a permanent nature. Glass-houses on market gardens were held to be 'buildings' (*Smith* v. *Richmond*, [1899] A.C. 448; *Meux* v. *Cobley*, [1892], 2 Ch. 253). But in *Mears* v. *Callender*, [1901] 2 Ch. 388, it was held that glass-houses erected by a tenant could be removed under his common law rights. It is thought that any engine, machinery, fencing, or temporary structure, which the tenant could remove at common law as not being in its nature a 'fixture', can be removed without regard to the provisions of this section. Section 100 seems to reserve the tenant's common law rights sufficiently in this connection. Section 14 is clearly intended to extend the rights of the tenant so far as regards fixtures and buildings erected by him. It is concerned with 'Tenant's property in fixtures and buildings'.

It is sometimes difficult to determine whether a thing falls within the category of fixture or not, and, where there is doubt, it will generally be safer to follow the procedure prescribed by this section. In determining, at common law, whether a thing is a fixture or not, the points for consideration are (1) whether the thing can be moved without injury to itself or to the subjects; and (2) the object for which the thing was attached to the soil, whether as a permanent improvement or for a merely temporary purpose.

There is this distinction, in the case of things which are not fixtures; they can generally only be removed before, or, at latest, at the termination of the tenancy (*Brand's Trustees* v. *Brand's Trustees*, 3 R. (H.L.) 16; *Miller* v. *Muirhead*, 1894, 21 R. 658), whereas in the case of 'fixtures' they may be removed within a reasonable time thereafter. What is a reasonable time depends on circumstances. For example, if there is a *bona fide* controversy as to the right of the tenant to remove the alleged fixtures, or if there is delay on the tenant's part in implementing an obligation under the lease, and he is acting *bona fide*, a reasonable period of delay might be of considerable extent.

[3] **'Termination of the tenancy'.** See definition in section 93 and notes. If the landlord should refuse to allow the removal of fixtures, this would confer a right of action by the tenant for their value (*Thomas* v. *Jennings*, 66 L.J. Q.B. 5). A renunciation of a lease, *prima facie*, includes all fixtures on the holding (*Leschallas* v. *Woolf*, [1908] 1 Ch. 641), but there seems no reason why the terms of this section should not apply in that case.

[4] Generally all claims at the landlord's instance must be settled before the removal of the fixtures, but this implies that the landlord does not unreasonably hold back his claims. In a case where a tenant was in arrears with his rent at the time of giving notice, it was held that this deferred his right to compensation but did not extinguish it. *Roberts* v. *Magor* (1953), 103 L.J. 703, C.C.

[5] See note to section 24 (6).

[6] **'Fair value to an incoming tenant'** seems to accord with the basis on which compensation is payable for permanent and other improvements. Any difference as to the value could be settled by arbitration under section 74. It will be noted that the fixtures 'become the property of the landlord'.

Compensation for damage by game.

15.[1]—(1) Subject to the provisions of this section, where the tenant of an agricultural holding has sustained damage to his crops from game,[2] the right to kill and take which is vested neither in him nor in anyone claiming under him other than the landlord, and which the tenant has not premission in writing to kill,[3] he shall be entitled to compensation from his landlord for the damage if it exceeds in amount the sum of one shilling per acre of the area over which it extends:[4]

Provided that compensation shall not be recoverable under this section unless—
  (a) notice in writing is given to the landlord as soon as may be after the damage was first observed by the tenant, and a reasonable opportunity is given to the landlord to inspect the damage[5]—
    (i) in the case of damage to a growing crop, before the crop is begun to be reaped, raised or consumed;[6] and
    (ii) in the case of damage to a crop reaped or raised, before the crop is begun to be removed from the land;[7] and
  (b) notice in writing of the claim, together with the particulars thereof, is given to the landlord within one month after the expiration of the calendar year, or such other period of twelve months as by agreement between the landlord and the tenant may be substituted therefor, in respect of which the claim is made.[8]

(2) The amount of compensation payable under this section shall, in default of agreement made after the damage has been suffered, be determined by arbitration.

(3) Where the right to kill and take the game is vested in some person other than the landlord, the landlord shall be entitled to be indemnified by that other person against all claims for compensation under this section;[9] and any question arising under the foregoing provisions of this sub-section shall be determined by arbitration.

(4) In this section the expression 'game' means deer, pheasants, partridges, grouse and black game.

[1] See Introduction.
[2] **'Game'**. For definition, see sub-section (4). Ground game is not included, because the tenant has statutory right to protect himself against damage by such game under the Ground Game Acts. Compensation is, however, payable under this section despite the tenant's right to kill deer under section 43 (1) of the Agriculture (Scotland) Act, 1948. *Lady Auckland* v. *Dowie*, 1964, S.L.T. (Land Ct.) 20; 1965 S.L.T. 76.
[3] In England an agricultural tenant is vested under common law in the right to kill 'game' unless the right is expressly excluded. The damage referred to in this section includes damage by game coming from an adjacent estate; and even during close time (*Thomson* v. *Earl of Galloway*, 1919 S.C. 611). It has been suggested that the permission to kill would require to be co-

extensive with the entire holding; at all events if the holding be in one 'unit'. But, on the contrary, it may be argued that, if the permission were confined to certain fields, that would bar a claim for damage in these fields. It is difficult to decide between these views, but the latter appears sounder, and would simply place the tenant in the same position, in relation to the excluded fields, as if they were in the occupation of another tenant. Obviously, the permission must be given in sufficient time to enable the tenant to have an opportunity of preventing the damage. At common law, a tenant has the right to kill ground game on his farm.

[4] Here the question is, does the damage over the area damaged (not over the whole field or over the farm) exceed 1s per acre of that area? If it does not, no compensation is due; if it does, the full compensation is due, without deduction of the one shilling per acre. The compensation is due, even though the landlord can prove that he took all reasonable steps to keep down the game.

[5] What is 'a reasonable opportunity' will be matter of opinion, and will depend on the circumstances. If for any reason the time is limited, the fact should be stated in the notice. For example, the tenant may desire to get on with his harvesting, and, if he were to delay long, he might thereby suffer greater loss than the amount of the game damage.

[6] This evidently means that the crop must be left growing, and it is a question whether it only applies to the area of the actual damage. It would be safer to leave the whole field uncut.

[7] This applies only to the case where the damage has been done after the crop was cut but before it was begun to be removed. In such a case the crop must be left standing on the ground. To remove it, say, into the stackyard, would be removing it from the land. It is a question if this applies only to the area of the actual damage. It would be safer to leave the whole crop in the particular field.

[8] The calendar year is from 1st January to 31st December. This proviso might result in a claim being made after the tenant has quitted possession. The particulars here required need only consist of a statement of the fact of the damage by game as defined in this section, of the approximate date when the damage was caused, of the crop damaged, of the locality (field), approximately the area of the damage, and the amount claimed.

[9] It seems strange that the game tenant should be liable, but there is nothing in this section to prevent him from contracting out with the landlord. Where the damage is caused not by game reared by the landlord but by game from coverts on an adjoining property, no right of indemnity is conferred by the Act, but it is thought that at common law the landlord would have a right of indemnity against the neighbouring proprietor who reared the game which caused the damage (*Farrer* v. *Nelson,* 15 Q.B.D. 258). The sporting tenant cannot be a party in his own name to an arbitration for game damage at the instance of the agricultural tenant against the landlord, nor can the agricultural tenant claim against the game tenant (*Inglis* v. *Moir's Tutors,* 1871, 10 M. 204), although the landlord has relief against the game tenant (*Kidd* v. *Byrne,* 1875, 3 R. 255). His right of relief is now to be determined by arbitration.

Restriction of landlord's right to penal rent or liquidated damages.

16.[1]—Notwithstanding any provision in a lease of an agricultural holding[2] making the tenant thereof liable to pay a higher rent or other liquidated damages in the event of a breach or non-fulfilment of any of the terms or conditions in the lease, the landlord shall not be entitled to recover any sum in consequence of any breach or non-fulfilment in excess of the damage actually suffered by him in consequence of the breach or non-fulfilment.³

¹ See Introduction.
² See definition, section 93.
³ This strikes at those clauses in leases, which require the tenant to pay a penalty on the basis of one, two, or more rents in cases where he infringes the conditions of his lease generally with reference to cultivation and cropping. A tenant who was bound not to sell hay or straw under a penalty of £3 a ton for every ton sold off was found liable only for the actual manurial value of the hay sold off (*Willson* v. *Love* [1896], 1 Q.B. 626).

17.¹—(1) The landlord or the tenant of an agricultural holding² may, at any time during the tenancy, require the making of a record³ of the condition of the fixed equipment on, and of the cultivation of, the holding; and the tenant may, at any time during the tenancy, require the making of a record of— [Making of record of condition, &c. of holding.]

(a) existing improvements carried out by him or in respect of the carrying out of which he has, with the consent in writing of his landlord, paid compensation to an outgoing tenant;⁴ and

(b) any fixtures or buildings which, under section *fourteen* of this Act, he is entitled to remove.

(2) Any record under this section shall be made by a person⁵ to be appointed by the Secretary of State, and shall be in such form as may be prescribed.

(3) The cost of making a record under this section⁸ shall, in default of agreement between the landlord and the tenant, be borne by them in equal shares.

(4) Any record made under this section shall show any consideration or allowances which have been made by the landlord to the tenant or by the tenant to the landlord.

(5) Subject to the provisions of section *five* of this Act, a record may, if the landlord or the tenant so requires, be made under this section relating to a part only of the holding or to the fixed equipment only.

(6) Any question or difference between the landlord and the tenant arising out of the making of a record under this section shall, on the application of the landlord or the tenant, as the case may be, be referred to the Land Court, and the Land Court shall determine such question or difference accordingly.

(7) The remuneration of the person appointed by the Secretary of State to make a record under this section shall be such amount as the Secretary of State may fix, and any other expenses of and

incidental to the making of the record shall be subject to taxation by the auditor of the Sheriff Court, but that taxation shall be subject to review by the Sheriff.

(8) The remuneration of the person appointed by the Secretary of State to make a record under this section shall be recoverable by that person from either the landlord or the tenant, but any amount paid by either of those parties in respect of that remuneration, or of any other expenses of and incidental to the making of the record, in excess of the share payable by him as aforesaid of the cost of making the record shall be recoverable by him from the other party.

[1] See Introduction and footnote 1 to section 5.
[2] See definitions, section 93.
[3] Such records are essential to claims for high farming (section 56) and to statutory claims for deterioration (sections 57 and 58). In any new lease there must be a record of the fixed equipment. It is suggested that photographs of defects in the condition of buildings, fences, &c., could be held part of a record. Such records are useful as evidence, especially in arbitrations at away going.
[4] See sections 46 and 55.
[5] Not necessarily a member of the panel of arbiters.

Power of landlord to enter on holding.

18. The landlord of an agricultural holding or any person authorised by him may at all reasonable times enter on the holding for any of the following purposes, that is to say—
   (a) viewing the state of the holding;[1]
   (b) fulfilling the landlord's responsibilities to manage the holding in accordance with the rules of good estate management;[2]
   (c) providing, improving, replacing or renewing fixed equipment on the holding otherwise than in fulfilment of his said responsibilities.

[1] Under common law, a landlord in Scotland can enter a holding for inspection or similar purposes. See also Agriculture (Safety, Health and Welfare Provisions) Act, 1956, section 25 (4).
[2] These rules are defined in the Fifth Schedule to the 1948 Act. See Introduction.

Removal of tenant for non-payment of rent.

19.—(1) When six months' rent of an agricultural holding is due and unpaid, the landlord[1] shall be entitled to raise an action of removing[2] in the Sheriff Court against the tenant, concluding for his removal from the holding at the term of Whitsunday or Martinmas[3] next ensuing after the action is raised, and the sheriff may, unless the arrears of rent then due are paid or caution is found to his satisfaction for them, and for one year's rent further, decern the tenant to remove, and may eject him at the said term in like manner as if the lease were determined and the tenant had been legally warned to remove.

(2) A tenant of a holding removed under the foregoing subsection shall have the rights of an outgoing tenant[4] to which he would have been entitled if his tenancy had terminated naturally at the term of Whitsunday or Martinmas aforesaid.

(3) The provisions of section five of chapter XV of Book L of the Codifying Act of Sederunt of the fourteenth day of June, nineteen hundred and thirteen, anent removings shall not apply in any case where the procedure under this section is competent.

[1] **'Landlord'**. See definition, section 93. The landlord in possession is entitled to take proceedings to recover rent although he may have to account therefor to his predecessors' representatives (*Lennox* v. *Reid*, 1893, 21 R. 77; *Lord Elibank* v. *Hay*, M. 13869; Hailes, 847).

[2] **'Action of removing'**. This action need not be raised forty days before the term of removal (*Ballantyne* v. *Brechin*, 1893, 1 S.L.T. 306). The action was formerly for immediate removal. An action was raised in the Sheriff Court for payment of arrears which failing for decree of removing. Held that the action was incompetent in respect that the Act gave the tenant right to prevent ejection by making payment of arrears at any time prior to extract of decree of removing, and the decree sought would deprive him of that right (*Fletcher* v. *Fletcher*, 1932 S.L.T. (Sh.Ct.) 10). Such an action can, of course, be met by a defence which will force both parties to arbitration—see *Brodie* v. *Ker*, 1952 S.C. 216. Held that an irritancy, being a conventional irritancy, e.g., for payment of rent, could not be purged (*M'Douall's Trustees* v. *MacLeod*, 1949 S.C. 593).

[3] **'Whitsunday'** and **'Martinmas'**. See Introduction and section 93 for definition.

[4] **'Rights of an outgoing tenant'**. This entitles the tenant to the usual common law away-going rights, implied or expressed by the terms of the lease, and including, it is thought, compensation for improvements under this Act, provided the provisions of the Act are complied with. Questions may arise with reference to the term of removal. Suppose a tenancy runs from Whitsunday as to houses and grass and separation of crop thereafter as regards land under crop, would the tenancy nevertheless expire as regards the whole at the one term, say Whitsunday, or would the Sheriff follow the away-going in terms of the lease? If he were not to do so the rights of the tenant would be prejudiced, apparently contrary to the intention of the statute. Under the Hypothec Abolition (Scotland) Act, 1880, the rights reserved to the tenant were such as he could have had at the term preceding his ejection, where he was ejected between terms.

20.[1]—[(1) Subject to the provisions of this section, the tenant of an agricultural holding[2] may, by will or other testamentary writing, bequeath[3] his lease of the holding to his son-in-law or daughter-in-law or any one of the persons who would be, or would in any circumstances have been, entitled to succeed to

Bequest of lease.

the estate on intestacy by virtue of the Succession (Scotland) Act, 1964.][4]

(2) A person to whom the lease of a holding is bequeathed as aforesaid (in this section referred to as 'the legatee') shall, if he accepts the bequest, give notice[5] of the bequest to the landlord of the holding within twenty-one days after the death of the tenant, or, if he is prevented by some unavoidable cause[6] from giving such notice within that period, as soon as possible thereafter. The giving of such notice shall import acceptance of the lease and, unless the landlord gives a counter-notice under the next following sub-section, the lease shall be binding on the landlord and on the legatee, as landlord and tenant respectively, as from the date of the death of the deceased tenant.

(3) Where notice as aforesaid has been given to the landlord he may within one month after the giving of the notice give to the legatee a counter-notice intimating that he objects to receive him as tenant under the lease.

(4) If the landlord gives a counter-notice under the last foregoing sub-section, the legatee may make application[7] to the Land Court for an order declaring him to be tenant under the lease as from the date of the death of the deceased tenant.

(5) If, on the hearing of such application, any reasonable ground of objection[8] stated by the landlord is established to the satisfaction of the Land Court, they shall declare the bequest to be null and void, but in any other case they shall make an order in terms of the application.

(6) Pending any proceedings under this section, the legatee [with the consent of the executor in whom the lease is vested under Section 14 of the Succession (Scotland) Act, 1964] shall, unless the Land Court on cause shown otherwise direct, have possession of the holding.[9]

[(7)[10] If the legatee does not accept the bequest, or if the bequest is declared null and void as aforesaid, the right to the lease shall be treated as intestate estate of the deceased tenant in accordance with Part I of the Succession (Scotland) Act, 1964.]

---

[1] This section must be read together with section 6 of the Agriculture Act, 1958, section 16 of the Succession (Scotland) Act, 1964, and Part III of the Agriculture (Miscellaneous Provisions) Act, 1968. The wording in square brackets was introduced by the Succession (Scotland) Act, 1964. See Appendix. Parties can contract out of this section. It has been held that where there was an express exclusion of the tenant's legatee in a lease, section

20 could not be invoked. *Kennedy* v. *Johnstone and Another*, 1956 S.C. 39. This decision is not affected by section 29 (2) of the Succession (Scotland) Act, 1964, which provides that an *implied* condition prohibiting assignation does not render a bequest of a lease invalid. Are testamentary trustees legatees within the meaning of this section? This was the view taken by the Land Court. *Andrew Linton's Trustees* v. *Wemyss Landed Estates Co. Ltd.*, 1953 S.L.C.R. 14. In *Kennedy* v. *Johnstone, supra*, the Lord President said *obiter* that it was inappropriate to describe trustees as legatees.

2 **'Agricultural holding'**. See definition, section 93.

3 It is prudent to make specific reference in the will to the lease.

4 In the 1949 Act the tenant's right of bequest extended to 'any person'. The Agriculture Act, 1958, section 6 (1) limited this right of bequest to 'any member of his family', as defined in section 6 (2). The present wording was introduced by section 16 of the Succession (Scotland) Act, 1964.

5 **'Notice'**. Although this is not expressly required to be in writing, it is clearly advisable, in order to avoid question both as to the fact and the date of the intimation. For intimation by telephone see *Irving* v. *Church of Scotland General Trustees*, 1960 S.L.C.R. 16.

6 **'Unavoidable cause'**. For example, the legatee might be abroad and he might not hear of the bequest in time to permit of the intimation being made within the twenty-one days. But see *Wight* v. *Marquis of Lothian's Trustees*, 1952 S.L.C.R. 25. Where a will was not found within 21 days of testator's death the legatee was held prevented by unavoidable cause: *Mackinnon* v. *Martin*, 1958 S.L.C.R. 19. For circumstances where a widow's exhaustion and inability to give instructions were held not to be an unavoidable cause see *Thomson* v. *Lyall*, 1966 S.L.C.R. App. 136.

7 **'Application'**. No time limit is specified for the application. If the tenant's legatee neglects to follow procedure in this section landlord may apply to have bequest declared null and void. *Wight* v. *Marquis of Lothian's Trustees Supra*.

8 **'Objection'**. The Land Court is given a wide discretion as to these objections. It is thought that they would consider that the landlord is entitled to have a reputable and satisfactory tenant. Objections must be on personal grounds affecting the heir. *Marquis of Lothian's Trustees* v. *Johnston*, 1952 Land Court, Roxburgh (R.N. 92). A person who has not a great amount of agricultural knowledge and skill would not necessarily be rejected, because he may be able to employ a thoroughly experienced grieve. Such a person might, however, be removed by notice to quit. Section 18 (2) of the Agriculture (Miscellaneous Provisions) Act, 1968, directs that the Land Court shall consent to the operation of notice to quit given to a 'near relative' successor if they are satisfied that he 'has neither sufficient training in agriculture nor sufficient experience in the farming of land to enable him to farm the holding . . . with reasonable efficiency'. If the legatee were a person of disreputable habits, or had insufficient capital objection on such grounds should succeed. In *Sloss* v. *Agnew*, 1923 S.L.T. (Sh.Ct.) 33, objection was sustained to a bequest to a daughter who did not possess sufficient skill, capital, or equipment. See also *Service* v. *Duke of Argyll*, 1951 S.L.T. (Sh.Ct.) 2, *Howie* v. *David Lowe & Sons Ltd.*, 1952 S.L.C.R. 14, *Fraser* v. *Murray's Trustees*, 1954 S.L.C.R. 10, *Reid* v. *Duffus Estate Ltd.*, 1955 S.L.C.R. 13.

9 Where the landlord attacks the validity of the bequest, that question falls to be dealt with by proceedings in the usual Court (*Mackenzie* v. *Cameron*, 1894, 21 R. 427).

10 If a legatee or the acquirer on intestacy succeeds to the tenancy a new lease is not thereby constituted.

[**21.**[1]—(1) The acquirer[2] of the lease of an agricultural holding shall give notice of the acquisition to the landlord of the holding within twenty-one days after the date of the acquisition,[3] or, if 'Right of landlord to object to acquirer of lease'.

he is prevented by some unavoidable cause from giving such notice within that period, as soon as possible thereafter, and unless the landlord gives a counter-notice under the next following subsection, the lease shall be binding on the landlord and on the acquirer, as landlord and tenant respectively, as from the date of the acquisition.

(2) Where notice as aforesaid has been given to the landlord he may, within one month after the giving of the notice, give to the acquirer a counter-notice intimating that he objects to receive him as tenant under the lease and not before the expiration of one month from the giving of the counter-notice the landlord may make application to the Land Court for an order terminating the lease.

(3) The Land Court, if they are satisfied that the landlord has established any reasonable ground of objection,[4] shall make such an order to take effect as from such term of Whitsunday or Martinmas as they may specify.

(4) Pending any proceedings under this section, the acquirer, with the consent of the executor in whom the lease is vested under section 14 of the Succession (Scotland) Act 1964, shall, unless the Land Court on cause shown otherwise direct, have possession of the holding.

(5) The termination of the lease under this section shall be treated, for the purposes of the provisions of this Act with respect to compensation, as the termination of the acquirer's tenancy of the holding; but nothing in this section shall be construed as entitling him to any compensation for disturbance.[5]

(6) In this section any reference in relation to the lease of an agricultural holding to an acquirer is a reference to any person to whom the lease is transferred under Section 16 of the Succession (Scotland) Act, 1964.]

[1] This section was substituted by the Succession (Scotland) Act, 1964, Schedule 2, para. 22, for the original section 21 which dealt with the landlord's right to object to the succession to the holding of the tenant's heir-at-law.

[2] **'Acquirer'**. This section provides for notice to be given by the 'acquirer' as defined in sub-section (6) in the same way as a legatee gives notice under section 20. The original section did not require the heir-at-law to give any notice to the landlord.

[3] **'Date of acquisition'**. This may be up to one year after the date of death of the tenant—see Succession (Scotland) Act, 1964, section 16. Delay by the deceased tenant's executor in transferring the lease to one of the persons entitled to succeed to the tenant's estate may thus have the effect of precluding

the landlord from giving the statutory notice to quit against a break in the lease or the conventional ish.

⁴ **'Reasonable ground of objection'.** See note 8 to preceding section.
⁵ He will be entitled to all usual way-going rights under the Act, but difficulties may arise if the occupation is terminated at other than the usual terms.

22.—(1) Where a tenant of an agricultural holding has entered into an agreement, or it is a term of the lease of the holding, that the tenant will, on quitting the holding, sell to the landlord or to the incoming tenant[1] any implements of husbandry, fixtures, farm produce or farm stock on, or used in connection with, the holding, it shall be deemed, notwithstanding anything in the agreement or in the lease to the contrary, to be a condition of the agreement or of the lease, as the case may be, that the property in the goods shall not pass to the buyer until the price is paid and that payment of the price shall be made within one month after the tenant has quitted the holding or, if the price of the goods is to be ascertained by a valuation, within one month after the delivery of the award in the valuation.[2]

(marginal note: Provisions as to payment for implements, &c., sold on quitting holding.)

(2) Where payment of the price is not made within one month as aforesaid the outgoing tenant shall be entitled to sell or remove the goods and to receive from the landlord or the incoming tenant, as the case may be, by whom the price was payable, compensation of an amount equal to any loss or expense unavoidably incurred by the outgoing tenant upon or in connection with such sale or removal, together with any expenses reasonably incurred by him in the preparation of his claim for compensation.[3]

(3) Any question arising as to the amount of compensation payable under the last foregoing sub-section shall be determined by arbitration.

[1] The outgoing tenant can elect with whom he will deal unless the terms of the lease are specific to the contrary. As a rule, he should deal with the landlord, whose interest is permanent.
[2] It is to the interest of the outgoing tenant that the valuation should be carried out and the award issued as soon as possible after the away-going and in the submission a date for the issue of the award might with advantage be fixed.
[3] This provision is difficult of application in the case of sheep stocks bound to the ground.

23.[1]—Where the tenant of an agricultural holding is liable in payment of the whole or any part of the premium due under a fire insurance policy in the name of the landlord over any buildings or other subjects included in the lease of the holding and the

(marginal note: Application of sums recovered under fire insurance policy.)

landlord recovers any sum under such policy in respect of the destruction of, or damage to, the buildings or other subjects by fire, he shall be bound, unless the tenant otherwise agrees, to expend such sum on the rebuilding, repair, or restoration of the buildings or subjects so destroyed or damaged in such manner as may be agreed or as may be determined, failing agreement, by the Secretary of State.

[1] This section can only affect leases entered into before 1st November, 1948, See section 5 (4).

PROVISIONS AS TO NOTICES TO QUIT

Provisions as to giving of notices to quit.

24.[1]—(1) Notwithstanding the termination of the stipulated[2] endurance of any lease[3] of an agricultural holding, the tenancy shall not come to an end unless, not less than one year nor more than two years before the termination of the lease, written notice has been given by either party to the other of his intention to bring the tenancy to an end.[4]

The provisions of this sub-section shall have effect notwithstanding any agreement[5] or any provision in the lease to the contrary.

(2) In the case of a lease continued in force by tacit relocation the period of notice required to terminate the tenancy shall be not less than one year nor more than two years.

(3) The provisions of the Sheriff Courts (Scotland) Act, 1907, relating to removings shall, in the case of an agricultural holding to which this section applies, have effect subject to the provisions of this section.[6]

(4) Notice by the landlord to the tenant under this section shall be given either—
    (*a*) in the same manner as notice of removal under section six of the Removal Terms (Scotland) Act, 1886; or
    (*b*) in the form and manner prescribed by the Sheriff Courts (Scotland) Act, 1907;
and such notice shall come in place of the notice required by the said Act of 1907.[7]

(5) Nothing in this section shall affect the right of the landlord of an agricultural holding to remove a tenant whose estate has been sequestrated under the Bankruptcy (Scotland) Act, 1913, or who by failure to pay rent or otherwise has incurred any irritancy of his lease or other liability to be removed.[8]

(6) The provisions of this section relative to notice shall not apply—

(a) to a notice given in pursuance of a stipulation in a lease entitling the landlord to resume land for building, planting, feuing or other purposes (not being agricultural purposes);[9] or

(b) to subjects let under a lease for any period less than a year, not being a lease which by virtue of section *two* of this Act takes effect as a lease from year to year.

[1] See Introduction. It is not competent to contract out of the requirements of this sub-section. *Duguid* v. *Muirhead*, 1926 S.C. 1078.

[2] **'Stipulated'**. If no period be 'stipulated', the period would be implied as for a year and thereafter from year to year. The 'expiration of the stipulated endurance' may be at a 'break' in the lease.

[3] **'Lease'**. See definition, section 93, and notes. **'Either party'**. See case of notice by one of two joint tenants (*Graham* v. *Stirling*, 1922 S.C. 90). (See also *Montgomerie* v. *Wilson*, 1924 S.L.T. (Sh.Ct.) 48, and *Wight* v. *Earl of Hopetoun*, 1864, 2 M. (H.L.) 35).

[4] Where there are several dates for ish, the notice must be given for the earliest of them. In a case where the termination was Martinmas from arable lands and Whitsunday following from grass lands, a notice a year before Whitsunday was held bad (*Montgomerie* v. *Wilson*, 1924 S.L.T. (Sh.Ct.) 48). It is thought that the section applies at a break as well as at the natural termination of the lease. This view is consistent with the opinion expressed in *Alston's Trustees* v. *Muir*, 1919, 2 S.L.T. 8. (See also *Strachan* v. *Hunter*, 1916 S.C. 901 (correspondence held not equivalent to notice); and *Edell* v. *Dulieu*, [1924] A.C. (H.L.) 38). In *Strachan's* case it was decided that the statutory notice to quit requires to be given even in order to take advantage of a conventional break, although there was a provision in the lease for a shorter notice. 'A notice to quit includes a notice to determine the tenancy.'

The Sheriff Courts (Scotland) Act, 1907, First Schedule, Rule 110 (b), places a tenancy, renewed on tacit relocation, in the same position as a tenancy from year to year as regards notice to quit. But this section provides that such notice shall be not less than one year and not more than two years. That is to say, the same length of notice must be given in the case where the tenancy is renewed on tacit relocation or for another year and thereafter from year to year, as is required in order to bring to an end a tenancy under a lease for a period of years.

Tacit relocation implies consent to a renewal, not strictly an extension, of the tenancy (Bell's Principles, 1265; *Forbes* v. *Lady Saltoun's Executors*, 1735 Elch. Cautioner 4; *M'Farlane* v. *Mitchell*, 1900, 2 F. 901; *Tayleur* v. *Wildin*, 1868, L.R. 3 Ex. 303; *Holme* v. *Brunskill*, [1878], 3 Q.B. 495; and *Inchiquin* v. *Lyons*, 1887, 20 Ir. L.R. 474). But in *Cowe* v. *Millar*, 1923 (unreported) the Lord President expressed the view that tacit relocation operated to extend the tenancy and not to create a new one. (See also *Donaldson* v. *Regent Photo Playhouse, Ltd.*, 1925 S.L.T. (Sh.Ct.) 92.)

[5] **'Agreement'**. Question reserved whether letter of removal is an agreement within meaning of this section. (*Cushnie* v. *Thomson*, 1954 S.L.C.R. 33.)

[6] and [7] See Appendix. It is vital to adhere to the statutory form of notice. The 1886 Act dealt only with *manner* of sending notice. The *form* of notice must be as nearly as possible that prescribed in form H annexed to the 1907 Act. The lease must be referred to. (*Rae and Cooper* v. *Davidson*, 1954

S.C. 361.) It was held by Sheriff Kidd in a stated case, in which 'Whitsunday' was not defined in the lease, reported in *Stirrat & Another* v. *Whyte*, 1968 S.L.T. 157 at p. 160 that in view of the terms of section 24 (3) notice to quit at Whitsunday must be served against 15th May notwithstanding the definition of Whitsunday in section 93 as 28th May in leases entered into after 1st November, 1948. The reasoning was that the sub-section states that the provisions of the Sheriff Courts (Scotland) Act, 1907, relating to removings are to apply and that Act provides that Whitsunday is 15th May. On the same reasoning a notice to quit at Martinmas, undefined in the lease, would require to be served against 11th November. A tenant of a cottage, garden, byre, and a five-acre field holding from year to year received notice to quit as follows: 'I beg to serve formal notice to quit at 28th May, 1918, as I shall be requiring the cottage for an employee'. This was held invalid for insufficient description of the subjects (*Scott* v. *Livingstone*, 1919 S.C. 1). A removal notice sent by a sheriff officer was held not invalid in respect that it did not state that it was given on behalf of the landlord (*Seggie* v. *Haggart*, 1926 S.L.T. (Sh.Ct.) 104). A notice not referring to the lease, but specifying subjects as 'presently possessed and leased by you,' and signed by a solicitor 'agent for M. W., by whom the said land and houses are sub-let to you', was held sufficient (*Watters* v. *Hunter*, 1927 S.C. 310). Correspondence was not held to be valid notice of intention to take advantage of a break (*Strachan* v. *Hunter*, 1916 S.C. 901). As to the arbiter's jurisdiction to decide whether a notice to quit was valid, see *Hoth* v. *Cowan*, 1926 S.C. 58. In this case the tenant was sitting, he said, on tacit relocation, and, he contended, at the old rent; the landlord contended he was sitting on under a new bargain subject to an increased rent. The tenant argued that the notice to quit which he had received was invalid, and that he was therefore sitting on under the old conditions, and the Board of Agriculture appointed an arbiter in virtue of an arbitration clause in the lease. The Court granted interdict against the arbiter proceeding, holding that the question at issue was not whether the notice to quit was invalid, but whether the tenant was sitting under the old lease or under a new bargain, and that, as the lease itself was in dispute, the arbitration clause in it did not apply. (See also *Bebington* v. *Wildman*, [1921], 1 Ch. 559). Where a farm was held at first by one owner, who sold part, and the proprietors of the two parts gave notice to quit the whole farm, and the tenant bought the part which belonged to one of the owners, and refused to give up possession of the other part, contending that under section 27 of the 1923 Act the contract of sale rendered the notice to quit void, it was held that, as the contract to sell was not made by the persons who gave the notice to quit, that section did not apply, and that the notice to quit was valid. (*Rochester and Chatham Joint Sewerage Board* v. *Clinch*, [1925] Ch. 753.) Where part of an agricultural holding is purchased, the purchaser cannot serve a valid notice to quit without the collaboration of the landlord of the remainder of the holding: *Secretary of State for Scotland* v. *Prentice*, 1963 S.L.T. (Sh.Ct.) 48. See also *Stewart* v. *Moir*, 1965 S.L.T. (Land Ct.) 11.

Service through the post, in a registered letter containing notice to quit, is sufficient. A letter sent by Recorded Delivery in terms of the Recorded Delivery Service Act, 1962, is equivalent to a registered letter but if this method is used the receipt issued by the Post Office must be initialled. It has been decided that notice sent by ordinary post is ineffective (*Department of Agriculture for Scotland* v. *Goodfellow*, 1931 S.L.T. 388). A tenant refused to sign a receipt for a registered letter which contained the notice which he therefore did not see. The letter was returned to the post office. The Court granted an order for ejectment (*Van Grutten* v. *Trevenen*, [1902], 2 K.B. 82). In England it has been held that a notice was effective if in fact delivered although not sent by registered post. *Re Poyser and Mills Arbitration*, [1964], 2 Q.B. 467. A notice to quit is valid though given on a Sunday (*Sangster* v. *Noy*, 1867, 16 L.T. 157). A notice given on a Monday would not be in time if Sunday was the last day (*Scott* v. *Scott*, 1927 S.L.T. (Sh.Ct.) 6).

For the purposes of this provision 'termination of the lease' is not necessarily the same as 'termination of the tenancy', to which reference has already been made. For example, in the case of an ish at Whitsunday as regards houses and pasture, and at separation of crop thereafter as regards the land in white crop, 'determination of the tenancy' has been held to be Martinmas after reaping, whereas in such a case the termination of the lease would be Whitsunday, and, therefore, it is thought that notice to quit would have to be given in writing for the required period counting back from the Whitsunday. Difficulty still arises as to the precise date against which a notice may with safety be served. See *Stirrat and Another* v. *Whyte*, 1968 S.L.T. 157.

It was held in an ordinary action of Removing brought against the tenant of a dwelling that a written notice in terms of section 37 of the Sheriff Courts Acts, 1907-1913, was not essential and that a verbal notice by the tenant, duly given, that he intended to flit at the term was sufficient. *Gillies* v. *Fairlie*, 36 Sh.Ct. Reports, 6. Approved in *Craighall Cast-Stone Co.* v. *Wood Brothers*, 1931 S.C. 66.

8 As regards a landlord's rights after irritating a lease on the tenant's bankruptcy, see *Chalmers' Trustee* v. *Dick's Trustee*, 1909 S.C. 761, and *McKinley* v. *Hutchison's Trustees*, 1935 S.L.T. 62. Where a lease was irritated for failure to pay the rent and there was an obligation to take over sheep stock at the expiry of the lease, it was held that there was no awaygoing within the meaning of the lease (*Marquis of Breadalbane* v. *Stewart*, 1904, 6 F. (H.L.) 23; *M'Douall's Trustees* v. *MacLeod*, 1949 S.C. 593. To irritate, landlord must discharge his own obligations—*Macnab of Macnab* v. *Willison*, 1960 S.L.T. (Notes) 25.

9 The words *'not being agricultural purposes'* were added by the Agriculture Act, 1958. The words 'other purposes' in the 1949 Act and in leases have given rise to difficulties of construction c.f. *Admiralty* v. *Burns*, 1910 S.C. 531, *Crichton Stuart* v. *Ogilvie*, 1914 S.C. 888, *Turner* v. *Wilson*, 1954 S.C. 296. *Pigott* v. *Robson*, 1958 S.L.T. 49. The argument is still open that too large a resumption for a non-agricultural purpose may be fraud on the lease—see the dicta of Lord Mackintosh in *Turner* v. *Wilson*, *supra*. *Edinburgh Corporation* v. *Gray*, 1948 S.C. 538. *Glencruitten Trustees* v. *Love*, 1966 S.L.T. (Land Ct.) 5 (resumption of buildings held to be contrary to good faith of lease—following *Trotter* v. *Torrance* (1891), 18 R. 848). On Resumption the tenant should be given such notice as will enable him to comply with the time limits in sections 14 and 56. See *Re Disraeli Agreement*, [1939] Ch. 382. There is authority to the effect that if the lease so provides the landlord may resume without any period of notice (*Kininmonth* v. *British Aluminium Company*, 1915 S.C. 271) but the safer course would be to give say two months' notice thus enabling the tenant to give the one month's notice required by sections 14 and 56. Additional compensation may be payable in terms of section 15 of the Agriculture (Miscellaneous Provisions) Act, 1968, in cases of early resumption. See Appendix.

It should be noted that this section applies to the case where notice is given to quit part of a holding held from year to year for the purposes specified in section 32. *Hamilton* v. *Lorimer*, 1959 S.L.C.R. 7. The full statutory notice is required.

Where there were two *pro indiviso* proprietors (husband and wife), and notice was served by the solicitors of the husband as 'proprietor of the farm', it was held that the notice was good subject to proof of the wife's authority. There was no infeftment when the notice was served (*Walker* v. *Hendry*, 1925 S.C. 855). The question was reserved whether notice by one party would stop tacit relocation. In *Hendry* v. *Walker*, 1927 S.L.T. 333, Lord Constable preferred the views expressed in *Mills* v. *Rose*, 1923 W.N. 330, to those in *Cave* v. *Page*, (1923), 67 S.J. 659, and held that the questions as to validity of notice to quit, &c., must be dealt with by the arbiter.

Restrictions on operation of notices to quit.

25.¹—(1) Where notice to quit an agricultural holding or part of an agricultural holding is given to the tenant thereof, and not later than one month from the giving of the notice to quit the tenant serves on the landlord a counter-notice in writing requiring that this sub-section shall apply to the notice to quit, then, subject to the provisions of the next following sub-section, the notice to quit shall not have effect unless the Land Court consents to the operation thereof.[1]

(2) The foregoing sub-section shall not apply where—

(a) Sub-section deleted by 1958 Act;

(b)[2] the notice to quit relates to land being permanent pasture which the landlord has been in the habit of letting annually for seasonal grazing or of keeping in his own occupation and which has been let to the tenant for a definite and limited period[3] for cultivation as arable land on the condition that he shall, along with the last or waygoing crop, sow permanent grass seeds, and it is stated in the notice that it is given by reason of the matter aforesaid;

(c)[2] the notice to quit is give on the ground that the land is required for a use, other than for agriculture, for which permission has been granted on an application made under the enactments relating to town and country planning, or for which (otherwise than by virtue of any provision of those enactments) such permission is not required, and that fact is stated in the notice;[4]

(d)[2] the Land Court, on an application in that behalf made not more than nine months before the giving of the notice to quit, was satisfied in relation to the holding that the tenant was not fulfilling his responsibilities to farm in accordance with the rules of good husbandry[5], and certified that it was so satisfied, and that fact is stated in the notice;

(e)[2] at the date of the giving of the notice to quit[6] the tenant had failed to comply with a demand in writing served on him by the landlord requiring him within two months from the service of the demand to pay any rent due in respect of the holding, or within a reasonable time[7] to remedy any breach[8] by the tenant, which was capable of being remedied, of any term or condition

PROVISIONS AS TO NOTICES TO QUIT

of his tenancy which was not inconsistent with the fulfilment of his responsibilities to farm in accordance with the rules of good husbandry,[5] and it is stated in the notice that it is given by reason of the matter aforesaid;

(f)[2] at the date of the giving of the notice to quit[6] the interest of the landlord in the holding had been materially prejudiced by the commission by the tenant of a breach, which was not capable of being remedied in reasonable time and at economic cost, of any term or condition of the tenancy which was not inconsistent with the fulfilment by the tenant of his responsibilities to farm in accordance with the rules of good husbandry,[5] and and it is stated in the notice that it is given by reason of the matter aforesaid;

(g)[2] at the date of the giving of the notice to quit[6] the tenant was a person who had become notour bankrupt[9] or had executed a trust deed for behoof of his creditors, and it is stated in the notice that it is given by reason of the matter aforesaid.

(3) This sub-section repealed by 1958 Act, sec. 10 (6).

[1] This section must be read together with section 27 which was introduced by the 1958 Act. The tenant's counter-notice should state simply that 'the tenant requires that section 25 (1) of the Agricultural Holdings (Scotland) Act, 1949, shall apply to the notice to quit, dated . . . . . , served upon him by . . . . in respect of (description of holding)'. If there is doubt as to the validity of the notice, then the tenant's right to contest should be expressly reserved, although this is not strictly necessary. An angry letter is not a counter notice (*Mountford* v. *Hodkinson*, [1956], 1 W.L.R. 422). Where a counter notice referred to a repealed statute held as referring to the corresponding sections in force (*Ward* v. *Scott*, 66 T.L.R. (Pt. 1) 340). But cf. *Secretary of State for Scotland* v. *Fraser*, 1954 S.L.C.R. 24. The counter-notice may be served upon the landlord's agents, *Hemington* v. *Walter* (1949), 100 L.J. 51. See also *Hewson* v. *Matthews* (1950), 100 L.J. 654. It is sufficient to serve upon the tenant in possession. *Egerton* v. *Rutter* [1951], 1 K.B. 472, *Wilbraham* v. *Colclough* [1952], 1 All E.R. 979. Where the landlord is not the legal owner at date when notice is given, see *G. C. Tebbs & Sons* v. *Edwards* (1950), 155 E.G. 335. For suspension of the operation of a notice to quit pending an arbitration, or decision of Land Court see section 27.

See also the Agriculture Act, 1967, section 29 (4) which provides that section 25 of the 1949 Act is not to apply to a notice to quit given by the Secretary of State to enable him to use or dispose of land to effect amalgamation or reshaping of uncommercial agricultural units, the tenant having agreed in the lease that the land might be so used. The Land Court is empowered to permit the serving of a counter-notice under this sub-section on behalf of a non-regular service man who is abroad—Reserve and Auxiliary Forces (Protection of Civil Interests) Act, 1951, and Agricultural Holdings (Service Men) (Scotland) Regulations, 1952 [1952/1338].

² This provision is of vital importance. Failure to state the reason in the notice to quit renders the notice invalid. Such an omission might be rectified by withdrawing the notice and serving a new one and stating therein the reason—this provided the second notice is otherwise in order and timeous. See, however, *Tayleur* v. *Wildin,* 1868 L.R., 3 Ex. 303, where it was held that a notice to quit could not be withdrawn except of consent. It has been decided in England that it is sufficient to refer to reasons as in paragraphs of the Act without setting out the reasons themselves (*in re Digby and Penny* [1932], 2 K.B. 491), also that the reason may be stated in a letter accompanying the notice to quit (*Turton* v. *Turnbull* [1934], 2 K.B. 197; *Selleck* v. *Hellens* (1928) L.J. K.B., 214). A statement of the reasons in a letter subsequent to the date of the notice to quit will not suffice. See also *Budge* v. *Hicks* [1951], 2 K.B. 335; Notice should identify paragraph relied on; *Cowan* v. *Wrayford* [1953], 2 All E.R. 1138; *Jones* v. *Gates* [1954], 1 All E.R. 158, and *Hammon* v. *Fairbrother* [1956], 2 All E.R. 108. Sub-sections (e) and (f) are mutually exclusive—*Macnabb* v. *Anderson*, 1957 S.C. 213. Where two reasons are given, or where two concurrent notices are given, see *French* v. *Elliot* [1960], 1 W.L.R. 40. Note that the sub-section does not authorise a landlord to serve notice to quit during the currency of a lease otherwise than against a break or the conventional ish. See *Macnabb* v. *Anderson*, 1955 S.C. 38 per Lord Patrick at p. 44.

³ Contrast 'definite and limited period' with the wording of the proviso to sub-section 1 of section 2.

'A definite and limited period' might even extend to seven years. A statement that the tenant should put the land through a rotation of cropping, and, at the termination of the tenancy, leave it 'in grass' would not, it is thought, meet this condition; it is necessary to stipulate expressly that the tenant shall, along with the last or away-going crop, sow permanent 'grass seeds'. See *Stirrat and Another* v. *Whyte*, 1968 S.L.T. 157.

⁴ In *Teignmouth U.D.C.* v. *Elliott*, 108 L.J. 204 it was held under the corresponding sub-section of the English Act that merely throwing land open to the public was a material change of use and required planning permission.

⁵ **'Rules of good husbandry'.** It is thought that there must be a material breach over a considerable extent of the farm. Some dirt in part of a field on a large farm would probably not suffice, more especially if a reasonable explanation is forthcoming. A tenant exercising rights of cropping and sale of produce could not be found in breach of the rules of good husbandry merely for doing so.

⁶ **'At the date of the giving of the notice'** is evidently not confined to acts in process of execution at precisely that date. It includes such acts and also prior acts, so far as these prior acts had results which were at the date of the notice contrary to the rules of good husbandry, and especially if the results were so serious that the tenant could not remedy them before the determination of the tenancy. That this is the intention seems to be obvious from the fact that it is a breach if there be failure to *maintain* the land clean and in a good state of cultivation and fertility, to have drains and ditches properly *maintained* and clear, and to have fences, &c., in a proper state of repair. On the other hand, acts done subsequent to the date of the notice, cannot afford justification for serving the notice. In the case of a tenancy for two years the arbiter took into consideration the shortness of the period and the condition of the farm at entry, and refused to find that the tenants were not cultivating according to the rules of good husbandry, although he found that at the date of the notice to quit, the holding was not clean and in a good state of cultivation and fertility and in good condition. No particular form of notice is required. Notice may be given by the agent of the landlord.

These notices requiring payment of rent, &c., may be given by an agent duly authorised. They should state the amount due. It is doubtful if the ordinary and usual notice to pay at the due date would suffice. The expression breach of the condition of the tenancy, &c., would probably apply to the breaking-up of permanent pasture which could only be restored over a period of years; that is assuming 'not capable of being remedied' means not capable of being remedied by the *tenant*, before the expiry of his lease.

Where landlord reserves power in lease to carry out maintenance at expense of tenant held he can nevertheless found on tenant's failure. *Halliday* v. *William Fergusson & Sons and Others*, 1961 S.C. 24, overruling *Forbes-Sempill's Trustees* v. *Brown*, 1954 S.L.C.R. 36. See also *Allan's Trustees* v. *Allan & Son*, 1891, 19 R. 215; Gloag on Contract, 2nd ed., 669.

7 '**Reasonable time**'. It is obviously convenient if the landlord can specify what he regards a reasonable time, but he need not so specify. *Morrison-Low* v. *Howison*, 1961 S.L.T. (Sh.Ct.) 53; *Stewart* v. *Brims*, 1969 S.L.T. (Sh.Ct.) 2. For circumstances where two months was not a reasonable time see *Pentland* v. *Hart*, 1967 S.L.T. (Land Ct.) 2.

8 '**Remedy any breach**'. Failure to comply with one of the requirements of a notice to remedy breaches is sufficient to found a notice to quit unless the breach is such that the rule *de minimis non curat lex* applies: *Price* v. *Romilly* [1960], 3 All E.R. 429; *Edmunds* v. *Woolacott*, 109 L.J. 204. Where certain breaches were not remedied because the landlord failed to supply materials the landlord was still entitled to rely on other failures by the tenant to remedy breaches where no materials had to be supplied: *Shepherd and Another* v. *Lomas* [1963], 2 All E.R. 902. It has been held that a notice to quit under section 25 (2) (e) was not invalid because the landlord was in breach of his obligations under the lease. *Wilson-Clarke* v. *Graham*, 1963 S.L.T. (Sh.Ct.) 2. The position is different where the landlord seeks to operate an irritancy clause. See note 8 to section 24.

The notice to remedy breaches must state clearly what is required of the tenant: otherwise a subsequent notice to quit may be invalid: *Morris and Another* v. *Muirhead Buchanan & Macpherson and Others*, 1969 S.L.T. 70. The observation by Lord Robertson at p. 74 however that 'section 25 (2) (e) is intended to apply to works of repair, maintenance and replacement rather than to general complaints of failure in husbandry,' must be regarded with reservation. See Muir Watt on *Agricultural Holdings* 12 Edition p. 57 where he deals with notices to remedy other than notices to do work.

9 This section would apply when tenant was 'notour bankrupt' at commencement of tenancy and subsequently received notice to quit bearing that it was given because of his bankruptcy (*Hart* v. *Cameron*, 1935 Sh.Ct. Rep. 166). Notour bankruptcy means practical insolvency not absolute insolvency: see Bankruptcy Act, 1913, section 5 (2); *Murray* v. *Nisbet*, 1967 S.L.C.R. App. 128.

26.—(1)[1] The Land Court shall consent under the last foregoing section to the operation of a notice to quit an agricultural holding or part of an agricultural holding if, but only if, they are satisfied as to one or more of the following matters, being a matter or matters specified by the landlord in his application for their consent,[2] that is to say—

Provisions as to consent for purposes of preceding section.

    (*a*) that the carrying out of the purpose for which the landlord proposes to terminate the tenancy is desirable in the interests of good husbandry[3] as respects the land to which the notice relates, treated as a separate unit;[4] or

(b) that the carrying out thereof is desirable in the interests of sound management[5] of the estates of which the land to which the notice relates forms part or which that land constitutes; or

(c) that the carrying out thereof is desirable for the purposes of agricultural research, education, experiment or demonstration, or for the purposes of the enactments relating to smallholdings or such holdings as are mentioned in section sixty-four of the Agriculture (Scotland) Act, 1948, or allotments; or

(d) that greater hardship[6] would be caused by withholding than by giving consent to the operation of the notice; or

(e) that the landlord proposes to terminate the tenancy for the purpose of the land being used for a use, other than for agriculture,[7] not falling within paragraph (c) of sub-section (2) of the last foregoing section:[8]

Provided that, notwithstanding that they are satisfied as aforesaid, the Land Court shall withhold consent to the operation of the notice to quit if in all the circumstances it appears to them that a fair and reasonable landlord would not insist on possession.[9]

(2)-(4) Repealed by Agriculture Act, 1958, Sch. 2, Pt. II.

(5) Where the Land Court consents under the last foregoing section to the operation of a notice to quit the Court may impose such conditions as appear to the Court requisite for securing that the land to which the notice relates will be used for the purpose for which the landlord proposes to terminate the tenancy.

(6) Where on an application by the landlord in that behalf, the Land Court is satisfied that by reason of any change of circumstances or otherwise any condition imposed under the last foregoing sub-section ought to be varied or revoked, the Court shall vary or revoke the condition accordingly.[10]

---

[1] This sub-section is enacted by the Agriculture Act, 1958, in substitution for the sub-section of the 1949 Act. See also the Opencast Coal Act, 1958, section 14 (6) (9). Agriculture (Miscellaneous Provisions) Act, 1968, section 19 (4).

² Note that it is necessary to specify in the application the reasons in respect of which consent is sought. The application should refer specifically to particular paragraphs of section 26 (1): *Benington Wood's Trustee* v. *Mackay*, 1969 S.L.T. (Land Ct.) 9.
³ **'Good husbandry'**—distinguished from 'rules of good husbandry'. (See *MacKenzie* v. *Tait*, 1951 S.L.C.R. 3; *Fenton* v. *Howie*, 1951 S.L.C.R. 7; *Young* v. *Steven*, 1951 S.L.C.R. 10.)
⁴ **'Treated as a separate unit'**. See *Yuill* v. *Semple*, 1954 S.L.C.R. 3; *Turner* v. *Wilson*, 1954 S.L.C.R. 7 and *ex parte Davies* [1953], 1 All E.R. 1182.
⁵ **'Sound management'**. This is not the same as good estate management. The effect on the whole estate is to be considered but not whether the tenant will suffer. This may be relevant under the proviso. *Evans* v. *Roper* [1960], 2 All E.R. 507.
⁶ **'Greater hardship'**. This is a practical question to be determined on the facts and circumstances of each case and a wide discretion is given to the Land Court. The personal circumstances of both parties must be carefully considered. See *Mackenzie* v. *Tait*, *supra*, also *Grant* v. *Murray*, 1950 S.L.C.R. 3. *Eastern Angus Properties Limited* v. *Chivers & Sons Ltd.*, 1960 S.L.C.R. 3. *Longair* v. *Reid*, 1960 S.L.C.R. 34. *Gibson* v. *McKechnie*, 1961 S.L.C.R. 11. *McBay* v. *Birse*, 1965 S.L.T. (Land Ct.) 10—(Consent granted to enable impoverished landlord to sell her holding). The onus of proof of greater hardship is on the landlord. *McLaren* v. *Lawrie*, 1964 S.L.T. (Land Ct.) 10. A bad investment is not hardship in this sense—*Crawford* v. *McKinlay*, 1954 S.L.C.R. 39; *Patrons of Cowane's Hospital* v. *Rennie*, 1966 S.L.C.R. App. 147; *Benington-Wood* v. *Mackay*, 1967 S.L.C.R. App. 135.
⁷ **'Other than for agriculture'**. The use of land for growing crops and weeds in order to test agricultural chemicals is a use othern than for agriculture. *Dow Agrochemicals Ltd.* v. *E. A. Lane (North Lynn) Ltd.* (1965), 115 L.J. 76.
⁸ It is to be noted that any one of the conditions is sufficient.
⁹ **'A fair and reasonable landlord'**. See *Evans* v. *Roper*, *supra*. In *Carnegie* v. *Davidson*, 1966 S.L.T. (Land Ct.) 3 consent was refused where the landlord sought possession for forestry, the Land Court finding that in the circumstances a fair and reasonable landlord would not insist on possession.
¹⁰ See *Burnett* v. *Gordon*, 1950 S.L.C.R. 9; *Cooper* v. *Muirden*, 1950 S.L.C.R. 45.
Land Court making order continuing appeal to allow landlords to provide another holding held competent. *University of Edinburgh* v. *Craik*, 1954 S.L.C.R. 16. See the Agriculture (Miscellaneous Provisions) Act, 1968, section 11, for circumstances in which additional payments may be due under section 9 of that Act if Land Court consents under this section to operation of notice to quit. See Introduction and forms of notice to quit.

27.—(1) An application by a landlord for the consent of the Land Court under section twenty-five of this Act to the operation of a notice to quit shall be made within one month after service on the landlord by the tenant of a counter-notice requiring that sub-section (1) of that section shall apply to the notice to quit.

(2) A tenant to whom has been given a notice to quit in connection with which any question arises under sub-section (2) of section twenty-five of this Act shall,² if he requires such question to be determined by arbitration under this Act, give notice to

Provisions supplementary to s. 25 and s. 26.

the landlord to that effect within one month after the notice to quit has been served on him; and where the award of the arbiter in an arbitration so required is such that the provisions of sub-section (1) of section twenty-five of this Act would have applied to the notice to quit if a counter-notice had been served within the period limited by that sub-section the period within which a counter-notice may be served under that sub-section shall be extended up to the expiration of one month from the issue of the arbiter's award.

(3) Where such an arbitration as is referred to in the last foregoing sub-section has been required by the tenant, or where an application has been made to the Land Court for their consent to the operation of a notice to quit, the operation of the notice to quit shall be suspended until the issue of the arbiter's award or of the decision of the Land Court, as the case may be.

(4) Where the decision of the Land Court giving its consent to the operation of a notice to quit, or the award of the arbiter in such an arbitration as is referred to in sub-section (2) of this section, is issued at a date later than six months before the date on which the notice to quit is expressed to take effect, the Land Court, on application made to them in that behalf at any time not later than one month after the issue of the decision or award aforesaid, may postpone the operation of the notice to quit for a period not exceeding twelve months.[3]

(5) If the tenant of an agricultural holding receives from the landlord notice to quit the holding or a part thereof and in consequence thereof gives to a sub-tenant notice to quit that holding or part, the provisions of sub-section (1) of section twenty-five of this Act shall not apply to the notice given to the sub-tenant; but if the notice to quit given to the tenant by the landlord does not have effect, the notice to quit given as aforesaid by the tenant to the sub-tenant shall not have effect.

For the purposes of this sub-section a notice to quit part of the holding which under the provisions of section thirty-three of this Act is accepted by the tenant as notice to quit the entire holding shall be treated as a notice to quit the holding.

(6) Where notice is served on the tenant of an agricultural holding to quit the holding or a part thereof, being a holding or part which is subject to a sub-tenancy, and the tenant serves on the landlord a counter-notice in accordance with the provisions of sub-section (1) of section twenty-five of this Act, the tenant shall also serve on the sub-tenant notice in writing that he has served such counter-notice on the landlord, and the sub-tenant shall be entitled to be a party to any proceedings before the Land Court for their consent to the notice to quit.

[1] In its present form this section was introduced by the 1958 Act, First Schedule, para. 37, in place of the 1949 Act provision. What is printed here is adapted to apply to Scotland. The words are not expressly as in the statute.
[2] This provision is imperative.
[3] See *Graham* v. *Wilson-Clarke*, 1962 S.L.C.R. 35.

28.[1]—For the purposes of paragraph (*d*) of sub-section (2) of section *twenty-five* of this Act, the landlord of an agricultural holding may apply to the Land Court for a certificate that the tenant is not fulfilling his responsibilities to farm in accordance with the rules of good husbandry, and the Land Court, if satisfied that the tenant is not fulfilling his said responsibilities, shall grant such a certificate.[3]

<small>Applications for certificates of bad husbandry.</small>

[1] This section has been substituted for the former section by the Agriculture Act, 1958, Schedule 1, para. 38.
[2] A notice given on the same day as the application is invalid. *Gilmour* v. *Osborne's Trustees*, 1951 S.L.C.R. 30.
[3] Under the new section application is made direct to the Land Court. Under the 1949 Act application was made to the Secretary of State who had delegated his powers to the A.E.Cs. The effect of this section is to save time since many applications went first to the A.E.Cs. and then to the Land Court by way of appeal.

29. * * * *

This section was repealed by Agriculture Act, 1958, Schedule 1, para. 39.

<small>Prevention of deterioration of holding after certificate of bad husbandry.</small>

30.[1]—(1) Where, on giving consent under section twenty-five of this Act to the operation of a notice to quit an agricultural holding or part of an agricultural holding, the Land Court

<small>Penalty for breach of condition accompanying consent to notice to quit.</small>

imposed a condition under section twenty-six of this Act for securing that the land to which the notice to quit related would be used for the purpose for which the landlord proposed to terminate the tenancy, and it is proved, on an application to the Land Court on behalf of the Crown—

    (a) that the landlord has failed to comply with the condition within the period allowed thereby, or

    (b) that the landlord has acted in contravention of the condition,

the Land Court may by order impose on the landlord a penalty of an amount not exceeding two years' rent of the holding at the rate at which rent was payable immediately before the termination of the tenancy, or, where the notice to quit related to a part only of the holding, of an amount not exceeding the proportion of the said two years' rent which it appears to the Land Court is attributable to that part.

(2) A penalty imposed under this section shall be a debt due to the Crown and shall, when recovered, be paid into the Exchequer.

    [1] This section is substituted by Agriculture Act, 1958, Schedule 1, Part II. The Secretary of State had power under section 30 of the 1949 Act to take possession of land where a condition imposed under section 26 had not been complied with. The power was seldom exercised and is now abolished.

    This section applies also to a condition imposed under section 18 of the Agriculture (Miscellaneous Provisions) Act, 1968.

Provisions as to notices to quit where holding agreed to be sold.

31.—(1) The provisions of the two following sub-sections shall have effect where, after the commencement of this Act, notice to quit land being or comprised in an agricultural holding has been given to the tenant and at any time while the notice is current a contract is made for the sale of the landlord's interest in the land or any part thereof.

(2) Unless within the period of three months ending with the making of the contract the landlord and the tenant have agreed in writing whether on the making of such a contract the notice shall continue in force or be of no effect—

    (a) the landlord shall, within the period of fourteen days from the making of the contract,[1] or, where the notice

to quit expires within the last-mentioned period, before the expiration of the notice to quit, give notice in writing to the tenant of the making of the contract, and
(b) the tenant may before the expiration of the notice to quit notify the landlord in writing that the tenant elects that the notice to quit shall continue in force, so however that the tenant shall not give a notification under this paragraph after the expiration of one month from the receipt by him of a notice under the last foregoing paragraph of the making of the contract.

(3) In default of any such agreement or notification as aforesaid the notice to quit shall be of no effect unless the landlord has failed duly to give notice of the making of the contract and the tenant quits the holding in consequence of the notice to quit.

(4) A notice to quit shall not be invalid by reason only that under any such agreement as aforesaid the operation of the notice is conditional.

[1] The landlord and tenant may come to an agreement but this will not be effective if it is entered into more than three months before the contract for sale.
[2] On corresponding section of English Act see *Blay* v. *Dadswell* [1922], 1 K.B. 632; *Rochester* v. *Chatham Joint Sewerage Board and Clinch* [1925] Ch. 753.

32.[1]—(1) A notice to quit part of an agricultural holding held on a tenancy from year to year given by the landlord of the holding shall not be invalid on the ground that it relates to part only of the holding if it is given— *Notices to quit part of holdings not to be invalid in certain cases.*
(a) for the purpose of adjusting the boundaries between agricultural units or amalgamating agricultural units or parts thereof, or
(b) with a view to the use of the land to which the notice relates for any of the purposes mentioned in the following sub-section,[2]
and the notice states that it is given for that purpose or with a view to any such use, as the case may be.

(2) The purposes referred to in paragraph (b) of the foregoing sub-section are the following, that is to say—
(a) the erection of farm labourers' cottages or other houses[3] with or without gardens;
(b) the provision of gardens for farm labourers' cottages or other houses;[3]

(c) the provision of allotments;[4]
(d) the provision of small holdings under the Small Landholders (Scotland) Acts, 1886 to 1931, or of such holdings as are mentioned in section sixty-four of the Agriculture (Scotland) Act, 1948;
(e) the planting of trees;
(f) the opening or working of any coal, ironstone, limestone, brick-earth, or other mineral, or of a stone quarry, clay, sand, or gravel pit, or the construction of any works or buildings to be used in connection therewith;
(g) the making of a watercourse or reservoir;
(h) the making of any road, railway, tramroad, siding, canal or basin, or any wharf, pier, or other work connected therewith.

[1] See Introduction. This section legalises a notice to quit part of a farm where such notice is not competent apart from this section. No special provision is made concerning the away-going tenant's rights apart from the statute. Apparently they would be enforced in full. Refer to section 60 dealing with rights to compensation.

[2] The notice to quit here is that required under section 24. The section gives no power to break the tenancy during its currency. Shortly, the section provides that a landlord of a holding, let from year to year (including a tenant sitting from year to year after renewal of the tenancy on tacit relocation), may give notice to the tenant to quit part only of the holding, when the object is to use the land for any of the purposes specified; but that the tenant may, within 28 days after service of the notice to quit, intimate to the landlord that he accepts the notice to quit the part as notice to quit the entire holding (section 33).

Where part of an agricultural holding is purchased, the purchaser cannot serve a valid notice to quit without the collaboration of the landlord of the remainder of the holding. *Secretary of State for Scotland* v. *Prentice*, 1963 S.L.T. (Sh.Ct.) 48.

[3] It is thought that this would not be confined to houses for purposes connected with agriculture, as power is given to terminate the tenancy for wider purposes.

[4] Not necessarily allotments under any statute.

Tenant's right to treat notice to quit part of holding as notice to quit entire holding.

33.—Where there is given to the tenant of an agricultural holding a notice to quit part of the holding, being such a notice as is rendered valid by the last foregoing section, then, if the tenant, within twenty-eight days after the giving of the notice, or, in a case where the operation of the notice depends on any proceedings under the foregoing provisions of this Act, within twenty-eight days after the time when it is determined that the notice has effect, gives to the landlord a counter-notice in writing to the effect that he accepts the notice to quit as a notice to quit the entire holding given by the landlord to take effect at the same time as the original notice, the notice to quit shall have effect accordingly.[1]

[1] This section gives the tenant right to terminate the tenancy at shorter notice than that required by section 24, in cases where the landlord's notice is given less than 28 days prior to the latest date at which the tenant could have given the notice. The tenant will be entitled to his usual way-going claims, including compensation for disturbance, subject to the provisions of section 35 (4).

34.—Where—   *Reduction of rent where tenant dispossessed of part of holding.*

(a) the tenancy of part of an agricultural holding[1] terminates by reason of such a notice to quit as is rendered valid by section *thirty-two* of this Act; or

(b) the landlord of an agricultural holding[1] resumes possession of part of the holding in pursuance of a provision in that behalf contained in the lease;[2]

the tenant shall be entitled to a reduction of rent, of an amount to be determined by arbitration, proportionate to that part of the holding and in respect of any depreciation of the value to him of the residue of the holding caused by the severance or by the use to be made of the part severed:

Provided that, in a case falling within paragraph (b) of this section, the arbiter, in determining the amount of the reduction, shall take into account any benefit or relief allowed to the tenant under the lease in respect of the land possession of which is resumed by the landlord.[3]

[1] See Interpretation, section 93.

[2] This relates to cases where there is provision in a lease, reserving right to the landlord to resume part of a holding for certain specific purposes, and generally on certain specific terms as to compensation or abatement of rent. (See Introduction.) The notice of resumption need not be in any particular form so long as it complies with the provisions in the lease.

[3] **'Benefit or relief'.** In some cases there is provision in the lease, by which the landlord undertakes to give the tenant an abatement of rent in respect of the land resumed, together with surface damages. Where there was power to resume on condition that the tenant should receive an abatement, he was held not entitled to claim for severance damages (*Robertson* v. *Ross & Co.*, 1892, 19 R. 967). 'Surface damages' includes damage to crops and plantations (*Galbraith's Trustee* v. *Eglinton Iron Co.*, 1868, 7 M. 167). For basis of a claim for loss of profits see *McIntyre* v. *Board of Agriculture*, 1916 S.C. 983; *Fleming* v. *Middle Ward of Lanark*, 23 R. 98. Loss of profits is only claimable when there is compulsory acquisition or a provision in the lease. The arbiter is directed to 'take into account any benefit or relief' allowed under the contract. (See Introduction under 'Ascertaining the Value of Improvements'.) It is thought that the arbiter must award in terms of this section and that in doing so he must have in view the terms of a contractual arrangement entered into between the parties. The intention is

not to have two awards, one under the lease and one under the statute. The tenant may, if he thinks proper, confine his claims to those in the lease. If he does so, he is not restricted by procedure under the 1949 Act, except that the arbitration must be under the Sixth Schedule. Refer to section 60 re compensation rights; also to sections 32, 33 and 35 (4).

COMPENSATION TO TENANT FOR DISTURBANCE

Right to, and measure of, compensation for disturbance

35.[1]—(1) Where the tenancy of an agricultural holding[2] terminates by reason either—

(a) of a notice to quit[3] the holding given by the landlord; or

(b) of a counter-notice given by the tenant under section *thirty-three* of this Act after the giving to him of such a notice to quit as is mentioned in that section;

and in consequence of the notice or counter-notice, as the case may be, the tenant quits the holding, then, subject to the provisions of this section, compensation for the disturbance shall be payable by the landlord to the tenant in accordance with the provisions of this section:

Provided that compensation shall not be payable under this sub-section where the operation of sub-section (1) of section *twenty-five* of this Act in relation to the notice to quit the holding or part, as the case may be, is excluded by virtue of paragraph (b), (d), (e), (f) or (g) of sub-section (2) of that section.[4]

(2) The amount of the compensation payable under this section shall be the amount of the loss or expense directly attributable to the quitting of the holding which is unavoidably incurred by the tenant upon or in connection with the sale or removal of his household goods, implements of husbandry, fixtures, farm produce or farm stock on or used in connection with the holding, and shall include any expenses reasonably incurred by him in the preparation of his claim for compensation (not being expenses of an arbitration to determine any question arising under this section):[5]

Provided that—

(a) the compensation payable under this section shall be an amount equal to one year's rent of the holding[7] at the rate at which rent was payable immediately before the termination of the tenancy without proof by the tenant of any such loss or expense as aforesaid;

(b) the tenant shall not be entitled to claim any greater amount than one year's rent of the holding unless he has given to the landlord not less than one month's notice of the sale of any such goods, implements, fixtures, produce or stock as aforesaid and has afforded him a reasonable opportunity of making a valuation thereof;[6]
(c) the tenant shall not in any case be entitled to compensation in excess of two years' rent of the holding.[7]

In this sub-section the expression 'rent' means the rent after deduction of such an amount as, failing agreement, the arbiter may find to be equivalent to the aggregate of the following amounts, that is to say—
- (i) the amount payable by the landlord in respect of the holding for the year in which the tenancy was terminated by way of any public rates, taxes or assessments or other public burdens, the charging of which on the landlord would entitle him to relief in respect of tax under Rule 4 of No. V of Schedule A to the Income Tax Act, 1918,[8] and
- (ii) the amount (if any) recovered in respect of that year from the landlord in pursuance of sub-section (1) of section forty-seven of the Local Government (Scotland) Act, 1929.

(3) Where the tenant of an agricultural holding has lawfully sub-let the whole or part of the holding, and in consequence of a notice to quit given by his landlord becomes liable to pay compensation under this section to the sub-tenant, the tenant shall not be debarred from recovering compensation under this section by reason only that, owing to not being in occupation of the holding or of part of the holding, on the termination of his tenancy he does not quit the holding or that part.[9]

(4) Where the tenancy of an agricultural holding terminates by virtue of such a counter-notice as is mentioned in paragraph (b) of sub-section (1) of this section and—
- (a) the part of the holding affected by the notice given by the landlord, together with any part of the holding affected by any such previous notice given by the landlord as is rendered valid by section *thirty-two* of this Act, is either less than one-fourth part of the area of the original holding or of a rental value less than one-fourth

part of the rental value of the original holding, and
(b) the holding as proposed to be diminished is reasonably capable of being farmed as a separate holding, compensation shall not be payable under this section except in respect of the part of the holding to which the notice to quit relates.[10]

(5) Compensation payable under this section shall be in addition to any compensation to which the tenant may be entitled apart from this section.

[1] See Introduction. The tenant must have quitted at the time required by the notice. See Note 2 to section 37. (*Roberts* v. *Magor* (1953), 103 L.J. 703.) The Court has held that quit 'in consequence of the notice' means quit at the time specified in the notice (*Hendry* v. *Walker*, 1927 S.L.T. 333). In that case a claim for compensation was rejected in respect that the tenant who quitted in July, 1925 (while he should have done so at Martinmas, 1923, and Whitsunday, 1924, in terms of the notice to quit), on being compelled to quit by order of the Court, did not quit in consequence of the notice to quit. The Lord Ordinary said that for more than a year prior to the quitting there had neither been a holding within the meaning of the Act nor a tenant thereof nor could it reasonably be said that the loss and expense then incurred was directly attributable to the quitting. 'I think that the Act plainly implies that the quitting of possession shall take place at or immediately after the termination of the tenancy'. In a Sheriff Court case a tenant was held to have left in consequence of a notice to quit and not in consequence of an obligation in his lease to grant a letter of removal and that the arbiter was entitled to award damages for breach of contract relying on the case of *Webster & Co.* v. *Cramond Iron Co.*, 2 R. 752. *Beveridge and Others* v. *McAdam*, 1925 Sh.Ct. Rep. 288. See also Part II of the Agriculture (Miscellaneous Provisions) Act, 1968, regarding additional payments for reorganisation of the tenant's affairs. (Where payable the additional sum is of four years' rent which together with compensation for disturbance gives a minimum of five years' and a maximum of six years' rent.) The additional payment is due only where the tenant is entitled to compensation for disturbance (1968 Act, section 9). It is not otherwise related to disturbance payments and is due irrespective of the actual loss or expense incurred by the tenant.

[2] '**Agricultural holding**'. See section 1 for definition.

[3] '**Notice to quit**'. Case in which it was questioned whether due notice to quit was given (*Forbes* v. *Pratt*, 1923 S.L.T. (Sh.Ct.) 91). '*Notice to quit*' is not defined, and it does not follow that the notice must be such as is required under section 24. Sheriff Brown (in *Earl of Galloway* v. *Elliot*, 1926 S.L.T. (Sh.Ct.) 123) decided that it did, in a case where the landlord gave notice more than two years before the termination of tenancy, but against that view there is the case of *Tidball* v. *Marshall*, 1922 C.A., Estates Gazette Digest, 1922 p. 396, to which he was referred (where six months' notice instead of 12 months' notice was given), and there is the more recent case of *Westlake* v. *Page* [1926], 1 K.B. 298, in which Lord Justice Banks, who gave the leading opinion, said: 'The first point taken by the landlord is a technical one, that the notice to quit was in law a bad notice, because it was conditional, &c., and in these circumstances the case does not come within the section at all, for that section, when speaking of a holding being terminated by a notice to quit, presumably refers to a termination by a valid notice. I cannot accept that contention. It seems to me that if a notice to quit is given, *whether it be a good or bad one*, and it is accepted as a good notice by the tenant who quits the holding in consequence of it, the case comes within the language of the

section. The technical point therefore fails'. In *Farrow* v. *Orttewell* [1933] Ch. 480, it was held that the notice to quit was invalid but that the landlord was estopped by his actings from denying the validity of the notice.

A tenant who, in expectation of receiving notice to quit, but before receiving such notice, made inquiries after another farm and left after getting a statutory notice, was held to have quitted in consequence of that notice (*Johnston* v. *Malcolm*, 1923 S.L.T. (Sh.Ct.) 81).

The notice to quit may be at a break or at the natural termination of the lease.

4 Compensation appears to be payable if the Land Court consents to the notice or if section 25 (2) (c) is applicable. Refer to *Kestell and another* v. *Langmaid*, 65 T.L.R. 699, *Dean* v. *Secretary of State for War* [1950], 1 All E.R. 344.

5 **'Directly attributable'** and **'unavoidably incurred'** are restrictive terms No indirect or avoidable loss or expense will be allowed. This would exclude loss or expense arising from the sale of stock, in so far as the stock might and should have been sold in the ordinary course, apart from the quitting. Again, if the tenant be going a long distance—say, to Canada—,. he would not be allowed to claim for the expense of removing his furniture &c., when it might be more economical to sell here and buy anew in Canada. The loss or expense of removal must not be too remote. As regards selling implements of husbandry, produce, and farm stock, it will sometimes be found that the chief item of expense is that attending the sale, because, in the general case, a tenant will get market value for what he sells. At the same time, there may be loss on the sale of immature stock, and on the dispersal of pedigree herds, and dairy and sheep stocks. It may no doubt be contended that the difference between going-concern value and break-up value is considerable, and that this difference represents the amount of the tenant's loss. Each case must, however, stand on its own merits. Where a sheep stock was sold by public roup, it was averred that it was sold at break-up value, whereas the claimant was entitled to going-concern value. The Court held that there was a relevant claim (*Keswick* v. *Wright*, 1924 S.C. 766). Where grain threshed and delivered showed less quantity than on valuation, it was held that the difference was not 'a loss directly attributable to the quitting of the holding' (*Macgregor* v. *Board of Agriculture*, 1925 S.C. 613).

Loss or expense, 'unavoidably' incurred in connection with the sale of the tenant's stock was held to include loss sustained in selling by public roup as against a sale under ordinary circumstances, and also the cost of supplying refreshments, but not a fee paid for valuing stock before the sale (*in re Evans and Glamorgan County Council* (1912), 28 T.L.R. 517). The compensation may include loss through deterioration of the stock on a sale (*Barbour* v. *M'Douall*, 1914 S.C. 844). A tenant cannot claim loss and expense through the removal of a fixture or building where he omits to give the notice required under the section relating to fixtures (*Harvey and Manns' Arbitration*, 89 L.J. K.B. 687).

'The expenses reasonably incurred in the preparation of the claim'. This is confined to the claim for compensation for disturbance. Though expenses of the arbitration are not included, it is still in the discretion of the arbiter to determine all questions as to such expenses and by whom and in what proportions they shall be paid.

6 This does not have effect where the tenant restricts his claim to a year's rent, as clearly it would serve no purpose in that case. What is a reasonable opportunity? Where a tenant gave notice in May, 1912, that he intended to claim compensation, and, without further notice, proceeded to sell stock and implements by public roup, it was held that he had given the landlord the reasonable opportunity here required, there being no duty on the tenant to give notice of such opportunity (*Barbour* v. *M'Douall*, 1914 S.C. 844). See also *Dale* v. *Hatfield Chase Corporation*, [1922], 2 K.B. 282, in which it was held that it was a question of fact for the arbiter to decide whether a reasonable opportunity has been given). A tenant may not know whether or not

he will desire to make a claim in excess of one year's rent, but where he is likely to do so it is a good precaution to have a valuation of stock made by an independent valuer and there should also be removal of any doubt if the landlord is notified of an opportunity of valuation.

[7] **'Rent'** here referred to is the net rent as defined. It is thought that the rent current for the last year of the tenancy should be taken as the basis both in the case of one year's rent and in the case of two years' rents. Where a tenant has been receiving abatements of rent, possibly only the sum actually paid can be considered, but the point is not free from doubt. There may arise questions as to whether for the purposes of this sub-section 'rent' should be held to include other payments by the tenant, such as insurance premiums on the farm buildings, interest on drainage, and other improvements. (See *Callander* v. *Smith*, 1900, 8 S.L.T. 109, as to interest being rent). In *Marquis of Breadalbane* v. *Robertson*, 1914 S.C. 215, the Court held that fire insurance premium, payable by the tenant on a policy effected by the landlord, was not 'present rent' within the meaning of the Small Landholders (Scotland) Act, 1911. But in *Duke of Hamilton's Trustees* v. *Fleming*, 1870, 9 M. 329; *Bennie* v. *Mack*, 1832, 10 S. 255; and *Clark* v. *Hume*, 1902, 5 F. 252, there is a good deal to suggest that such premiums are 'rent' and even more can be said in favour of the inclusion of interest on improvement expenditure. On the other hand, where a landlord, in implement of an obligation in a lease, repaid the tenant half the cost of lime put upon the land, the Court held that, in entering the farm in the valuation roll, the assessor was bound to allow a deduction from the rent specified in the lease of the sum so repaid. The deduction amounted to £29 5s. (*Miller's Trustees* v. *Berwickshire Assessor*, 1911 S.C. 908.) It is unfortunate that the statute does not give a definition of 'rent', apart from the deductions to be made from gross rent for the purposes of this sub-section. Either landlord or tenant may make a demand for arbitration as to 'rent' to be paid for the holding. In fixing the rent, the arbiter would require to take into account the landlord's improvements in respect of which interest was being paid; in other words, he would have to fix a rent for the holding in its then improved state (subject, however, to section 7 (2)). That being so, practically the same result is reached as by including interest as part of the rent. Although the word 'taxes' is used, this cannot be held to include Imperial taxes, which are levied on the same basis in England as in Scotland and it is thought that the word is used in the popular sense, as we frequently use 'local taxation' for 'local rating' (*Lord Provost, &c., Edinburgh* v. *The Lord Advocate*, 1923 S.L.T. 14).

Shortly stated, the 'year's rent' in Scotland is subject to deduction of such of the local rates paid by occupiers in England as are paid by owners in Scotland. Owners' rates having been abolished, reference thereto in this section was deleted by the Valuation and Rating (Scotland) Act, 1956, Schedule 7, Part III. No deduction falls to be made in respect of stipend.

'Two years' rent of the holding' would be twice the net rent for the last year. There may be difficulty in determining what is the year's rent in cases where the rent is fixed, according to fiars prices, but that is a matter of fact for the arbiter, though a question of law might arise.

[8] Re-enacted with minor alteration by Income Tax Act, 1952, section 95.

[9] See Agriculture (Miscellaneous Provisions) Act, 1968, section 11 (5).

[10] This sentence means that compensation for disturbance is confined to the part to which the notice to quit applied, except where the area included in that notice, and in any previous notice to quit, amounts to or exceeds one-fourth part of the area (not rent) of the original holding, or where the holding, as proposed to be diminished, is not reasonably capable of being cultivated as a separate holding. In such cases compensation for disturbance may be claimed in respect of the entire holding, as it was at the time when the last notice to quit was served.

COMPENSATION TO TENANT, ON TERMINATION OF TENANCY, FOR IMPROVEMENTS BEGUN BEFORE 1ST NOVEMBER, 1948

36.—(1) The provisions of the ten next following sections shall have effect with respect to the rights of the tenant[1] of an agricultural holding[2] with respect to compensation for an improvement specified in the Second Schedule to this Act carried out on the holding, being an improvement begun before the thirty-first day of July, nineteen hundred and thirty-one (in this Act referred to as 'a 1923 Act improvement'), or for an improvement specified in the Third Schedule to this Act so carried out, being an improvement begun on or after that date and before the first day of November, nineteen hundred and forty-eight (in this Act referred to as 'a 1931 Act improvement'). _{Application of sections 37 to 46.}

(2) An improvement being a 1923 Act improvement or a 1931 Act improvement is in this Act referred to as 'an old improvement'.

[1] **'Tenant'.** See definition, section 93.
[2] **'Holding'.** See definition, section 93, and notes. A holding is not a holding within the meaning of the Act unless the tenancy under which it is held is not less than from year to year or falls within the requirements of section 2 (1). (See under 'Lease' section 93.)

37.—(1) The tenant shall, subject to the provisions of this Act, be entitled, at the termination of the tenancy,[1] on quitting[2] the holding, to obtain from the landlord[3] compensation for an old improvement[4] carried out by the tenant: _{Right of tenant to compensation for old improvements.}

Provided that where the lease was entered into before the first day of January, nineteen hundred and twenty-one, the tenant shall not be entitled to compensation under this section for an improvement which he was required to carry out by the terms of his tenancy.[5]

(2) Nothing in this section shall prejudice the right of a tenant to claim any compensation to which he may be entitled under custom, agreement or otherwise, in lieu of any compensation provided by this section.[6]

[1] **'Termination of the tenancy'.** See definition, section 93 and notes.
[2] **'Quitting'.** If the tenant does not go out on the expiry of a notice to quit but stays on in spite of the notice, he is generally not entitled to compensation. *Cave* v. *Page*, 1923, 67 S.J. 659; *Hendry* v. *Walker*, 1927 S.L.T. 333. Case in which tenant left the holding but not the farmhouse (owing to illness of wife)—held that, if the tenant was ejected, the ejectment was in consequence of the landlord's notice to quit, and that therefore the tenant had quitted the

farm in consequence of the notice to quit, and that he was entitled to compensation for disturbance. *Mills* v. *Rose*, 1923, 68 S.J. 420.

³ **'Landlord'**. See definition, section 93. The seller, and not the purchaser with entry at same time as the tenant quits, is the person against whom a claim for compensation falls to be made in spite of the fact that the tenant may claim within two months after termination of the tenancy. *Waddell* v. *Howat*, 1925 S.C. 484; *Tombs* v. *Turvey*, 1924, 68 S.J. 385.—C.A.

⁴ **'Old Improvement'**. See section 36 (2); definition, section 93.

⁵ This excludes compensation for improvements which the tenant contracted to execute under lease entered into prior to 1st January, 1921. But where a lease permitted alternative methods of cropping, one of which would result in an improvement and the other would not, and the tenant chose the former method, it was held that, as the tenant was not contractually bound to choose that method, he was entitled to compensation. *Gibson* v. *Sherret*, 1928 S.C. 493.

⁶ Customary compensation has been more in vogue in England than in Scotland, but the effect of custom on liability may be seen in *Stewart* v. *Maclaine*, 1899 (H.L.) 37 S.L.R. 623.

This sub-section (section 37 (2)) reserves the right of a tenant to claim compensation for improvements where, apart from the Act, he is entitled to claim under local custom or custom of the country. But in any case such a claim must be dealt with by arbitration under the Act. Under section 42 there is provision in certain cases only for substituting compensation under 'agreement' for compensation under the Act. There is no similar provision expressly relating to compensation under 'custom', but if the arbiter awards under custom, he cannot, of course, also award 'under the Act' for the same improvement.

Section 42 only allows substituted compensation under agreement as regards improvements specified in Part III of the Second and Third Schedules where the claim arises under leases entered into before 1st January, 1921.

Amount of compensation for old improvements.

38.—The amount of any compensation under this Act for an old improvement shall be such sum as fairly represents the value of the improvement to an incoming tenant.¹

¹ **'To an incoming tenant'**. The indefinite, not the definite, article is deliberately used. '*An*' incoming tenant is not necessarily '*the*' incoming tenant. There may be no incoming tenant, or the particular incoming tenant may intend to use the farm in such a way that the 'improvements', e.g., certain buildings, would be of little or no value to him. These special conditions cannot affect the right of the quitting tenant to compensation, nor the measure of the compensation. It is thought that, in the event of dispute, the arbiter would be entitled to fix the compensation at the value to a hypothetical new tenant on the supposition that that tenant had been taken bound to carry on the farm on the same lines substantially as his predecessor had done. (See *MacMaster* v. *Esson*, 1921, 9 S.L.C.R. 18.)

Compensation for certain old improvements conditional on consent of landlord.

39.¹—(1) Compensation under this Act shall not be payable for a 1923 Act improvement specified in Part I of the Second Schedule to this Act or for a 1931 Act improvement specified in Part I of the Third Schedule thereto unless, before the carrying out thereof, the landlord consented in writing² (whether unconditionally³ or upon terms as to compensation or otherwise agreed on between him and the tenant) to the carrying out thereof.

(2) Where the consent was given upon terms as to compensation agreed on as aforesaid, the compensation payable under the agreement shall be substituted for compensation under this Act.

[1] See Introduction. This section relates to permanent (or landlord's) improvements, which are affected generally by the provisions of the immediately preceding sections. Being so affected, is compensation payable for improvements under this section, even when the tenant contracted to make the improvements, where the lease was entered into on or after 1st January, 1921? It is thought not, because otherwise the landlord could derive no advantage by binding his tenant to effect such improvements except that he would get the improvement effected without having to pay for it immediately. If the landlord gives his consent—the tenant being free to make or not to make the improvements—the tenant would have to be satisfied with what compensation he had agreed to accept as a condition of the consent.

[2] The consent in writing may be embodied in the lease or separately. Agreements giving a conditional right to compensation and giving no right were both held legal in *Turnbull* v. *Millar*, 1942 S.C. 521. The English case of *Mears* v. *Callender* [1901], 2 Ch. 388 was not followed. An estate factor would probably be held impliedly to have authority to consent on behalf of his principal. It was held to be within the power of a factor to consent to the conversion of a farm into a market garden (re *Pearson and I'Anson* [1899], 2 Q.B. 618; *Turner* v. *Hutchinson*, 2 F. & F. 185).

The consent need not be expressed in a formal or probative deed (sec. 85).

The question whether consent had been granted or not is a dispute which must go to arbitration, under section 74 of the 1949 Act. *Sinclair* v. *Clyne's Trustee*, 1887, 15 R. 185, has in effect been superseded.

The landlord may give consent on such terms as to compensation or otherwise as may be agreed. It may be argued that 'or otherwise' would allow the landlord to attach the condition that no compensation should be paid. That is at least doubtful, and in *Mears* v. *Callender* [1901], 2 Ch. 388, compensation was found to be due, though consent was given to the tenant to make the improvement 'at his own cost'. It is thought that an agreement for compensation, though small, if not illusory, would be competent.

[3] If consent be given unconditionally, the compensation would be the value to an incoming tenant. If a definite sum is agreed on, and the improvement destroyed, say, by fire, before the termination of the tenancy, a question may arise whether the sum is nevertheless payable (*Duke of Hamilton's Trustees* v. *Fleming*, 1870, 9 M. 329; where the lease terminated by renunciation, see *Walker's Trustees* v. *Manson*, 1886, 13 R. 1198).

40.[1]—(1) Compensation under this Act shall not be payable for a 1923 Act improvement specified in Part II of the Second Schedule to this Act unless the tenant, not more than three nor less than two months before he began to carry out the improvement, gave to the landlord notice in writing under section three of the Agricultural Holdings (Scotland) Act, 1923, of his intention to carry out the improvement and of the manner in which he proposed to carry it out,[2] and either— {Compensation for certain old improvements conditional on notice to landlord.}

    (*a*) the landlord and the tenant agreed on the terms as to compensation or otherwise on which the improvement was to be carried out;[3] or

(b) where no such agreement was made and the tenant did not withdraw the notice, the landlord failed to exercise the right conferred on him by that section to carry out the improvement himself within a reasonable time:

Provided that this sub-section shall not have effect if the landlord and the tenant agreed, by the lease or otherwise, to dispense with notice under the said section three.

(2) Compensation under this Act shall not be payable for a 1931 Act improvement specified in Part II of the Third Schedule to this Act unless the tenant, not more than six months nor less than three months before he began to carry out the improvement, gave to the landlord notice in writing under section three of the Agricultural Holdings (Scotland) Act, 1923, of his intention to carry out the improvement and of the manner in which he proposed to carry it out and either—

(a) the landlord and the tenant agreed on the terms as to compensation or otherwise on which the improvement was to be carried out; or

(b) where no such agreement was made and the tenant did not withdraw the notice, the landlord failed to exercise the right conferred on him by that section to carry out the improvement himself within a reasonable time;[4] or

(c) in a case where the landlord gave notice of objection and the matter was, in pursuance of sub-section (2) of section twenty-eight of the Small Landholders and Agricultural Holdings (Scotland) Act, 1931, referred for determination to the appropriate authority,[5] that authority was satisfied that the improvement ought to be carried out and the improvement was carried out in accordance with the directions (if any) given by that authority as to manner in which the improvement was to be carried out:

Provided that this sub-section shall not have effect—

(i) if the landlord and the tenant agreed, by the lease or otherwise, to dispense with notice under the said section three;[6] or

(ii) where the improvement consists of drainage which was carried out by the tenant for the purpose of complying

COMPENSATION: BEFORE 1st NOVEMBER, 1948   159

with the directions given under Defence Regulations, but which he was not required to carry out by the term of the tenancy.

(3) If the landlord and the tenant agreed (whether after notice was given under the said section three or by an agreement to dispense with notice under that section) on the terms as to compensation on which the improvement was to be carried out, the compensation payable under the agreement shall be substituted for compensation under this Act.[7]

(4) In this section the expression 'the appropriate authority' means, in relation to the period before the fourth day of September, nineteen hundred and thirty-nine, the Department of Agriculture for Scotland, and in relation to the period commencing on that day, the Secretary of State.

[1] See Introduction.
This category of improvements under the 1923 Act was confined to 'drainage', but the 1931 Act added further items (see Schedule), most of which were transferred from Part I. Further alterations were made by the 1948 Act in respect of what are now known as 'new improvements'. It is thought that the section would apply to the substantially complete renewal of worn-out drains of considerable extent, but not to mere repairs. It applies to all kinds of agricultural drainage. A test to be applied to all improvements proposed to be executed under this section will be whether they are those which the tenant himself should carry out in terms of the obligations of his lease.

[2] The notice must contain a statement of the manner in which the tenant proposes to do the work (*Hamilton Ogilvy* v. *Elliot*, 1905, 7 F. 1115). In the case of drainage this can generally be done most effectively by giving a reference to the field, with a statement of the direction, length, and depth of the proposed drains and particulars of the kind and diameter of pipes to be used. A simple tracing plan showing the direction of the drains is an advantage. Knowledge on the part of the landlord or his factor, or the fact that either agreed to the tenant's scheme of drainage, would not be held to imply that the landlord had dispensed with notice (*Barbour* v. *M'Douall*, 1914 S.C. 844).

[3] The agreement here contemplated may provide much less than full compensation and could not be repudiated on that ground.

[4] If the landlord failed to do the work within a reasonable time, the tenant could do so and claim compensation. What is a reasonable time is a question of fact for the arbiter to decide.

[5] See sub-section (4) *infra*.

[6] It is not expressly provided that this agreement must be in writing. Lord Kincairney held the contrary in *Hamilton Ogilvy* v. *Elliot*, 1904, 7 F. 1115. While writing may not be essential under the proviso to this sub-section it is in the interests of both parties that an agreement to dispense with notice should be in writing.

[7] See section 42 regarding substituted compensation.

Conditions attaching to right to compensation for repairs to buildings.

41.[1]— Compensation under this Act shall not be payable in respect of any such repairs as are specified in paragraph 29 of the Second Schedule to this Act or in paragraph 29 of the Third Schedule thereto unless, before beginning to execute any such repairs, the tenant gave to the landlord notice in writing under paragraph (29) of the First Schedule to the Agricultural Holdings (Scotland) Act, 1923, or under paragraph (30) of the First Schedule to the Small Landholders and Agricultural Holdings (Scotland) Act, 1931, of his intention to execute the repairs, together with particulars thereof, and the landlord failed to exercise the right conferred on him by the said paragraph (29) or the said paragraph (30) to execute the repairs himself within a reasonable time after receiving the notice.

[1] See Introduction.

Agreements as to compensation for old improvements specified in Part III of Second or Third Schedule.

42.[1]—Where an agreement in writing[2] entered into before the first day of January, nineteen hundred and twenty-one, secures to the tenant for an old improvement specified in Part III of the Second Schedule to this Act or in Part III of the Third Schedule thereto fair and reasonable compensation, having regard to the circumstances existing at the time of the making of the agreement,[3] the compensation so secured shall as respects that improvement be substituted for compensation under this Act.

[1] See Introduction. This section is confined, in its effect, to the temporary improvements comprised in Part III of the Second and Third Schedules and it applies to all agricultural holdings within the meaning of the Act, including market gardens, regarding which see the special provisions of sections 65-67.

[2] **'Agreement in writing'** need not be formal or probative (see section 85), but it should bear an agreement stamp (*Gardiner* v. *Abercromby*, 1893, 9 Sh.Ct.Rep. 33). It will generally be a lease, but in any case it is thought that the agreement would fall to be interpreted in accordance with section 37 (2), and 'custom of the country' would accordingly have effect so far as applicable, and the section would apply alike to compensation payable under agreement and 'custom of the country'. There remains, however, the difference regarding local custom. It is doubtful if local custom is reserved under section 37 (2), which depends on whether it is included under the words, 'or otherwise'. If there is an agreement in writing dated prior to 1st January, 1921, and it provides more liberal compensation than is 'fair and reasonable', it would appear that the arbiter should award in terms of the agreement.

[3] Where an arbiter is required to assess compensation in terms of an agreement in writing, if the agreement is objected to as not providing 'fair and reasonable compensation' as under this section, the arbiter has the duty to determine the point. But his determination may not be final and may be set aside by the Court (*Bell* v. *Graham*, 1908 S.C. 1060). What is 'fair and reasonable compensation having regard to the circumstances existing at the time of making the agreement' is, of course, a question of fact, but there may also be involved a question of law. It may be suggested that the landlord should be entitled to object to the agreement where it gives more than fair

COMPENSATION: BEFORE 1st NOVEMBER, 1948

and reasonable compensation. It is thought, however, that the Court would hold that the section means that the agreement must have effect so long as it secures *at least* fair and reasonable compensation, the Act not aiming at interfering with contracts except to the extent of ensuring that the minimum compensation, viz., what is fair and reasonable, shall be secured. It has hitherto been usual to embody such agreement in leases—the agreements referring to a scale or scales annexed thereto. In *Bell* v. *Graham (supra)* it was laid down by Lord Dunedin that the agreement being signed by both parties, it is necessary for the party seeking to repudiate it to condescend specifically on the provisions objected to, and to the reasons for maintaining that the provisions are not fair and reasonable.

43.[1]—The tenant shall be entitled to compensation under this Act in respect of the 1931 Act improvement specified in paragraph 28 of the Third Schedule to this Act, being the laying down of temporary pasture in accordance with that paragraph, notwithstanding that the laying down or the leaving at the termination of the tenancy of such pasture was in contravention of the terms of the lease or of any agreement made by the tenant respecting the method of cropping the arable lands; but in ascertaining the amount of the compensation the arbiter shall take into account any injury to, or deterioration of, the holding due to the contravention except in so far as the landlord has recovered damages in respect of such injury or deterioration.

Compensation in respect of temporary pasture.

[1] It is to be noted that the section is affected by the terms of section 40 of the same (1931) Act and accordingly only applies to pasture laid down after the passing of that Act. See *Blair* v. *Meikle*, 1934 Sh.Ct.Rep. 224. It applies whether the lease was entered into before or after the passing of that Act.

44.—(1) In the ascertainment of the amount of the compensation payable under this Act to the tenant in respect of an old improvement, there shall be taken into account[1]—

Reduction in amount of, or exclusion of right to, compensation for old improvements in certain cases.

(*a*) any benefit which the landlord has given or allowed to the tenant in consideration of the tenant carrying out the improvement, whether expressly stated in the lease to be so given or allowed or not;[2] and

(*b*) as respects manuring, the value of the manure required by the lease or by custom to be returned to the holding in respect of any crops grown on and sold off or removed from the holding within the last two years of the tenancy or other less time for which the tenancy has endured, not exceeding the value of the manure which would have been produced by the consumption on the holding of the crops so sold off or removed.[3]

(2) In assessing the amount of any compensation payable to the tenant, whether under this Act or under custom or agreement, by reason of the improvement of the holding by the addition thereto of lime in respect of which a contribution has been made under Part I of the Agriculture Act, 1937, the contribution shall be taken into account as if it had been a benefit allowed to the tenant in consideration of his carrying out the improvement, and the compensation shall be reduced accordingly.

(3) In assessing the amount of any compensation payable under this Act to the tenant in respect of such an improvement as is mentioned in paragraph (ii) of the proviso to sub-section (2) of section *forty* of this Act, if it is shown to the satisfaction of the person assessing the compensation that the improvement consisted of, or was wholly or in part the result of or incidental to, operations in respect of which any grant has been or is to be made to the tenant out of moneys provided by Parliament, the grant shall be taken into account as if it had been a benefit allowed to the tenant in consideration of his carrying out the improvement, and the compensation shall be reduced to such extent as that person considers appropriate.

(4) Notwithstanding anything in the foregoing provisions of this Act, the tenant shall not be entitled to compensation thereunder for an old improvement carried out on land which, at the time the improvement was begun, was not a holding within the meaning of the Agricultural Holdings (Scotland) Act, 1923, as originally enacted, and would not have fallen to be treated as such a holding by virtue of section thirty-three of that Act.

(5) In this section the expression 'manuring' means any of the improvements specified in paragraphs 25 to 27 of the Second Schedule to this Act or in paragraphs 25 to 27 of the Third Schedule thereto.

[1] **'Shall be taken into account'**. This is imperative—not optional, and throws on the arbiter the duty to inquire into the relevant facts and apply this provision to them.

[2] **'Any benefit'**. See Introduction. This is very wide, and would embrace not money payments only, but also anything given or allowed which has a value to the tenant, e.g., agreeing to accept a low rent, providing drain pipes for drainage to be executed by the tenant, or the providing of timber or other material for permanent improvements, or fencing. Clearly 'benefit' cannot take the form of a relaxation of cropping conditions which, in respect of the statutory right to freedom of cropping, the tenant is entitled, subject to maintaining fertility, to ignore. A 'benefit' could, however, be conferred by allowing the tenant to break up permanent pasture which by contract

was to remain unbroken up, or to exercise freedom of cropping or sale of crops during the last year of the tenancy. The mere non-termination of a tenancy by the landlord is not a 'benefit' under the Act (*Mackenzie* v. *McGillivray*, 1921 S.C. 722).

The question of 'benefit' has been raised with reference to claims for compensation for unexhausted manures and temporary pasture (*McQuater* v. *Fergusson*, 1911 S.C. 640; and *Mackenzie* v. *McGillivray*, *supra*).

It does not matter whether the benefit be mentioned in the lease or not. In order that 'benefit' may be taken into account, the arbiter must be satisfied that (1) benefit of actual value to the tenant was actually given, and (2) that it was given or allowed in consideration of the tenant making the improvement in question. No writing is required. Proof that a benefit was given is not enough of itself. There must be a clear connection between the benefit and the improvement, viz., agreement between the parties of intention to set the former against the latter. There must be a bilateral agreement of which the benefit is a counterpart for the improvement.

³ What is here meant is that, if the tenant has removed from the holding within the last two years of his tenancy any crops grown on the land, the money value of the manure, which by the lease or by custom is required to be returned to the holding in respect of these crops, must be taken into account, but only to an extent *not exceeding* the value of the manure which would have been produced by the consumption of the crops on the holding. This does not affect crops destroyed, say, by extraordinary flood or by fire.

It must not be supposed that this in any way restricts the landlord from claiming for deterioration through the tenant's selling off crops in previous years, and, in this connection, the obligations of the tenant, where he exercises his statutory rights to freedom of cropping and sale of produce, should be kept in view. It would appear that if, in the exercise of these statutory rights, the tenant sells off and removes, say, turnips or straw, within the penultimate year of his tenancy, the value of the manure that would otherwise have been made from such produce would have to be taken into account.

If the tenant has brought in and applied manure equal in value to that which would have been made by consuming the crops in question the one thing would simply be set against the other; and if the value brought in be short, the arbiter would only make allowance in respect of the deficiency.

The direction to take into account . . . 'not exceeding the value of the manure which would have been produced . . .' suggests that the arbiter is not bound to deduct the whole value of the manure that might have been made by consuming the crops sold off. The arbiter has to consider to what extent deduction should be made in all the circumstances. He has to take the value in question into account but not necessarily to deduct it in full.

'Grain', apart from straw, does not generally require to be taken into account, because neither by lease nor by custom has it been required to return manure in respect of grain. The same remark applies to straw of the away-going crop, which it is not practicable for the tenant to consume on the holding. At common law a tenant has right to sell off or remove his away-going crop (*Hamilton* v. *Reid's Trustees*, 1824, 2 S. 611; *Miln* v. *Earl of Dalhousie*, 1869 S.L.R. 689). But this is now qualified by section 13, which gives the landlord or incoming tenant right to purchase manure, hay, straw, roots, &c.

45.—Where the tenant[1] has remained in the holding during two or more tenancies, he shall not be deprived of his right to compensation under this Act for old improvements by reason only that the improvements were not carried out during the tenancy on the termination of which he quits the holding.[2]

Provision as to change of tenancy.

¹ By section 93 'tenant' is defined as including 'executors, administrators, &c.', and accordingly the section may enable a tenant's successors to antedate their claims by a considerable number of years.

² Where a tenant has sat on under successive leases or for a period from year to year, he may claim compensation when he actually quits even for the unexhausted value of improvements effected during the leases or years preceding those at the expiry of which he quits. This requires to be read along with section 96, which relates to improvements under repealed statutes. As to whether a tenancy continued on tacit relocation is a new tenancy or a continuation of the old one, see *Mackenzie v. McGillivray*, 1921 S.C. 722. The sitting tenant constructively takes over improvements from himself each time the tenancy is renewed (*Earl of Galloway v. McClelland*, 1915 S.C. 1062). See also section 3.

It is thought that section 45 would not apply where there was a material reduction in the area of the subjects let, but it probably would so far as the improvements related to the part remaining. Can it be said that a tenant remains in his holding if he only remains in, say, half of it? Section 61 does not apply to the case. Again, if there be any change in the personnel of the 'tenant' the section would not apply, e.g., the original tenant may have been 'A', and he, along with 'B', becomes a joint tenant. A mere change in the rent or conditions of tenancy would probably make no difference, nor would the fact that part of the farm had been resumed under the Act or under the lease.

Right to compensation for old improvements of tenant who has paid compensation therefor to outgoing tenant.

46.¹—Where, on entering into occupation of the holding, the tenant, with the consent in writing of the landlord and in pursuance of an agreement made before the first day of November, nineteen hundred and forty-eight, paid to an outgoing tenant any compensation¹ payable under or in pursuance of this Act² or the Agricultural Holdings (Scotland) Acts, 1923 to 1948, in respect of the whole or part of an old improvement, or, with the like consent and in pursuance of an agreement in writing made after that day, paid to an outgoing tenant any compensation payable as aforesaid in respect of the whole or part of an old improvement of the kind specified in Part III of the First Schedule to this Act, he shall be entitled, on quitting the holding, to claim compensation for the improvement or part in like manner, if at all, as the outgoing tenant would have been entitled if the outgoing tenant had remained tenant of the holding and quitted it at the time at which the tenant quits it.³

¹ See Introduction. The consent in writing need not be holograph or tested (section 85). It is necessary, however, if this section is to be founded on, that the written evidence of consent and the vouchers for the sum paid be carefully preserved. Vouchers are useful as evidence not to enable the value of the improvement to be determined, but to prove that the consent was actually obtained. If the consent were given in general terms, the incoming tenant would not be in safety to agree with the away-going tenant regarding the amount of compensation payable to the latter. In such circumstances, the safe course is to let an arbiter fix the amount, and to preserve his award. A common arrangement has been to bind the incoming tenant in the lease to relieve the landlord of any claim by the away-going

tenant for compensation for improvements. In this connection see section 11 which, with exceptions, makes such agreements invalid after the passing of the Act. It is not clear whether this would base a claim for the improvements at the termination of the tenancy. It may be argued that the obligation imposed on the tenant implies the landlord's consent. The obligation may, however, be imposed by the landlord without his being aware of what improvements had in fact been claimed. In any event, the cases to which the section may apply will mainly refer to temporary improvements which will be exhausted before a claim arises. In respect of improvements upon land agreed to be let or treated as a market garden, the consent of the landlord to the purchase by the incoming tenant is not necessary. See also Opencast Coal Act, 1958, section 24.

[2] Evidently this includes such substituted compensation as is recognised under the Act.

[3] Improvements, for which a tenant has paid in terms of this section, may be included in a record of the holding under section 17.

COMPENSATION TO TENANT, ON TERMINATION OF TENANCY, FOR IMPROVEMENTS BEGUN ON OR AFTER 1ST NOVEMBER, 1948

47.[1]—(1) The provisions of the eight next following sections shall have effect with respect to the rights of the tenant of an agricultural holding with respect to compensation for improvements specified in the First Schedule to this Act carried out on the holding, being improvements begun on or after the first day of November, nineteen hundred and forty-eight; and the said provisions shall have effect whether the tenant entered into occupation of the holding before or on or after the said first day of November. *Application of sections 48 to 55.*

(2) An improvement falling within the foregoing sub-section is in this Act referred to as 'a new improvement'.

[1] See notes to section 36.

48.[1]—(1) The tenant shall, subject to the provisions of this Act, be entitled at the termination of the tenancy, on quitting the holding, to obtain from the landlord compensation for a new improvement carried out by the tenant: *Tenant's right to compensation for new improvements.*

Provided that where the lease was entered into before the first day of January, nineteen hundred and twenty-one, the tenant shall not be entitled to compensation under this section for an improvement which he was required to carry out by the terms of his tenancy.

(2) Nothing in this section shall prejudice the right of a tenant

to claim any compensation to which he may be entitled under an agreement in writing in lieu of any compensation provided by this section.

[1] See notes to section 37.

Amount of compensation for new improvements.

49.[1]—(1) The amount of any compensation under this Act for a new improvement shall be such sum as fairly represents the value of the improvement to an incoming tenant.

(2) In the ascertainment of the amount of the compensation payable under this Act for a new improvement there shall be taken into account—

(a) any benefit which under an agreement in writing the landlord has given or allowed to the tenant in consideration of the tenant carrying out the improvement; and

(b) any grant out of moneys provided by Parliament which has been or will be made to the tenant in respect of the improvement.

[1] See note to section 38.

Compensation for Sch. I. Pt. I, improvements conditional on consent of landlord.

50.[1]—(1) Compensation under this Act shall not be payable for a new improvement specified in Part I of the First Schedule to this Act unless, before the carrying out thereof, the landlord has consented in writing (whether unconditionally or upon terms as to compensation or otherwise agreed on between him and the tenant) to the carrying out thereof.

(2) Where the consent is given upon terms as to compensation agreed on as aforesaid, the compensation payable under the agreement shall be substituted for compensation under this Act.

[1] See notes to section 39.

Compensation for Sch. I, Pt. II, improvements conditional on notice to landlord.

51.[1]—(1) Compensation under this Act shall not be payable for a new improvement specified in Part II of the First Schedule to this Act unless the tenant has, not less than three months before he began to carry out the improvement, given to the landlord notice in writing of his intention to carry out the improvement and of the manner in which he proposes to carry it out.

(2) On such notice being given, the landlord and the tenant may enter into an agreement in writing with respect to the

terms as to compensation or otherwise on which the improvement is to be carried out, and if any such agreement is entered into, the compensation payable under the agreement shall be substituted for compensation under this Act.

(3) The landlord and the tenant may, by the lease or otherwise, enter into an agreement in writing to dispense with any notice under sub-section (1) of this section; and an agreement so entered into may provide for anything for which an agreement entered into under the last foregoing sub-section may provide, and in such case shall be of the like validity and effect as such last-mentioned agreement.

[1] See notes to section 40. See also Housing (Scotland) Act 1964 section 38(2) and Housing (Scotland) Act 1966 section 80(2).

52.—(1) Subject to the provisions of this section, compensation under this Act shall not be payable in respect of a new improvement specified in Part II of the First Schedule to this Act if, within one month after receiving notice under sub-section (1) of the last foregoing section from the tenant of his intention to carry out the improvement, the landlord gives notice in writing to the tenant that he objects to the carrying out of the improvement or to the manner in which the tenant proposes to carry it out. *Compensation for Sch. I, Pt. II, improvements conditional on approval of Land Court in certain cases.*

(2) Where notice of objection has been given as aforesaid, the tenant may apply to the Land Court for approval of the carrying out of the improvement, and on any such application the Land Court may approve the carrying out of the improvement either unconditionally or upon such terms, whether as to reduction of the compensation which would be payable if the Land Court approved unconditionally or as to other matters, as appear to the Land Court to be just, or may withhold its approval.

(3) If, on an application under the last foregoing sub-section, the Land Court grants its approval, the landlord may, within one month after receiving notice of the decision of the Land Court, serve notice in writing on the tenant undertaking to carry out the improvement himself.

(4) Where the Land Court grants its approval, then if either—
    (*a*) no notice is served by the landlord under the last foregoing sub-section, or

(b) such a notice is served but, on an application made by the tenant in that behalf, the Land Court determines that the landlord has failed to carry out the improvement within a reasonable time,

the tenant may carry out the improvement and shall be entitled to compensation under this Act in respect thereof as if notice of objection had not been given by the landlord and any terms subject to which the approval was given shall have effect as if they were contained in an agreement in writing between the landlord and the tenant.

[1] This section was considerably amended by the Agriculture Act, 1958, Schedule 1, Part II. Under the 1949 Act application for approval of the carrying out of an improvement was made to the Secretary of State who delegated his functions under the section to the Agricultural Executive Committees.

Approval by the Land Court under this section must be limited to such improvements as are required to enable the tenant to carry on the type of farming specified in the lease: *Taylor* v. *Burnett's Trs* 1966 S.L.C.R. App. 139.

See also Housing (Scotland) Act 1964 section 38(2) and Housing (Scotland) Act 1966 section 80(2).

Compensation in respect of temporary pasture.

53.—The tenant shall be entitled to compensation under this Act in respect of the new improvement specified in paragraph 33 of the First Schedule to this Act, being the laying down of temporary pasture[1] in accordance with that paragraph notwithstanding that the laying down or the leaving at the termination of the tenancy of such pasture[2] is in contravention of the terms of the lease or of any agreement made by the tenant respecting the method of cropping the arable lands; but in ascertaining the amount of the compensation the arbiter shall take into account any injury to, or deterioration of, the holding due to the contravention except in so far as the landlord has recovered damages in respect of such injury or deterioration.

[1] See Introduction.
[2] The onus of showing the extent of the pasture at his ingoing probably lies on the outgoing tenant making a claim. This section applies only to pasture laid down after 1st November, 1948, and seems to apply whether the lease was entered into before or after that date.

Provision as to change of tenancy.

54.[1]—Where the tenant has remained in the holding during two or more tenancies, he shall not be deprived of his right to compensation under this Act for new improvements by reason only that the improvements were not carried out during the tenancy on the termination of which he quits the holding.

[1] See Introduction. See Opencast Coal Act, 1958, section 24.

COMPENSATION: AFTER 1st NOVEMBER, 1948

55.[1]—(1) Where, on entering into occupation of the holding the tenant, in pursuance of such an agreement as is mentioned in sub-section (2) of section *eleven* of this Act, paid to an outgoing tenant or refunded to his landlord any compensation payable by the landlord under or in pursuance of this Act or the Agricultural Holdings (Scotland) Acts, 1923 to 1948, in respect of the whole or part of a new improvement, he shall be entitled on quitting the holding to claim compensation in respect of the improvement or part in like manner, if at all, as the outgoing tenant would have been entitled if he had remained tenant of the holding and quitted it at the time at which the tenant quits it. Right to compensation for new improvements of tenant who has paid compensation therefor to outgoing tenant.

(2) Where, in a case not falling within the foregoing sub-section or section *eleven* of this Act, the tenant, on entering into occupation of the holding, paid to his landlord any amount[2] in respect of the whole or part of a new improvement, he shall, subject to any agreement in writing between the landlord and the tenant, be entitled on quitting the holding to claim compensation in respect of the improvement or part in like manner, if at all, as he would have been entitled if he had been tenant of the holding at the time when the improvement was carried out and the improvement or part had been carried out by him.

[1] See Opencast Coal Act, 1958, section 24.

[2] It is important to preserve evidence of amount paid on entry of tenant with details—arbiter's award, receipts, &c.

COMPENSATION TO TENANT, ON TERMINATION OF TENANCY, FOR CONTINUOUS ADOPTION OF SPECIAL STANDARD OF FARMING

56.[1]—(1) Where the tenant of an agricultural holding proves that the value of the holding[2] to an incoming tenant has been increased during the tenancy[3] by the continuous adoption of a standard of farming or a system of farming[4] which has been more beneficial to the holding than the standard or system required by the lease,[5] or, in so far as no system of farming is so required, than the system of farming normally practised on comparable holdings in the district, the tenant shall be entitled, on quitting the holding, to obtain from the landlord such compensation as represents the value to an incoming tenant of the adoption of that standard or system:[5] Compensation for continuous adoption of special standard of farming.

Provided that compensation shall not be recoverable under this section unless—

(i) the tenant has, not later than one month before the termination of the tenancy, given to the landlord notice in writing of his intention to claim such compensation;[6] and

(ii) a record of the condition of the fixed equipment on, and the cultivation of, the holding has been made under section *seventeen* of this Act;[7]

and shall not be so recoverable in respect of any matter arising before the date of the record so made or, where more than one such record has been made during the tenancy, before the date of the first such record.

(2) In assessing the compensation to be paid under this section due allowance shall be made for any compensation agreed or awarded to be paid to the tenant for any old or new improvement which has caused or contributed to the benefit.[8]

(3) Nothing in this section shall entitle a tenant to recover, for an old or a new improvement or an improvement to which the provisions of this Act relating to market gardens apply, any compensation which he would not be entitled to recover apart from this section.[9]

[1] See Introduction.
For convenience and brevity this improvement may be named 'high farming'. In some respects it resembles the improvement which has sometimes been claimed under the name of 'cumulative or accumulated fertility'. A tenant would probably not be safe to confine himself to the name 'high farming' or to the name 'cumulative fertility', in claiming under this section. It has been held that an arbiter who allowed a claim for 'continuous high farming' was acting *ultra vires* (*Brodie-Innes* v. *Brown*, 1917, 1 S.L.T. 49). It is thought that the same objection could not be taken to a claim if stated thus—'for improvement resulting from the long and extensive use of purchased manures and feeding-stuffs during the years. . . .' A claim thus stated appears to be competent independently of the provisions of this section.

[2] Here it will be noted that the increase in the value 'of the holding' is referred to. In the case of all other improvements it is the 'value of the improvement'.

[3] **'During the tenancy'** is apparently not confined to the tenancy under the lease at the termination of which the claim is made, but this is not made quite clear. (Sections 45 and 54 refer to 'two or more tenancies'.) *Findlay* v. *Munro*, 1917 S.C. 419. See Muir Watt on Agricultural Holdings 12th Edition p. 314.

[4] **'Standard or system of farming'.**—The interpretation of these terms is left to the arbiter, but his interpretation might be subject to review by the Court on a stated case. Standard or system required by the lease does not necessarily refer to a specific standard or system in a written lease, for under section 93 a lease may be verbal. One essential is that the substituted standard or system must have been more beneficial to the holding than the standard or system required by the contract of tenancy. If the tenant were continuously to have land much longer in grass than his contract requires, if he were continuously to follow a five-course shift while he was at liberty to follow a

COMPENSATION: SPECIAL STANDARD OF FARMING

four-course, if he were greatly to improve his pastures by sowing down and specially nurturing wild white clover, if, by the bringing in of exceptionally large quantities of dung, he were to improve the mechanical quality of the soil as well as its fertility—in any of these cases it is thought a valid claim might be made under this section, always, of course, keeping in view the provisos.

[5] Here we find the arbiter is directed to award compensation representing the value to an incoming tenant 'of the adoption of that standard or system'. What appears to be intended is that the arbiter is to fix the value to an incoming tenant of the improvement which has resulted from the adoption by the away-going tenant of the standard or system.

[6] This, as regards notice, differs from compensation for other improvements which may be claimed, so long as intimation of intention to claim in terms of section 68 is given within two months after the termination of the tenancy. On a resumption sufficient notice should be given to enable the tenant to give the notice required by this section. See note 1 to section 24(6).

[7] Presumably, the record may have been made under the Act of 1920, or under the Acts 1923-1949.

[8] This is intended to prevent the tenant from getting payment twice for the same improvement. A claim might be made under this section for special treatment of pastures (say, by liberal sowing of wild white clover) which had been improved by the liberal consumption on them of purchased feedingstuffs which may be claimed for separately. The arbiter must be careful to discriminate between the improvement resulting from the respective operations so as to ensure that double compensation is not allowed. (See note 4.)

[9] This seems to mean that, so far as regards the improvements in the First, Second, Third and Fourth Schedules, the arbiter shall award no more compensation than he would and could have awarded had this section not been passed.

See Opencast Coal Act, 1958 section 25.

COMPENSATION TO LANDLORD, ON TERMINATION OF TENANCY, FOR DETERIORIATION OF HOLDING

57.[1]—(1) The landlord of an agricultural holding shall be entitled to recover from the tenant of the holding, on the tenant's quitting the holding on the termination of the tenancy, compensation in respect of any dilapidation or deterioration of, or damage to, any part of the holding or anything in or on the holding, caused by non-fulfilment by the tenant of his responsibilities to farm in accordance with the rules of good husbandry.[2]

Compensation to landlord for deterioration, &c., of particular parts of holding.

(2) The amount of the compensation payable under the foregoing sub-section shall be the cost, as at the date of the tenant's quitting the holding, of making good the dilapidation, deterioration or damage.[3]

(3) Notwithstanding anything in this Act, the landlord may, in lieu of claiming compensation under sub-section (1) of this section, claim compensation in respect of matters specified therein under and in accordance with a lease in writing,[4] so however that—

(a) compensation shall be so claimed only on the tenant's quitting the holding on the termination of the tenancy;

(b) compensation shall not be claimed in respect of any one holding both under such a lease and under the said sub-section (1);[5]

and for the purposes of paragraph (b) of this sub-section any claim under sub-section (1) of section *six* of this Act shall be disregarded.

[1] A record is essential to a claim under this section.

[2] See Introduction. The English equivalent of this section has been held not to bar action for damages for breach of covenant to repair. *Gulliver* v. *Catt*, [1952] 2 Q.B. 308 and *Kent* v. *Conniff*, [1953] 1 Q.B. 361.

[3] For matters to be considered by an arbiter in determining a claim under this section see *Barrow Green Estate Co.* v. *Walkers' Executors*, [1954] 1 All E.R. 204. Claims for rehabilitation of fields have been held to fall within the English equivalent of this provision. *Evans* v. *Jones*, [1955] 2 Q.B. 58. A claim for damages during the currency of tenancy if made under the lease is competent. *Kent* v. *Conniff and Another, supra.*

[4] It should be borne in mind that a tenant has freedom of cropping, except perhaps in the last year of his tenancy.

[5] This refers to rights and not to methods of enforcing them—a landlord who gave notice of claims under the lease and under the Act could abandon one and validly pursue the other: *Boyd* v. *Wilton* [1957] 2 Q.B. 277.

See Opencast Coal Act, 1958, section 25.

Compensation to landlord for general deterioration of holding.

58.[1]—Where, on the quitting of an agricultural holding by a tenant thereof on the termination of the tenancy, the landlord shows that the value[2] of the holding generally has been reduced, whether by reason of any such dilapidation, deterioration or damage as is mentioned in sub-section (1) of the last foregoing section or otherwise by non-fulfilment by the tenant of his responsibilities to farm in accordance with the rules of good husbandry, the landlord shall be entitled to recover from the tenant compensation therefor, in so far as the landlord is not compensated therefor under sub-section (1) of that section or in accordance with sub-section (3) thereof, of an amount equal to the decrease attributable thereto in the value[1] of the holding.

[1] As to the scope of section 58 see *Evans* v. *Jones, supra.*

[2] It will be noted that here the reference is to the value of the holding. The damage will depend on the expenditure required to bring the land, &c., back to proper condition and the loss incurred during the period when the subjects are not earning full rent.

See Opencast Coal Act, 1958, section 25.

Provisions supplementary to s. 57 and s. 58.

59.—(1) Compensation shall not be recoverable under sub-section (1) of section *fifty-seven* of this Act or under section

COMPENSATION: FOR DETERIORATION OF HOLDING

fifty-eight thereof unless the landlord has, not later than three months before the termination of the tenancy, given notice in writing to the tenant of his intention to claim compensation thereunder.

(2) Compensation shall not be recoverable—
  (a) under sub-section (1) of section *fifty-seven* of this Act or under section *fifty-eight* thereof in any case where the lease was entered into after the thirty-first day of July, nineteen hundred and thirty-one, or
  (b) under and in accordance with any lease entered into on or after the first day of November, nineteen hundred and forty-eight,

unless during the occupancy of the tenant a record of the condition of the fixed equipment on, and the cultivation of, the holding has been made under section *seventeen* of this Act, or in respect of any matter arising before the date of the record so made, or, where more than one such record has been made during his occupancy, before the date of the first such record:

Provided that if the landlord and the tenant enter into an agreement in writing in that behalf, a record of the condition of the holding shall, notwithstanding that it was made during the occupancy of a previous tenant, be deemed, for the purposes of this sub-section, to have been made during the occupancy of the tenant and on such date as may be specified in the agreement and shall have effect subject to such modifications (if any) as may be so specified.

(3) Where the tenant has remained in his holding during two or more tenancies, his landlord shall not be deprived of his right to compensation under section *fifty-seven* or section *fifty-eight* of this Act in respect of any dilapidation, deterioration or damage by reason only that the tenancy during which an act or omission occurred which in whole or in part caused the dilapidation, deterioration or damage was a tenancy other than the tenancy at the termination of which the tenant quits the holding.

SUPPLEMENTARY PROVISIONS WITH RESPECT TO COMPENSATION

60.—(1) Where—
  (a) the tenancy of part of an agricultural holding terminates by reason of such a notice to quit as is rendered valid by section *thirty-two* of this Act; or

Compensation provisions of this Act to apply to parts of holdings in certain cases.

(b) the landlord of an agricultural holding resumes possession of part of the holding in pursuance of a provision in that behalf contained in the lease;

the provisions of this Act with respect to compensation shall apply as if that part of the holding were a separate holding which the tenant had quitted in consequence of a notice to quit:

Provided that, in a case falling within paragraph (b) of this section, the arbiter, in assessing the amount of compensation payable to the tenant, shall take into account any benefit or relief[1] allowed to the tenant under the lease in respect of the land possession of which is resumed by the landlord.

(2)[2] Where any land comprised in a lease is not an agricultural holding[3] within the meaning of this Act by reason only that the land so comprised includes land (in this sub-section referred to as 'non-statutory land') which, owing to the nature of the buildings thereon or the use to which it is put, would not, if it had been separately let, be an agricultural holding within the meaning of this Act, the provisions of this Act with respect to compensation for improvements and for disturbance shall, unless it is otherwise agreed in writing, apply to the part of the land exclusive of the non-statutory land as if that part were a separate agricultural holding.

[1] 'Benefit or relief'. See note to section 34.
[2] This sub-section extends the provisions for compensation for improvements and disturbance to the holdings referred to. The effect is to allow such compensation to the holding, excluding the 'non-statutory' land, which would also include 'non-statutory' buildings thereon, but contracting out is permissible.
[3] See Introduction and definition of 'agricultural holding' section 93. A number of decisions excluded subjects from the definition of 'holding' under the previous Acts for the reasons here stated (see notes to section 93). Apparently an arbiter will have to determine, in connection with a claim for compensation for disturbance, what is the rent of the part to which this section applies. Must the notice prescribed under section 24 be given as regards the entire subject, including the non-statutory land, or only as regards the part which becomes 'a holding' under this section? The safest course is to give the notice under that section in respect of all the subjects.

Determination of claims for compensation where holding is divided.

61.[1] Where an agricultural holding has become vested in more than one person in several parts and the rent payable by the tenant of the holding has not been apportioned with his consent or under any statute, the tenant shall be entitled to require that any compensation payable to him under this Act shall be determined as if the holding had not been divided; and the arbiter shall, where necessary, apportion the amount awarded between the persons who for the purposes of this Act together constitute the landlord of the holding, and any additional expenses of the award caused by the apportionment shall be directed by the arbiter to be paid by those persons in such proportions as he shall determine.

[1] This provision should be useful in cases where estates have been put on the market for sale, and it has been found expedient to sell part of a particular farm to one purchaser and part to another. Where that occurs it is necessary for the tenant to comply with the provisions of the statute in relation to both proprietors. In the case of two purchasers of separate parts of the same farm, and where there has been no apportionment of rent, the tenant should duplicate all his notices, claims, &c., and send, in proper time, one to each of the owners. He should make no attempt to allocate between them, but leave the arbiter to deal with the claim under this section. In an English case two purchasers individually gave notice to the tenant to quit; both notices were served timeously, so as to expire at the same time. But they were independent notices, which made no reference to each other; they were served at different dates, and each was limited to a particular part of the farm. The Court held that the notices were bad (*Bebington* v. *Wildman*, [1921], 1 Ch. 559). It has also been decided in England that the liability is not joint and several (*Weston* v. *Duke of Devonshire*, [1923], 12 L.J.C.C.R. 74).

This section applies to sums payable under section 9 of the Agriculture (Miscellaneous Provisions) Act 1968. See section 11(5) of 1968 Act.

The compensation under this section would include compensation for disturbance where applicable. The arbiter would be entitled to apportion rent in order to ascertain the proportion of compensation for disturbance payable by each proprietor.

62. In assessing the amount of compensation payable, whether under this Act or under custom or agreement, to the tenant of an agricultural holding comprising land in respect of which a payment in respect of a ploughing grant under Part IV of the Agricultural Development Act, 1939,[1] has been made to the tenant, or has been or is to be applied for by him, if it is shown to the satisfaction of the person assessing the compensation that the improvement or cultivations in respect of which the compensation is claimed was wholly or in part the result of or incidental to the operations by virtue of which the land became eligible for the grant, the grant shall be taken into account as if it had been a benefit allowed to the tenant in consideration of his carrying out the improvement or cultivations, and the compensation shall be reduced to such extent as that person considers appropriate. *Adjustment of compensation in respect of ploughing grants.*

[1] Repealed by the Statute Law Revision Act 1953. This section is obsolescent.

63.—(1) Notwithstanding anything in the foregoing provisions of this Act or any custom or agreement— *Compensation not to be payable for things done in compliance with this Act.*

(a) no compensation shall be payable to the tenant of an agricultural holding in respect of anything done in pursuance of an order under paragraph (b) of sub-section (2) of section *nine*[1] of this Act;

(b) in assessing compensation to an outgoing tenant of an agricultural holding where land has been ploughed up in pursuance of a direction under that section, the value per acre of any tenant's pasture comprised in the holding shall be taken not to exceed the average value per acre of the whole of the tenant's pasture comprised in the holding on the termination of the tenancy.[2]

In this sub-section the expression 'tenant's pasture' means pasture laid down at the expense of the tenant or paid for by the tenant on entering the holding.

(2) The tenant of an agricultural holding shall not be entitled to any compensation for an old improvement specified in Part III of the Second Schedule to this Act or in Part III of the Third Schedule thereto or a new improvement specified in Part III of the First Schedule thereto, being an improvement carried out for the purposes of the proviso to sub-section (1) of section thirty-five of the Agricultural Holdings (Scotland) Act, 1923, or of the proviso to sub-section (1) of section *twelve* of this Act.[3]

[1] This section makes provision for the ploughing up of permanent pasture.
[2] See section 9(2)(b).
[3] Providing for restoration of fertility.

Extent to which compensation recoverable under agreements.

64.—(1) Save as expressly provided[1] in this Act, in any case for which, apart from this section, the provisions of this Act provide for compensation a tenant or a landlord shall be entitled to compensation in accordance with these provisions and not otherwise, and shall be so entitled notwithstanding any agreement to the contrary:[2]

Provided that where the landlord and the tenant of an agricultural holding enter into an agreement in writing for any such variation of the terms of the lease as could be made by direction or order under section *nine* of this Act, the agreement may provide for the exclusion of compensation in like manner as under sub-section (1) of section *sixty-three* of this Act.

(2) Nothing in the said provisions, apart from this section,

SUPPLEMENTARY PROVISIONS TO COMPENSATION 177

shall be construed as disentitling a tenant or a landlord to compensation in any case for which the said provisions do not provide for compensation, but a claim for compensation in any such case aforesaid shall not be enforceable except under an agreement in writing.

[1] 'Save as expressly provided'. E.g. the provisions of sections 37 (2), 39 (2), 42, 48 (2), 50 (2), 51 (2), 57 (3) and 67 (2).
[2] Held that a limitation in a lease of the right to compensation for disturbance was void: *Coates* v. *Diment*, [1951] 1 All E.R. 890.

SPECIAL PROVISIONS AFFECTING MARKET GARDENS AS REGARDS COMPENSATION AND FIXTURES

65.[1]—(1) In the case of an agricultural holding in respect of which it is agreed by an agreement in writing made on or after the first day of January, eighteen hundred and ninety-eight, that the holding shall be let or treated as a market garden[2]— *Effect of agreement to let or treat an agricultural holding as a market garden.*

(a) the provisions of this Act shall apply as if improvements of a kind specified in the Fourth Schedule thereto begun before the thirty-first day of July, nineteen hundred and thirty-one, were included among the improvements specified in Part III of the Second Schedule thereto, as if improvements of such a kind begun on or after that day and before the first day of November, nineteen hundred and forty-eight were included among the improvements specified in Part III of the Third Schedule thereto, and as if improvements of such a kind begun on or after the said first day of November were included among the improvements specified in Part III of the First Schedule thereto;[3]

(b) section *fourteen* of this Act shall extend to every fixture or building affixed or erected by the tenant to or upon the holding or acquired by him since the thirty-first day of December, nineteen hundred, for the purposes of his trade or business as a market gardener;[4]

(c) it shall be lawful for the tenant to remove all fruit trees and fruit bushes planted by him on the holding and not permanently set out, but if the tenant does not remove such fruit trees and fruit bushes before the termination of his tenancy they shall remain the property of the landlord and the tenant shall not be entitled to any compensation in respect thereof;[5] and

(d) the right of an incoming tenant to claim compensation in respect of the whole or part of an improvement which he has purchased may be exercised although the landlord has not consented in writing to the purchase.[6]

(2) Where under a lease current on the first day of January, eighteen hundred and ninety-eight, an agricultural holding was at that date in use or cultivation as a market garden with the knowledge of the landlord and the tenant thereof had then carried out thereon, without having received previously to the carrying out thereof a written notice of dissent from the landlord an improvement of the kind specified in the Fourth Schedule to this Act (other than one consisting of such an alteration of a building as did not constitute an enlargement thereof) the provisions of this section shall apply in respect of the holding as if it had been agreed in writing after that date that the holding should be let or treated as a market garden, so however that the improvements in respect of which compensation is payable under those provisions as so applied shall include improvements carried out before as well as improvements carried out after that date:[7]

Provided that where such tenancy was a tenancy from year to year, the compensation payable in respect of such an improvement as aforesaid shall be such (if any) as could have been claimed if this Act had not been passed.[8]

(3) Where the land to which such agreement relates or so used and cultivated consists of part of an agricultural holding only, this section shall apply as if that part were a separate holding.

(4) Nothing in this section shall confer a right to compensation for the alteration of a building (not being an alteration constituting an enlargement of the building) where the alteration was begun before the first day of November nineteen hundred and forty-eight.

[1] See Introduction and section 66.
[2] For definition, see section 93. Under sub-section (3), part of a farm may be a market garden as regards compensation (*Callander* v. *Smith*, 1900, 2 F. 1140; *Taylor* v. *Steel Maitland*, 1913 S.C. 562), and, under section 66 there is provision whereby the tenant of a farm may, with the authority of the Land Court, use part thereof as a market garden. An estate factor has implied authority to bind the landlord by an agreement with the tenant authorising the latter to change the cultivation of the land from ordinary agriculture to market gardening (*re Pearson and I'Anson*, [1899], 2 Q.B. 618). Where there was a lease for 28 years of a residence, garden, pleasure grounds, and 27 acres,

and the tenant was bound not to use the subjects for any trade except that of nurseryman, market gardener, or florist, held that there was an agreement within the section although there was no other reference to market garden cultivation (*Saunders-Jacobs* v. *Yates*, [1933], 2 K.B. 240). Where a holding was let on condition that it would not be deemed 'to be let or treated as a market garden' within the meaning of the Act of 1908, it was decided that the tenant could not claim compensation. (*Masters* v. *Duveen*, [1923], 2 K.B. 729). Land used as an orchard, with rhubarb and other crops grown underneath the trees, the fruit and crops being sold, is 'a market garden' (*Lowther* v. *Clifford*, [1927], 1 K.B. 130). See also *Grewar* v. *Moncur's curator bonis*, 1916 S.C. 764; *Bickerdike* v. *Lucy*, [1920], 1 K.B. 707; *Watters* v. *Hunter*, 1927 S.C. 310).

3 This is intended to enable the tenant of a market garden to effect the the improvements comprised in the First, Second and Third Schedules without the consent of, or notice to, the landlord. As these improvements include the erection or enlargement of buildings for the purpose of the trade or business of a market gardener, the compensation payable by a landlord at the termination of a tenancy, might be large in amount. This consideration operated to deter some landlords from freely letting land for market gardens, and to increase the rents of market garden land, a fact which led to the passing of section 10 (3) of the Act of 1920 (section 66 of the 1949 Act), which aimed at introducing the Evesham custom into Scotland.

Tenants of market gardens should be careful not to erect expensive buildings without first coming to an arrangement with the landlord as to the amount of the compensation which would be paid therefor at quitting. Compensation is only payable for buildings required for the purpose of the trade or business of a market gardener, and is limited to such sum as 'fairly represents the value of the improvement to an incoming tenant'. As an incoming tenant can only, at the best, have a temporary interest, it would appear that the compensation would be on a lower scale than if it had been based on the more permanent interest of the landlord. Thus there is some risk of an improving tenant failing to recover what he may consider to be adequate compensation.

4 See Introduction, and section 14, relating to fixtures and buildings. It is thought that the tenant is entitled to treat buildings erected under the First, Second and Third Schedules either as fixtures subject to the conditions under section 14, or as improvements for which he would be entitled to claim compensation provided he complies with the provisions of the Act for claiming such compensation. As, however, there may be doubts on this point, the latter alternative should be preferred. A lease of a market garden contained a reservation in favour of the landlord of right to work minerals. Glasshouses, which were tenant's fixtures, and were in existence at the commencement of the lease, were destroyed by mineral workings sanctioned by the landlord. Held that the landlord was bound to compensate the tenant (*Gibson* v. *Farie*, 1918, 1 S.L.T. 404).

5 As this relates to fruit trees and fruit bushes *not* permanently set out, this part of the sub-section seems merely to state the common law (*Wardell* v. *Usher*, (1841), 3 Scott, (N.R.) 508). Generally, trees are held to be permanently set out when they are in a position from which it is not intended to transplant them.

6 This refers to section 46.

7 This sub-section places land which, under lease current on 1st January, 1898, was being used as a market garden with the knowledge of the landlord, and on which the tenant had 'then' (interpreted by the decision as 'thereafter') executed thereon, without previous notice of dissent from the landlord, any improvement comprised in the Fourth Schedule, in the same position as land agreed in writing, on or after that date, to be treated as a market garden. See also *Morse* v. *Dixon*, (1917), 87 L.J. K.B. 1, in which it was held that a term in a lease under which the tenant who planted fruit trees was entitled to

remove them, was not an agreement within the meaning of the sub-section.

[8] This differentiates a year-to-year market garden tenancy from a year-to-year farm tenancy. In the former case the compensation in respect of an improvement comprised in the First, Second or Third Schedule is such (if any) as could have been claimed had this Act not been passed; that is to say, it is necessary to fall back on the provisions of the previous statutes (section 96). Section 42 of the Act of 1883 had the effect of interrupting the currency of the lease when neither party had right to terminate the tenancy after 1st January, 1898 (*in re Kedwell* v. *Flint & Co.*, [1911], 1 K.B. 797). In the case, therefore, of a year-to-year tenancy, a new lease would be held to have commenced at the end of each year, with the result that an agreement in writing would be required under sub-section (1) to entitle the tenant to compensation.

Power of the Land Court in default of agreement to treat an agricultural holding as a market garden.

66.[1]—(1) Subject to the provisions of this section[2] where the tenant of an agricultural holding intimates to the landlord in writing his desire to carry out on the holding or any part thereof an improvement specified in the Fourth Schedule to this Act and the landlord refuses or within a reasonable time fails, to agree in writing that the holding, or that part thereof, shall be treated as a market garden, the Land Court may, on the application of the tenant, and after being satisfied that the holding or that part thereof is suitable for the purposes of market gardening, direct that the last foregoing section shall, either in respect of all the improvements specified in the said Fourth Schedule or in respect of some only of those improvements, apply to the holding or to that part thereof, and the said section shall apply accordingly as respects any improvement carried out after the date on which the direction is given.[3]

(2) Where a direction is given under the foregoing sub-section, then, if the tenancy is terminated by notice to quit given by the tenant or by reason of the tenant becoming notour bankrupt or executing a trust deed for behoof of his creditors, the tenant shall not be entitled to compensation in respect of improvements specified in the direction unless the tenant not later than one month after the date on which the notice to quit is given or the date of the bankruptcy or the execution of the trust deed, as the case may be, or such later date as may be agreed,[4] produces to the landlord an offer in writing by a substantial and otherwise suitable person[5] (being an offer which is to hold good for a period of three months from the date on which it is produced) to accept a tenancy of the holding from the termination of the existing tenancy thereof, and on the terms and conditions of that tenancy so far as applicable, and, subject as hereinafter provided, to

COMPENSATION AND FIXTURES

pay to the outgoing tenant all compensation[6] payable under this Act or under the lease, and the landlord fails to accept the offer within three months after the production thereof.[7]

(3) If the landlord accepts any such offer as aforesaid, the incoming tenant shall pay to the landlord on demand all sums payable to him by the outgoing tenant on the termination of the tenancy in respect of rent or breach of contract or otherwise in respect of the holding, and any amount so paid may, subject to any agreement between the outgoing tenant and incoming tenant, be deducted by the incoming tenant from any compensation payable by him to the outgoing tenant.

(4) A direction under sub-section (1) of this section may be given subject to such conditions, if any, for the protection of the landlord[8] as the Land Court may think fit to attach to the direction, and, without prejudice to the generality of this subsection, where the direction relates to part only of the holding, the direction may, on the application of the landlord, be given subject to the condition that the tenant shall consent to the division of the holding into two parts (one such part being the part to which the direction relates) to be held at rents agreed by the landlord and tenant or in default of agreement determined by arbitration, but otherwise on the same terms and conditions (so far as applicable) as those on which the holding is held.[9]

(5) A new tenancy created by the acceptance of a tenant in accordance with the provisions of this section on the terms and conditions of the existing tenancy shall be deemed for the purposes of section *seven* of this Act not to be a new tenancy.[10]

[1] See Introduction. This important provision is substantially based on the Evesham custom. But this section goes further by putting it in the power of any farmer, with the approval of the Land Court, to have part of his land treated as a market garden subject to conditions for the protection of the landlord.

[2] This section is as amended by the Agriculture Act, 1958, Schedule 1, Part II, and Schedule II, Part II. The functions which now belong to the Land Court under the section formerly were fulfilled by the Secretary of State who delegated them to Agricultural Executive Committees.

[3] It may be directed that all or only certain of the improvements in the Fourth Schedule shall apply to the holding or to the part, and, of course, any claim for compensation would be confined to these particular improvements. For example, 'Erection or enlargement of buildings for the purpose of the trade or business of a market gardener' might be excluded. What is 'a reasonable time' is a question of fact.

[4] The date of bankruptcy is the date of the first deliverance on the petition.

[5] **'A substantial and otherwise suitable person'**.—Apparently this person must have sufficient capital, and he must be otherwise capable from experience

and character of undertaking the obligations of the tenancy. Probably objections similar to those that may be stated against a legatee under section 20 relating to bequest of a lease may be taken here. A dispute as to whether a 'substantial and otherwise suitable person' has been put forward would be settled by arbitration under the Act. Where the holding is not divided, as provided for under (c), the provisions of (a) and (b) appear to apply to the *whole* farm, including the part which has been turned into a market garden as a result of a 'direction' under the section.

6 'All compensation'—i.e., in respect of the whole subject where it is undivided, or in respect of the part in question, where it is divided. The compensation embraces all the compensation payable under the Act, and not merely the compensation for market garden improvements, as 'directed' by the Land Court.

7 There does not appear to be any express reference to arbitration regarding the questions which may arise under this sub-section. It is thought, however, that in view of the wide terms of section 74, these questions would fall to be determined by an arbiter.

8 **'For the protection of the landlord'.**—One can imagine that the effect of putting this section into operation would be prejudicial to the landlord, and that the only or the main way of protecting him might be to require payment of an increased rent. It might not be fair to other market gardeners who generally pay higher than the ordinary agricultural rent to let the tenant who converts the whole or part of his land into market garden under the Act, escape with an abnormally low rent. The conditions would doubtless include such obligations for cultivation, cleaning, manuring, &c., as are usual on market garden land.

9 The effect of this clause is simply to make the one holding into two separate holdings—separate as regards improvements, compensation for disturbance, &c. One of them will remain under the same conditions as the original lease (subject, it may be, to an alteration in rent). The other will also remain under these conditions, plus the conditions attached by the Land Court and by this section. The apportionment of rent falls to be settled by arbitration failing mutual agreement. It seems to be in the power of an arbiter to increase the rent so that joint rents of the two holdings would exceed the rent of the holding as undivided. Further, it could be stipulated that the use of the buildings would be available for both holdings, and that, should the tenant give up one of the holdings, the landlord would be entitled to insist on his giving up the other at the same time. In some cases it would, evidently, be very awkward if two different tenants had joint use of the same buildings for separate holdings.

10 Section 7 *inter alia* prevents an increase of rent (under demand) from having effect until the lapse of five years from 'the commencement of the tenancy'. In a case to which this sub-section applies the five years would apparently be counted as if the original tenant had sat on under tacit relocation, and there had been no change of tenant.

Agreements as to compensation relating to market gardens.

67.—(1) Where an agreement in writing secures to the tenant of an agricultural holding for an improvement for which compensation is payable by virtue of either of the two last foregoing sections fair and reasonable compensation, having regard to the circumstances existing at the time of making the agreement, the compensation so secured shall, as respects that improvement, be substituted for compensation under this Act.[1]

(2) The landlord and the tenant of an agricultural holding how

COMPENSATION AND FIXTURES

have agreed that the holding shall be let or treated as a market garden may by agreement in writing substitute, for the provisions as to compensation which would otherwise be applicable to the holding, the provisions as to compensation set out in subsections (2) and (3) of the last foregoing section.

[1] The effect of this sub-section is that, in regard to market garden improvements (in the Fourth Schedule), substituted compensation may be provided as hitherto, while substituted compensation can only be provided now for the other improvements embraced in Part III of the Schedule where the lease was entered into prior to 1st January, 1921.

SETTLEMENT OF CLAIMS BETWEEN LANDLORD AND TENANT ON TERMINATION OF TENANCY

68.—(1) Without prejudice to any other provision of this Act, any claim of whatever nature by the tenant or the landlord of an agricultural holding against his landlord or his tenant, being a claim which arises— *Settlement of claims by arbitration.*

(a) under this Act or any custom or agreement, and

(b) on or out of the termination of the tenancy of the holding or part thereof,[1]

shall, subject to the provisions of this section, be determined by arbitration.[2]

(2) Without prejudice to any other provision of this Act, no such claim as aforesaid shall be enforceable unless before the expiration of two months from the termination of the tenancy the claimant has served notice in writing on his landlord or his tenant, as the case may be, of his intention to make the claim.[3]

A notice under this sub-section shall specify the nature of the claim, and it shall be a sufficient specification thereof if the notice refers to the statutory provision, custom, or term of an agreement under which the claim is made.

(3) The landlord and the tenant may within the period of four months from the termination of the tenancy by agreement in writing settle any such claim as aforesaid,[4] and the Secretary of State may upon the application of the landlord or the tenant made within that period extend the said period by two months and, on a second such application made during these two months, by a further two months.[5]

(4) Where before the expiration of the said period and any extension thereof under the last foregoing sub-section any such claim as aforesaid has not been settled, the claim shall cease to

184 AGRICULTURAL HOLDINGS (SCOTLAND) ACTS

be enforceable unless before the expiration of one month from the end of the said period and any such extension, or within such longer time as the Secretary of State may in special circumstances allow, an arbiter has been appointed[6] by agreement between the landlord and the tenant under the provisions of this Act in that behalf or an application for the appointment of an arbiter under those provisions has been made by the landlord or the tenant.[7]

(5) Where a tenant lawfully remains in occupation of part of an agricultural holding after the termination of a tenancy, references in sub-sections (2) and (3) of this section to the termination thereof shall be construed as references to the termination of the occupation.[8]

(6) This section shall not apply to a claim arising on or out of the termination of a tenancy before the first day of November, nineteen hundred and forty-eight.

[1] Both conditions must be fulfilled. Refer to section 93 for definition of 'termination of tenancy'.
[2] See also the general arbitration section (section 74). This section (68) (1) (b) appears to apply to waygoing valuations; although this results in inconsistencies in that the Secretary of State does not appoint two arbiters.
[3] At this stage only a general intimation is required. Where land is resumed the claim should be intimated two months from the date of resumption. The provisions appear to apply when the tenancy is renounced. In a notice under the corresponding section of the English Act the landlord's son's name was inserted in place of the landlord's in error. The notice was held valid as it was received by the landlord and he could not reasonably mistake its intentions: *Frankland* v. *Capstick*, [1959], 1 All E.R. 209. It would be unwise to assume that this decision would be followed in Scotland.
[4] This envisages settlement without detailed particulars.
[5] These provisions give scope for delaying tactics which the general scheme of the Act seeks to prevent.
[6] Mere agreement to appoint is not enough (*Chalmers Property Investment Co. Ltd.* v. *MacColl*, 1951 S.C. 24). Paragraph 5 of the 6th Schedule to the Act lays down the manner in which the parties must present their claims to the arbiter and failure to comply therewith renders the claim invalid. *Robertson's Trustees* v. *Cunningham*, 1951 S.L.T. (Sh.Ct.) 89.
[7] It is important to note that the provision is one month or longer.
[8] It has been judicially observed that section 68 (5) applies only to the case where the landlord permitted the tenant to retain possession of part of the holding after the ish in terms of a new agreement. *Coutts* v. *Barclay-Harvey*, 1956 S.L.T. (Sh.Ct.) 54.
See Opencast Coal Act, 1958, section 24.
See also Agriculture (Miscellaneous Provisions) Act, 1968 section 11 (5) which applies this section to sums payable under section 9 of that Act.

RECOVERY OF SUMS DUE UNDER THIS ACT

Recovery of compensation and other sums due.

69. Any award or agreement under this Act as to compensation, expenses or otherwise may, if any sum payable thereunder is not payed within one month after the date on which it becomes payable, be recorded for execution in the books of council and session or in the sheriff court books, and shall be enforceable in like manner as a recorded decree arbitral.[1]

[1] It is thought that the submission or award would not require to contain a warrant of registration for execution.
See Agriculture (Miscellaneous Provisions) Act, 1968 section 11 (5).

70.[1]—(1) Where on or after the first day of November, nineteen hundred and forty-eight, any sum has become payable to the tenant of an agricultural holding in respect of compensation by the landlord and the landlord has failed to discharge his liability therefor within one month after the date on which the sum became payable, the Secretary of State may, on the application of the tenant and after giving not less than fourteen days' notice of his intention so to do to the landlord, create, where the landlord is the absolute owner of the holding, a charge on the holding, or where the landlord is the lessee of the holding under a lease recorded under the Registration of Leases (Scotland) Act, 1857, a charge on the lease for the payment of the sum due. Power of tenant to obtain charge on holding in respect of compensation.

(2) For the purpose of creating a charge under this section for the payment of any sum due, the Secretary of State may make in favour of the tenant a charging order charging and burdening the holding or the lease, as the case may be, with an annuity to repay the sum due together with the expenses of obtaining the charging order and recording it in the appropriate Register of Sasines; and the provisions of sub-section (2) and sub-sections (4) to (10) of section fifty-five of the Water (Scotland) Act, 1946, shall, with the following and any other necessary modifications, apply to any such charging order—
  (a) for any reference to the local authority there shall be substituted a reference to the Secretary of State;
  (b) for any reference to the period of thirty years there shall be substituted a reference to such period (not exceeding thirty years) as the Secretary of State may determine;
  (c) for references to Part III of the said Act of 1946 there shall be substituted references to this Act.

(3) The creation of a charge on a holding or the lease of a

holding under this section shall not be deemed to be a contravention of any prohibition against charging or burdening contained in the deed or instrument under which the holding is held.

[1] See Agriculture (Miscellaneous Provisions) Act, 1968 section 11 (5).

SUPPLEMENTARY PROVISIONS

71.—Sections 71 and 72 of the 1949 Act were repealed by the 1958 Act, Second Schedule, Part II.

Proceedings of the Land Court.

73.—The provisions of the Small Landholders (Scotland) Acts, 1886 to 1931, with regard to the Land Court shall, with any necessary modifications, apply for the purpose of the determination of any matter which they are required by or under this Act to determine, in like manner as those provisions apply for the purpose of the determination by the Land Court of matters referred to them under those Acts.[1]

[1] There is an appeal by way of stated case on a question of law to the Court of Session.
This section applies also for purpose of determination of any matter under section 11 (7) of the Agriculture (Miscellaneous Provisions) Act, 1968—(1968 Act section 11 (8)).

Matters to be referred to arbitration.

74.[1]—Save as otherwise expressly provided in this Act, any question or difference of any kind whatsoever between the landlord and the tenant of an agricultural holding arising out of the tenancy or in connection with the holding (not being a question or difference as to liability for rent[2]) shall, whether such question or difference arises during the currency or on the termination of the tenancy, be determined by arbitration.

[1] In *Houison-Craufurd's Trs. v. Davies*, 1951 S.L.T. 25, landlords served on a tenant a notice under a resumption clause in a lease of their intention to resume certain parts of land and called upon the tenant to remove. The tenant refused and the landlords raised an action of removing in the Sheriff Court. The Sheriff held the defences irrelevant and granted decree to the landlords. The tenant appealed to the Court of Session and took the plea, which he had not taken in the inferior Court, that the question was one for arbitration in terms of section 74. It was held (Lord Patrick dissenting) that the question fell to be determined by arbitration under section 74 and that as section 74 is imperative parties cannot prorogate the jurisdiction of the Courts. The decision in *Houison-Craufurd's Trs. v. Davies* was affirmed by a Court of seven judges in *Brodie v. Ker* and *McCallum v. Macnair*, 1952 S.C. 216. The Agriculture (Miscellaneous Provisions) Act, 1968 section 11 (7) expressly excludes from arbitration and refers to the Land Court any

SUPPLEMENTARY PROVISIONS

question as to the purpose for which the landlord is terminating the tenancy under section 6 (3) of the Agriculture Act, 1958.
² See *Budge* v. *Mackenzie*, 1934 Sh.Ct.Rep. 168; *Ross* v. *Macdonald*, 1934 Sh.Ct.Rep. 201.
See *Goldsack* v. *Shore*, [1950], 1 K.B. 708; *Davidson* v. *Chiskan Estate Co. Ltd.*, 1952 S.L.C.R. 41; *Galbraith and Ors.* v. *Ardnacross Farming Co.*, 1953 S.L.T. Notes 30; *Gulliver* v. *Catt*, [1952], 2 Q.B. 308.

75.—(1) Any matter which by or under this Act, or by regulations made thereunder, or under the lease of an agricultural holding is required to be determined by arbitration[1] shall, whether the matter arose before or after the passing of this Act, be determined, notwithstanding any agreement under the lease or otherwise providing for a different method of arbitration, by a single arbiter in accordance with the provisions of the Sixth Schedule to this Act,[2] and the Arbitration (Scotland) Act, 1894, shall not apply to any such arbitration.

Provisions as to arbitrations.

(2) The Secretary of State may by rules make such provision as he thinks desirable for expediting, or reducing the expenses of proceedings on arbitrations under this Act:[3]

Provided that the Secretary of State shall not make rules inconsistent with the provisions of the Sixth Schedule to this Act.

(3) The power conferred by the last foregoing sub-section on the Secretary of State shall be exercisable by statutory instrument which shall be subject to annulment in pursuance of a resolution of either House of Parliament.

(4) This section and the last foregoing section shall not apply to valuations of sheep stocks, dung, fallow, straw, crops, fences and other specific things the property of an out-going tenant, agreed under a lease to be taken over from him at the termination of a tenancy by the landlord or the incoming tenant, or to any questions which it may be necessary to determine in order to ascertain the sum to be paid in pursuance of such agreement, and that whether such valuations and questions are referred to arbitration under the lease or not.[4]

[1] See Arbitration—Introduction. Practically every question arising between landlord and tenant in relation to the holding comes under the arbitration procedure of the Act. See also Agriculture Act 1967 Schedule 3 paragraph (5).
[2] The arbiter has not exclusive jurisdiction to determine whether a tenancy has terminated. It was held that the question whether a tenancy had terminated was a question precedent to the existence of a statutory claim. *Donald-*

son's Hospital v. Esslemont, 1926 S.C. (H.L.) 68. See Cowdray v. Ferries, 1919 S.C. (H.L.) 27, where it was held that the arbiter had jurisdiction to decide whether notice to quit was valid.

[3] No such directions have been yet issued.

[4] See note 1 and Introduction. It is unfortunate that 'out-going tenant' is not stated to include the landlord where the landlord is also the occupier. A landlord who was in possession agreed with an incoming tenant for valuation of the crops, the property of the landlord, by two arbiters and an oversman. The tenant refused to pay, as the valuation had been by more than one arbiter, and it was held that section 1 of the Agricultural Holdings Act, 1910, did not apply to a reference between a landlord in personal occupation and the tenant in-going (Methven v. Burn, 1923 S.L.T. (Sh.Ct.) 25).

Constitution of panel of arbiters, and provisions as to remuneration of arbiter.

76.—(1) Such number of persons as may be appointed by the Lord President of the Court of Session, after consultation with the Secretary of State, shall form a panel of persons from whom any arbiter appointed, otherwise than by agreement, for the purposes of an arbitration under and in accordance with the provisions of the Sixth Schedule to this Act shall be selected.

(2) The panel of arbiters constituted under the foregoing sub-section shall be subject to revision by the Lord President of the Court of Session, after such consultation as aforesaid, at such intervals not exceeding five years, as the Lord President and the Secretary of State may from time to time agree.

(3) The remuneration of an arbiter appointed as aforesaid by the Secretary of State shall be such amount as is fixed by the Secretary of State, and the remuneration of an arbiter appointed by the parties to any such arbitration shall, in default of agreement between those parties and the arbiter, be such amount as, on the application of the arbiter or of either of the parties, is fixed by the auditor of the sheriff court, subject to appeal to the sheriff.[1]

(4) The remuneration of an arbiter, when agreed or fixed under this section, shall be recoverable by the arbiter as a debt due from either of the parties to the arbitration, and any amount paid in respect of the remuneration of the arbiter by either of those parties in excess of the amount (if any) directed by the award to be paid by him in respect of the expenses of the award shall be recoverable from the other party to the arbitration.

[1] It is for the arbiter to apply to the Secretary of State to fix his fee, which is over and above his out-of-pocket expenses. The application should be accompanied by a detailed note of all the work done and time occupied by the arbiter in connection with the arbitration. In important cases the process may with advantage be placed before the Secretary of State.

SUPPLEMENTARY PROVISIONS 189

77.[1]—Where the Secretary of State is a party to any question or difference which under this Act is to be determined by arbitration or by an arbiter appointed in accordance with the provisions of this Act, the arbiter shall, in lieu of being appointed by the Secretary of State, be appointed by the Land Court, and the remuneration of the arbiter so appointed shall be such amount as may be fixed by the Land Court.

Appointment of arbiter in cases to which the Secretary of State is a party.

[1] *Secretary of State* v. *John Jaffray and others*, 1957 S.L.C.R. 27. *Commissioners of Crown Lands* v. *Grant*, 1955 S.L.C.R. 25.
See Agriculture Act, 1967 Schedule 3 paragraph (5).

78.[1]—Any question or difference between the landlord and the tenant of an agricultural holding which by or under this Act or under the lease is required to be determined by arbitration may, if the landlord and the tenant so agree, in lieu of being determined in pursuance of sub-section (1) of section *seventy-five* of this Act be determined by the Land Court, and the Land Court shall, on the joint application of the landlord and the tenant, determine such question or difference accordingly.

Determination of questions by Land Court in lieu of arbitration.

[1] This procedure is now being invoked to some extent. In disputes sent to the Land Court, the decision of that Court is not subject to a stated case to the Sheriff or on appeal to the Court of Session. There is, however, provision in the Land Court Rules for a special case stated on any question of law 'for the opinion of either Division of the Court of Session'—Rules 102-108.
See Agriculture Act, 1967 Schedule 3 paragraph (5).

79.[1]—(1) The Secretary of State may, after consultation with persons appearing to him to represent the interests of landlords and tenants of agricultural holdings, by order vary the provisions of the First and Fourth Schedules to this Act.

Power of Secretary of State to vary First and Fourth Schedules to this Act.

(2) An order under this section may make such provision as to the operation of this Act in relation to tenancies current when the order takes effect as appears to the Secretary of State to be just having regard to the variation of the said Schedules effected by the order:

Provided that nothing in any order made under this section shall affect the right of a tenant to claim, in respect of an improvement made or begun before the date on which such order takes effect, any compensation to which, but for the making of the order, he would have been entitled.

(3) An order under this section shall be embodied in a

statutory instrument which shall be of no effect unless approved by resolution of each House of Parliament.

¹ See Opencast Coal Act, 1958, sections 26 and 28.

Power of limited owners to give consents, &c.

80.—The landlord¹ of an agricultural holding,¹ whatever may be his estate or interest in the holding, may for the purposes of this Act give any consent, make any agreement, or do or have done to him any act which he might give or make or do or have done to him if he were absolute owner of the holding.

¹ See section 93 for definitions.
See Agriculture (Miscellaneous Provisions) Act, 1968 section 11 (5).

Power of heir to entail to apply entailed moneys for improvements.

81.—The price of any entailed land sold under the provisions of the Entail Acts, when such price is entailed estate within the meaning of those Acts, may be applied by the heir of entail in respect of the remaining portion of the entailed estate, or in respect of any other estate belonging to him and entailed upon the same series of heirs, in payment of any expenditure and expenses incurred by him in pursuance of this Act in carrying out or paying compensation for any old improvement specified in Part I or Part II of the Second or Third Schedule to this Act or any new improvement specified in Part I or Part II of the First Schedule thereto, or in discharge of any charge with which the estate is burdened in pursuance of this Act in respect of such an improvement.¹

¹ See Introduction. It is doubtful if this provision would apply to cases where the parties agree on the amount of compensation. In any event it is wiser to let an arbiter fix the amount where it is intended to charge the estate therewith.
An absolute owner is entitled at common law to charge his property as he may decide. This section is, therefore, of avail only to limited owners.

Power of landlord to obtain charge on holding in respect of compensation, &c., paid by him.

82.¹—(1) Where on or after the first day of November, nineteen hundred and forty-eight, the landlord of an agricultural holding, not being the absolute owner of the holding, has paid to the tenant of the holding the amount due to him under this Act, or under custom or agreement, or otherwise, in respect of compensation for an old or a new improvement or in respect of compensation for disturbance, or has himself defrayed the cost of an improvement proposed to be executed by the tenant, the Secretary of State may, on the application of the landlord and after giving not less than fourteen days' notice to the absolute owner of the holding, make in favour of the landlord a charging order charging and burdening the holding with an annuity to repay the amount of the compensation or the cost of the improvement, as the case may be, together with the expenses of obtaining the charging order and recording it in the appropriate Register of Sasines and the provisions of sub-sections (2) and (4) and of sub-sections (6) to (10) of section fifty-five of the Water (Scotland) Act, 1946, shall, with the following and any other necessary modifications, apply to any such charging order—

(a) for any reference to the local authority there shall be substituted a reference to the Secretary of State;

(b) for any reference to the period of thirty years there shall be substituted in the case of a charging order made in respect of compensation for, or of the cost of, an improvement a reference to the period within which the improvement will, in the opinion of the Secretary of State, have become exhausted;

(c) for references to Part III of the said Act of 1946 there shall be substituted references to this Act.

(2) An annuity constituted a charge by a charging order recorded in the appropriate Register of Sasines shall be a charge on the holding specified in the order and shall rank after all prior charges heritably secured thereon.

(3) The creation of a charge on a holding under this section shall not be deemed to be a contravention of any prohibition against charging or burdening contained in the deed or instrument under which the holding is held.

[1] See Agriculture (Miscellaneous Provisions) Act, 1968 section 11 (5).

83.[1]—Any company now or hereafter incorporated by Parliament or incorporated under the Companies Act, 1948, and having power to advance money for the improvement of land, or for the cultivation and farming of land, may make an advance of money upon a charging order duly made and recorded under this Act, on such terms and conditions as may be agreed upon between the company and the person entitled to the order.

Power of land improvement companies to advance money.

[1] See Agriculture (Miscellaneous Provisions) Act, 1968 section 11 (5).

84.—Where the landlord[1] or the tenant[1] of an agricultural holding is a pupil or a minor or is of unsound mind, not having

Appointment of guardian to landlord or tenant in certain cases.

a tutor, curator or other guardian, the sheriff, on the application of any person interested, may appoint to him, for the purposes of this Act, a tutor or a curator, and may recall the appointment and appoint another tutor or curator if and as occasion requires.[2]

[1] See section 93 for definitions.
[2] See Agriculture (Miscellaneous Provisions) Act, 1968 section 11 (5).

Validity of consents, &c.

85.—It shall be no objection to any consent in writing or agreement in writing under this Act signed by the parties thereto or by any persons authorised by them that the consent or agreement has not been executed in accordance with the enactments regulating the execution of deeds in Scotland.[1]

[1] This practically places 'consent in writing' or 'agreement in writing' under the Act, among the class of privileged writing such as those *in re mercatoria*, which 'are effectual although neither attested (by witnesses) nor holograph', on account of the rapidity which may be necessary in preparing them.

Provisions as to Crown Land

Application of Act to Crown land.

86.—(1) This Act shall apply to land belonging to His Majesty in right of the Crown, subject to such modifications as may be prescribed; and for the purposes of this Act the Commissioners of Crown Lands or other the proper officer or body having charge of the land for the time being, or if there is no such office or body, such person as His Majesty may appoint in writing under the Royal Sign Manual, shall represent His Majesty and shall be deemed to be the landlord.

(2) Without prejudice to the provisions of the foregoing sub-section it is hereby declared that the provisions of this Act apply to land notwithstanding that the interest of the landlord or the tenant thereof belongs to a government department or is held on behalf of His Majesty for the purposes of any government department; but in their application to any land belonging, or an interest in which is held as aforesaid, the said provisions shall have effect subject to such modifications as may be prescribed.

(3) Repealed by Crown Estate Act, 1961, Third Schedule.

Determination of matters relating to holdings of which the Secretary of State is landlord or tenant.

87.—(1) Any section of this Act under which any matter is referred to the decision of the Secretary of State shall, in its application to an agricultural holding of which the Secretary of State is himself the landlord or the tenant,[1] have effect with the substitution of the Land Court for the Secretary of State, and any provision in any such section for an appeal to an arbiter from the decision of the Secretary of State shall not apply.

(2) The provisions of this Act shall, in their application to any arbitration with regard to an agricultural holding of which the Secretary of State is himself the landlord or the tenant, have effect with the substitution of the Land Court for the Secretary of State.[2]

[1] See observations in *Commissioners for Crown Lands* v. *Grant*, 1955 S.L.C.R. 25.
[2] See Agriculture Act, 1967, Schedule 3 paragraph (5).

General

88.—(1) All expenses incurred by the Secretary of State under this Act shall be defrayed out of moneys provided by Parliament.

Expenses and receipts.

(2) All sums received by the Secretary of State under this Act, including sums received on his behalf by any person or body of persons exercising functions on behalf of the Secretary of State, shall be paid into the Exchequer.

89.—(1) Any person authorised by the Secretary of State in that behalf shall have power at all reasonable times to enter on and inspect any land for the purpose of determining whether, and if so in what manner, any of the powers conferred on the Secretary of State by this Act are to be exercised in relation to the land, or whether, and if so in what manner, any direction given under any such power has been complied with.

Provisions as to entry and inspection.

(2) Any person authorised by the Secretary of State who proposes to exercise any power of entry or inspection conferred by this Act shall, if so required, produce some duly authenticated document showing his authority to exercise the power.

(3) Admission to any land used for residential purposes shall not be demanded as of right in the exercise of any such power as aforesaid unless twenty-four hours' notice of the intended entry has been given to the occupier of the land.

(4) Save as provided by the last foregoing sub-section, admission to any land shall not be demanded as of right in the

exercise of any such power as aforesaid unless notice has been given to the occupier of the land that it is proposed to enter during a period, specified in the notice, not exceeding fourteen days and beginning at least twenty-four hours after the giving of the notice and the entry is made on the land during the period specified in the notice.

(5) Any person who obstructs any person authorised by the Secretary of State exercising any such power as aforesaid shall be guilty of an offence and shall be liable on summary conviction to a fine not exceeding five pounds in the case of a first offence or twenty pounds in the case of a second or any subsequent offence.

Service of notices, &c.

90.—(1)[1] Any notice or other document required or authorised by or under this Act to be given to or served on any person shall be duly given or served if it is delivered to him, or left at his proper address, or sent to him by post in a registered letter[2].

(2) Any such document required or authorised to be given to or served on an incorporated company or body shall be duly given or served if given to or served on the secretary or clerk of the company or body.

(3) For the purposes of this section and of section twenty-six of the Interpretation Act, 1889, the proper address of any person to or on whom any such document as aforesaid is to be given or served shall, in the case of the secretary or clerk of any incorporated company or body, be that of the registered or principal office of the company or body, and in any other case be the last known address of the person in question.

(4) Unless or until the tenant of an agricultural holding shall have received notice that the person theretofore entitled to receive the rents and profits of the holding (hereinafter referred to as 'the original landlord') has ceased to be so entitled, and also notice of the name and address of the person who has become entitled to receive such rents and profits, any notice or other document served on or delivered to the original landlord by the tenant shall be deemed to have been served on or delivered to the landlord of the holding.

[1] This sub-section is concerned with proof of service, and may raise questions of corroboration. Service by ordinary post would in any event raise difficulties. Service on person in occupation held valid. *Wilbraham* v. *Colclough and Ors.*, [1952], 1 All E.R. 979.

² Includes letter sent by recorded delivery service: Recorded Delivery Service Act, 1952 sections 1 and 2. Notice sent by recorded delivery before the passing of the 1952 Act was held duly served—*re Poyser and Mills Arbitration* [1964], 2 Q.B. 467.
See Agrilculture (Miscellaneous Provisions) Act, 1968 section 11 (5).

91.—Where any jurisdiction committed by this Act to the sheriff is exercised by the sheriff-substitute, there shall be no appeal to the sheriff.¹ <small>Prohibition of appeal from sheriff substitute.</small>

¹ Either the Sheriff or the Sheriff-substitute is entitled to exercise jurisdiction. The section applies only to jurisdiction conferred by the Act, not to the common law jurisdiction. (*Cameron* v. *Ferrier*, 1912, 28 Sh.Ct.Rep. 220).

92.—Any power conferred on the Secretary of State by this Act to make an order shall include a power, exercisable in the like manner and subject to the like conditions, to revoke or vary the order. <small>Revocation and variation of orders.</small>

93.—(1) In this Act, unless the context otherwise requires, the following expressions have the meanings hereby respectively assigned to them, that is to say— <small>Interpretation.</small>

    'absolute owner' means the owner or person capable of disposing by disposition or otherwise of the fee simple or dominium utile of the whole interest of or in land, although the land, or his interest therein, is burdened, charged, or encumbered;

    'agricultural holding'¹ has the meaning assigned to it by section one of this Act;

    'agricultural unit' means land which is an agricultural unit for the purposes of the Agriculture (Scotland) Act, 1948;

    'agriculture' includes horticulture, fruit growing, seed growing, dairy farming and livestock breeding and keeping, the use of land as grazing land, meadow land, osier land, market gardens and nursery grounds, and the use of land for woodlands where that use is ancillary to the farming of land for other agricultural purposes, and 'agricultural' shall be construed accordingly;

    'building' includes any part of a building;

    'Defence Regulations' means Regulations made under the Emergency Powers (Defence) Acts, 1939 and 1940;

'fixed equipment' includes any building or structure affixed to land and any works on, in, over or under land, and also includes anything grown on land for a purpose other than use after severance from the land, consumption of the thing grown or of produce thereof, or amenity, and, without prejudice to the foregoing generality, includes the following things, that is to say—

(*a*) all permanent buildings, including farm houses and farm cottages, necessary for the proper conduct of the agricultural holding;

(*b*) all permanent fences, including hedges, stone dykes, gate posts and gates;

(*c*) all ditches, open drains and tile drains, conduits and culverts, ponds, sluices, flood banks and main water courses;

(*d*) stells, fanks, folds, dippers, pens and bughts necessary for the proper conduct of the holding;

(*e*) farm access or service roads, bridges and fords;

(*f*) water and sewerage systems;

(*g*) electrical installations including generating plant, fixed motors, wiring systems, switches and plug sockets;

(*h*) shelter belts;

and references to fixed equipment on land shall be construed accordingly;

'former enactment relating to agricultural holdings' means Part I of the Agriculture (Scotland) Act, 1948, Part II of the Small Landholders and Agricultural Holdings (Scotland) Act, 1931, the Agricultural Holdings (Scotland) Act, 1923, and any enactment repealed by the last-mentioned Act;

'Land Court' means the Scottish Land Court;

'landlord' means any person for the time being entitled to receive the rents and profits or to take possession of any agricultural holding, and includes the executor, administrator, assignee, heir-at-law, legatee, disponee, next-of-kin, guardian, curator bonis or trustee in bankruptcy, of a landlord;[2]

'lease' means a letting of land for a term of years, or for

lives, or for lives and years, or from year to year;[1]
'livestock' includes any creature kept for the production of food, wool, skins or fur, or for the purpose of its use in the farming of land;
'market garden' means a holding, cultivated, wholly or mainly, for the purpose of the trade or business of market gardening;[3]
'new improvement' has the meaning assigned to it by sub-section (2) of section *forty-seven* of this Act;
'old improvement' has the meaning assigned to it by sub-section (2) of section *thirty-six* of this Act;
'1923 Act improvement' and '1931 Act improvement' have the meanings respectively assigned to them by sub-section (1) of section *thirty-six* of this Act;
'pasture' includes meadow;
'prescribed' means prescribed by the Secretary of State by regulations made by statutory instrument which shall be subject to annulment in pursuance of a resolution of either House of Parliament;
'produce' includes anything (whether live or dead) produced in the course of agriculture;
'tenant' means the holder of land under a lease and includes the executor, administrator, assignee, heir-at-law, legatee, disponee, next-of-kin, guardian, curator bonis, or trustee in bankruptcy, of a tenant;[4]
'termination,' in relation to a tenancy, means the termination of the lease by reason of effluxion of time or from any other cause;[5]
'Whitsunday' and 'Martinmas' in relation to any lease entered into on or after the first day of November, ninteen hundred and forty-eight, mean respectively the twenty-eighth day of May and the twenty-eighth day of November.[6]

(2) The provisions of the Fifth and Sixth Schedules to the Agriculture (Scotland) Act, 1948 (which have effect respectively for the purpose of determining for the purposes of that Act whether the owner of agricultural land is fulfilling his responsibilities to manage it in accordance with the rules of good estate management and whether the occupier of such land is fulfilling his responsibilities to farm it in accordance with the rules of good husbandry)

shall have effect for the purposes of this Act as they have effect for the purposes of that Act.

(3) References in this Act to the farming of land include references to the carrying on in relation to the land of any agricultural activity.

(4) References in this Act to the use of land for agriculture include, in relation to land forming part of an agricultural unit, references to any use of the land in connection with the farming of the unit.

(5) References to the terms, conditions, or requirements of a lease of or of an agreement relating to an agricultural holding shall be construed as including references to any obligations, conditions or liabilities implied by the custom of the country in respect of the holding.[7]

(6) The designations of landlord and tenant shall continue to apply to the parties until the conclusion of any proceedings taken under or in pursuance of this Act in respect of compensation for improvements or under any agreement made in pursuance of this Act.[8]

(7) Anything which by or under this Act is required or authorised to be done by, to or in respect of the landlord or the tenant of an agricultural holding may be done by, to or in respect of any agent of the landlord or of the tenant.[9]

[1] In England it has been held that a tenancy for one year certain was an interest less than a tenancy from year-to-year. *Bernays* v. *Prosser*, [1963], 2 Q.B. 592. There must be very few cases in Scotland of lets for a single year, except perhaps with reference to grass parks which are usually let for the season only, and do not, therefore, fall within the Act.

A let of land 'for a course of cropping for four years' comes under the Act (*M'Kenzie* v. *Buchan*, 1889, 5 Sh.Ct.Rep. 40), but a let for a 'rotation of cropping' subject to termination in the event of sale was not within the Act. (*Stirrar and Anor.* v. *Whyte*, 1968 S.L.T. 157).

Notice served under the Act must generally apply to the entire 'holding'. (But see sections 32, 33, 34, 60 and 61) (*Gates* v. *Blair*, 1923 S.C. 430).

The mere fact that a farmhouse on a large farm has been let to a sub-tenant, with approval of the landlord, for the purpose of taking in paying guests does not prevent the farm from being a 'holding' as defined in this section (*in re Russell & Harding's Arbitration*, (1922), 67 S.J. 123). In *Salmon* v. *Eastern Regional Hospital Board*, it was held by Sheriff Ford at Forfar early in 1960 that although the defenders used an agricultural unit to grow produce for consumption in the hospital and had substantial sales the unit was not a holding to which the Act applied.

A 'holding' may be even quarter of an acre in extent (*Malcolm* v. *M'Dougall*, 1916 S.C. 283).

House and grounds principal subjects, and land accessory (*Taylor* v. *Earl of Moray*, 1892, 19 R. 399). A farmhouse, outbuildings and garden ground held not to be a holding (*Barr* v. *Strang and Gardner*, 1935 S.L.T.

(Sh.Ct.) 10). Where dwelling-house of greater value than holding (*Hamilton v. Duke of Hamilton's Trustees*, 1918 S.C. 282). Inclusion of non-agricultural subjects in holding (*Stormonth-Darling v. Young*, 1915 S.C. 44—blacksmith's shop; *Yool v. Shepherd*, 1914 S.C. 689—spinning mill). Small non-agricultural element immaterial (*Taylor v. Fordyce*, 1918 S.C. 824—porter and ale licence; *in re Russell & Harding, supra*—dwelling-house sub-let). Hotel and about 30 acres land (*Mackintosh v. Lord Lovat*, 1886, 14 R. 282). (See also *in re Lancaster & Macnamara*, [1918], 2 K.B. 472; a 'stud-farm' is not a holding (*Berwick v. Baird*, [1930], 2 Ch. 359). Poultry farm (*Lean v. Inland Revenue*, 1926 S.C. 15). *Robinson v. Rowbottom*, 1942 S.L.T. (Sh.Ct.) 43. Piggery (*Inland Revenue v. Assessor for Lanarkshire*, 1930 S.L.T. 164).

For case in which subjects let in part remuneration for services, see *Dunbar's Trustees v. Bruce*, 1900, 3 F. 137.

² **'Landlord'.** See note 9, *infra*. 'The primary object of this definition was to extend the meaning of the word "landlord" so as to include heritable creditors in possession, liferenters, &c'. (*Waddell v. Howat*, 1925 S.L.T. 403). A purchaser having given notice of termination of a tenancy is impliedly liable to pay compensation to the tenant. The vendor has no duty to disclose a sale to the tenant (*re Lord Derby and Fergusson's contract*, [1912], 1 Ch. 479). But the Agricultural Holdings Acts (Amendment) Act, 1923, provided that, 'where a seller does not intimate to his tenant that the sale has taken place, the notice to the seller will be deemed to be notice to the landlord'. Where landlord not legal owner at date when notice given see *Tebbs & Sons v. Edwards*, 1950 E.G. 335 (see also section 90 (4) of the 1949 Act and note 9, *infra*). The provisions referred to protect the tenant, but do not interfere with the liability of seller and purchaser *inter se*. A tenant who sub-lets stands in the relation of proprietor to his sub-tenant. As to sub-tenancy under the Crofters Act, see *Livingstone v. Beattie*, 1891, 18 R. 735, and *Dalgleish v. Livingston*, 1895, 22 R. 646. Where farm sold and notice to quit given by seller to tenants between date of contract and date of completion thereof, held that completion meant 'actual completion', and that seller being person entitled to receive rents and profits was the landlord and liable for disturbance (*Richards v. Pryse*, [1927], 2 K.B. 76). For meaning of 'assignee' see *Cunningham v. Fife C.C.*, 1948 S.C. 439. 'Landlord' includes an uninfeft proprietor (*Alexander Black & Sons v. Paterson*, 1968 S.L.T. Sh.Ct.) 64).

Executors of a lessor whose estate was entailed were held bound by a clause in a lease under which the sheep stock was to be taken over by the proprietor or incoming tenant (*Riddel's Executors v. Milligan's Executors*, 1909 S.C. 1137).

³ **'Market garden'.** A farmer, who was in the habit of growing, in open fields, a quantity of peas and young potatoes as a fallow crop, and sending them to London for sale, was held not to be a market gardener within 5 *and* 6 Vict. c. 122, *section* 10. *Hammond, ex parte*, [1844], De G. 93; cp. *in re Wallis*, 14 Q.B.D. 950.

As to definition of 'market gardens', see *Cooper v. Pearce*, [1896], 1 Q.B. 562; *Purser v. Worthing Local Board*, 18 Q.B. 818 (whole under glass). Holding held a market garden where wholly under raspberries (*Grewar v. Moncur's Curator Bonis*, 1916 S.C. 764). Lease of buildings and ground for two years for the purpose of growing bulbs held not a 'holding' (*Watters v. Hunter*, 1927 S.C. 310). Where under flowers only (*Drummond v. Thomson*, 1921, 37 Sh.Ct.Rep. 180). Where occupier of a residence sold surplus of his requirements—not a market garden (*Bickerdike v. Lucy*, [1920], 1 K.B. 707); a piece of ground for growing raspberries for jam held not to be a market garden, but a fruit farm is probably a market garden.

⁴ **'Tenant'** includes sub-tenants, and as below. A sub-tenant cannot impugn the title of the principal tenant (*Dunlop & Co. v. Meiklem*, 1876, 4 R. 11). A sub-tenant falls within the definition of a 'tenant'. The principal tenant is the landlord in relation to the sub-tenant. Where there is a sub-tenant, and the principal tenant has no property on the holding, the latter

would not be entitled to claim compensation for disturbance (*Ministry of Agriculture* v. *Dean*, [1924], 1 K.B. 851).

A general conveyance, by will, of all estate, heritable and moveable, may carry a lease (*Gordon (Lindsay's Trustees)* v. *Welsh*, 1890; Rankine on Leases, p. 164).

Bankruptcy does not in the absence of a condition in the lease terminate a tenancy; the trustee in bankruptcy is entitled to carry on the lease in virtue of the effect of the act and warrant as an assignation in his favour, assuming he is not barred from doing so by the lease. So also in the case of all the other successors of a tenant above enumerated; if they are entitled to take up, and do take up the lease, they become tenants, subject to all a tenant's rights and liabilities, and may be entitled to claim compensation under the Act.

A trustee in bankruptcy, who had taken up a tenancy, was held entitled to claim under the Act of 1883 (*Sinclair* v. *Clyne's Tr.*, 1887, 15 R. 185). On the other hand, it was held that the bankrupt could not claim where the lease was prematurely terminated by bankruptcy (*Walker, &c.* v. *M'Knights*, 1886, 13 R. 599; *Haden and Others (Scott's Executors)* v. *Hepburn*, 3 R. 816; *ex parte Dyke, in re Morrish*, 22 Ch. 410, *Schofield* v. *Hincks*, 58 L.J. Q.B. 147). Held that a trustee for creditors was not entitled to claim (*Christison's Trustees* v. *Callender Brodie*, 1905, 8 F. 928). See also *Egerton* v. *Rutter*, [1951], 1 K.B. 472.

Again, a tenant who renounces his lease, his renunciation being accepted by the landlord, does not thereby lose his right to claim compensation (*Strang* v. *Stuart*, 1887, 14 R. 637).

Where there is an express exclusion of the tenant's legatee in the lease, section 20 cannot be invoked. *Kennedy* v. *Johnstone*, 1956 S.C. 39 and see notes to section 20. Waiving right to exclude assignees (*Lord Elphinstone* v. *Monkland Iron Co.*, 1886, 13 R. (H.L.) 98). Abandonment of lease by tenant's heir (*Forbes* v. *Ure*, 1856, 18 D. 577; *Duff* v. *Keith*, 1857, 19 D. 713). Heir's failure to vindicate right (*Gray* v. *Low*, 1859, 21 D. 293; *Wilson* v. *Stewart*, 1853, 16 D. 106; *M'Iver* v. *M'Iver*, 1909 S.C. 639). Cases in which trustee in bankruptcy not held to have taken over lease, notwithstanding that he cut and harvested crops, &c. (*M'Gavin* v. *Sturrock's Trustee*, 1891, 18 R. 576; *Taylor's Trustee* v. *Paul*, 1888, 15 R. 313; *Imrie's Trustee* v. *Calder*, 1897, 25 R. 15). Cases *contra* to above (*Ross* v. *Monteith*, 1786, M. 15290; *Kirkland* v. *Gibson*, 1831, 9 S. 596; *Dundas* v. *Morison*, 1875, 20 D. 225; *M'Lean's Trustee* v. *M'Lean*, 1850, 13 D. 90; *Moncrieffe* v. *Ferguson*, 1890, 24 R. 47; *Mackessack & Son* v. *Molleson*, 1886, 13 R. 445).

[5] See Introduction. The interpretation of 'determination of tenancy' has been given by the Court of Session in the following decisions:—

(a) *Strang* v. *Stuart*, 1887, 14 R. 637. Ish at Martinmas as to arable land, and at Whitsunday following as to the houses and grass. Held that termination of tenancy was the latter term.

(b) *Black* v. *Clay*, 1894, 21 R. (H.L.) 72. Ish Whitsunday as to houses (except as after), grass and fallow; separation of crop thereafter as to arable land under crop, and Whitsunday following as to barns, barn yard, and cot houses. Held 'determination of tenancy' was at 'separation of crop' (and notice of intention to claim given before Martinmas of that year held in time).

(c) *Waldie* v. *Mungall*, 1896, 23 R. 792. A ten years' lease of a farm, with entry 'at Martinmas, 1885', bound the tenant to consume all turnips on the farm and apply the dung to the lands. On the other hand, he was entitled to retain the use of a barn and threshing mill 'till the term of Whitsunday after the termination of the lease' for threshing his away-going crop. Held that the determination of tenancy was Martinmass 1895, although the tenant had a limited right of occupation beyond that date.

(d) *Todd* v. *Bowie*, 1902, 4 F. 435. A tenant is not entitled to renounce his lease because the fences are in bad order, contrary to stipulation. Accordingly, if on such a ground he abandons his lease, there has been no 'determination of the lease' in the sense of section 42 of the Agricultural Holdings (Scotland) Act, 1883.

(e) Whitsunday and separation of crop waygoing. *Coutts* v. *Barclay-Harvey*, 1956 S.L.T. (Sh.Ct.) 54.

(f) In *Breadalbane* v. *Stewart*, 1904, 6 F. (H.L.) 23, the expression 'at my away-going' meant the expiry of the lease through the effluxion of time, and did not apply where the landlord had irritated the lease on the tenant being in arrear with his rent. (See also *Pendreigh's Trustee* v. *Dewar*, 1871, 9 M. 1037). Where a tenant took advantage of a break at the end of fourteen years (in a twenty-one years' lease), the landlord agreeing to pay the tenant compensation for seeds, gooseberry and currant bushes, 'at the end of the term', it was held that the tenant was entitled to compensation for improvements under the English Act (*Bevan* v. *Chambers*, 12 T.L.R. 417). In England, under a yearly tenancy, where the tenant entered to the main portion of a holding on 6th April and to the farmhouse, buildings, and remainder of land on 13th May, subject to the express condition that 'on termination of tenancy' he should give up possession of different portions on these respective dates—(one rent being payable for the whole farm)—held that the termination of tenancy for the purpose of section 10, sub-section (7), of the Act of 1920, was 13th May, when the tenant had to quit the boozy pasture ('an outlet or outrun adjoining the homestead') notwithstanding that the main portion of the subjects let had to be surrendered at the earlier date (*Swinburne* v. *Andrews*, [1923], 2 K.B. 483). (See also *Russell* v. *Freen*, 1835, 13 S. 752).

Termination of the tenancy by voluntary renunciation of lease does not bar a claim for compensation (*Strang* v. *Stuart*, 1887, 14 R. 637; *Gardiner* v. *Lord Abercromby*, 1893, 9 Sh.Ct.Rep. 33; and *Osler* v. *Lansdowne*, 1885, 1 Sh.Ct.Rep. 48). In the last of these cases not only the lease, but also all claims in the tenancy, were expressly discharged in the renunciation. A tenancy from year-to-year, on the death of the tenant, vests in England in his personal representatives (*Shaw* v. *Porter*, 3 T.R. 13; *James* v. *Dean*, 15 Ves. 241).

A tenant under an agreement for a lease for fourteen years was to receive compensation on a valuation basis for the tillages and improvements he might leave on the farm. A dispute arose between landlord and tenant. The tenant said he would quit at the end of the year and the landlord agreed. He left accordingly, and on his claiming compensation the Court held that such a quitting was not a quitting under the terms of the tenancy, but in reality a running away, and, therefore, they rejected the claim (*Whittaker* v. *Barker*, 1 C. & M. 113).

'**Termination of the lease**' may not in all cases be the same as 'termination of the tenancy'. (See section 24, note 4).

6 '**Whitsunday' and 'Martinmas'**. For the meaning of these words in leases, see Introduction, and sections 93 (1) but note that despite the definition in section 93 (1) notice to quit in terms of section 24 may require to be served against 15th May or 11th November. See notes 6 and 7 to section 24. See also *Hunter* v. *Barron's Trustees*, 1886, 13 R. 883; and *Fraser's Trustees* v. *Maule & Son*, 1904, 6 F. 819. In an improbative lease, where no term of entry was specified, it was held that any legal presumption which there might be as to the implied term of entry might be displaced by evidence and proof as to the actual entry to, and possession of, the subjects allowed (*Watters* v. *Hunter*, 1927 S.L.T. 232; *Macleod* v. *Urquhart*, 1808, Hume, 840; *Russell* v. *Freen*, 1835, 13 S. 752).

7 Always in so far as the 'custom of the country' is not inconsistent with the express terms of the contract of tenancy.

8 '**Proceedings**' here referred to are 'judicial proceedings', and do not include negotiations with regard to the question of compensation (*Waddell* v. *Howat*, 1924 S.L.T. 684).

This clause might be read as suggesting that, where a claim is made against the person who was proprietor at the time, he would continue under obligation to meet the claim even after he has sold the farm to another who has taken possession prior to the settlement of the claim. The sounder position appears to be that the person who is owner at the quitting terms must meet the claim even though the notice to quit was given to or by a previous owner. The reason is that the Act bears that the tenant shall be entitled to compensation from the landlord *on quitting*. Apparently it is the person who is then the owner who must pay (*Bradshaw* v. *Bird*, [1920], 3 K.B. 144). This was decided mainly on the definition of landlord as the person for the time being entitled to receive the rents, &c. This decision was followed in the case of *Dale* v. *Hatfield Chase Corporation*, [1922], 2 K.B. 282. In *Bennett* v. *Stone*, [1920], 1 Ch. 226, where a tenant gave notice after the date of an agreement to sell by the landlord, and where the notice expired before the completion of the purchase, the vendor was obliged to pay compensation for improvements to the tenant, but it was held that he was entitled to recover the compensation from the purchaser (see also *re Lord Derby and Fergusson's Contract*, [1912], 1 Ch. 479). In *Bradshaw* v. *Bird*, [1920], 3 K.B. 144, the seller gave notice in 1917 to terminate the tenancy at 29th September, 1918. In October, 1917, the purchaser agreed to purchase the farm, 'and from that date became the person entitled to the rents and profits of the land'. The conveyance in his favour was dated 18th July, 1918. Claim for compensation was made on 10th December, 1918, against the purchaser, and the tenants quitted at Michaelmas, 1918, in accordance with the notice to quit. Held that he was liable, in respect that he was the person who was entitled to the rents and profits of the land for the time being. Scrutton, L.J., said: 'So that the purchaser of the land at the termination of the tenancy, *when the tenant has quitted the holding*, is the person to pay compensation'; while Atkin, L.J., said: 'The person entitled to receive the rents and profits of the land at the termination of the tenancy, or when payment is to be made, is the person to pay compensation. Where the landlord sold the farm with entry at the same term as the tenant quitted, the Board of Agriculture appointed an arbiter to deal with a claim for compensation for improvements by the tenant. Lord Constable interdicted the arbiter from proceeding, as he found that the purchaser was the party liable. The First Division reversed, holding that the seller was liable (*Waddell* v. *Howat*, 1925 S.C. 484).

It was held in England that purchaser who got entry after the date when he contracted to get entry, which was also the date of termination of tenancy, was not, at the material time (viz., 'the expiration of the tenancy'), the landlord within the meaning of the Act (*Tombs* v. *Turvey*, 93 L.J. K.B. 785).

See Agriculture (Miscellaneous Provisions) Act, 1968 section 11 (5).

[9] Notice to landlord's agent is *prima facie* notice to the landlord *Ingham* v. *Fenton*, 1893, 10 T.L.R. 113; *Hemington* v. *Walter*, 1950, 100 L.J. 51.

Amendments of other Acts.

94.—The enactments specified in the Seventh Schedule to this Act shall have effect subject to the amendments specified in that Schedule.

Construction of references in other Acts to holdings as defined by the Agricultural Holdings (Scotland) Act, 1923.

95.—(1) References, in whatever terms, in any enactment, other than an enactment contained in this Act, in the Agricultural Holdings (Scotland) Acts, 1923 and 1931, or in Part I of the Agriculture (Scotland) Act, 1948, to a holding within the meaning of the Agricultural Holdings (Scotland) Act, 1923, or of the Agricultural Holdings (Scotland) Acts, 1923 to 1948, shall be construed as references to an agricultural holding as defined by section one of this Act.

GENERAL 203

(2) The foregoing sub-section shall not apply to an enactment in so far as its operation is material for the purposes of the provisions of the said Acts of 1923 and 1931 or the said Act of 1948 to the extent to which they are expected from the repeal of enactments effected by this Act.

96.—The compensation in respect of an improvement made or begun before the first day of January, nineteen hundred and nine (being the date of the commencement of the Agricultural Holdings (Scotland) Act, 1908), or made upon an agricultural holding held under a lease, other than a lease from year to year, current on the first day of January, eighteen hundred and eighty-four, shall be such (if any) as could have been claimed if the Agricultural Holdings (Scotland) Acts, 1923 to 1948, and this Act had not passed, but the procedure for the ascertainment and recovery thereof shall be such as is provided by this Act, and the amount so ascertained shall be payable, recoverable and chargeable as if it were compensation under this Act. Improvements carried out before 1909.

97.—Subject to the provisions of the next following section, the enactments specified in the first and second columns of the Eighth Schedule to this Act are hereby repealed to the extent specified in the third column of that Schedule. Repeal of enactments.

98.—In a case where the tenant of an agricultural holding has quitted the holding before the commencement of this Act, or quits it after the commencement of this Act in consequence of a notice to quit given (whether by him or his landlord) before the first day of November, nineteen hundred and forty-eight, or in consequence of a renunciation of the tenancy in pursuance of an agreement in writing made before that day, the provisions of this Act, so far as relating to the rights of landlords and tenants to compensation (including the provisions relating to the determination of compensation where a holding is divided, the apportionment of compensation in such a case and the payment of expenses caused by such an apportionment), and the payment and recovery of compensation shall not apply, and in lieu thereof the enactments specified in the Eighth Schedule to this Act, so far as relating to the matters Provisions as to tenants quitting before commencement of this Act, or thereafter in consequence of notice given, &c., before 1st November, 1948.

aforesaid, shall continue to apply and shall accordingly be excepted from the operation of the last foregoing section.

General Savings.

99.—(1) Nothing in this Act shall affect any order, rule, regulation, record, appointment, application or complaint made, approval, consent or direction given, proceeding or assignation taken, notice served or given, certificate issued, condition imposed or thing done under a former enactment relating to agricultural holdings but any such order, rule, regulation, record, appointment, application, complaint, approval, consent, direction, proceeding, assignation, notice, certificate, condition, or thing which is in force at the commencement of this Act, shall continue in force, and so far as it could have been made, given, taken, served, issued, imposed or done under the corresponding provision of this Act shall (save where it is material only for the purposes of the enactments specified in the Eighth Schedule to this Act so far as continued in force by virtue of the last foregoing section) have effect as if it had been made, given, taken, served, issued, imposed or done under that corresponding provision:

Provided that this sub-section shall not apply to any such regulations or directions as are mentioned in the two next following sub-sections.

(2) Nothing in this Act shall affect any regulations having effect for the purposes of section seventy-one or eighty of the Agriculture (Scotland) Act, 1948, which are in force at the commencement of this Act, but any such regulations shall continue to have effect for those purposes and shall also have effect for the purposes of section *seventy-two* or *seventy-one* of this Act, as the case may be, as if they had been made by virtue of those sections respectively.

(3) Nothing in this Act shall affect any direction given under sub-section (1) of section five of the Agricultural Holdings (Scotland) Act, 1923, or the corresponding provision of an enactment repealed by that Act, by the Board or the Department of Agriculture for Scotland or by the Secretary of State, but any such direction which is in force at the commencement of this Act shall continue in force and shall (save where it is material only for the purposes of the enactments specified in the Eighth Schedule to this Act so far as continued in force by virtue of

GENERAL 205

the last foregoing section) have effect as if it were a direction of the Secretary of State under sub-section (1) of section *sixty-six* of this Act.

(4) Any notice deemed to have been given by the Secretary of State under the Second Schedule to the Agriculture (Scotland) Act, 1948, shall be deemed to have been given under sub-section (2) of section *twenty-eight* of this Act.

(5) Any provision of the Agricultural Holdings (Scotland) Act, 1923, or Part I of the Agriculture (Scotland) Act, 1948, or of any other enactment which (whatever its terms) has the effect of requiring a matter to be determined by arbitration under the said Act of 1923, shall be construed as having the effect of requiring that matter to be determined by arbitration under this Act, and an arbitration under the said Act of 1923 uncompleted at the commencement of this Act may be carried on and completed as if it had been begun under this Act:

Provided that, in the application of the Sixth Schedule to this Act to an arbitration for the purposes of the enactments specified in the Eighth Schedule to this Act so far as continued in force by virtue of the last foregoing section, paragraph 12 of the said Sixth Schedule shall have effect with the substitution, for references to this Act, of references to those enactments.

(6) Notwithstanding sub-section (1) of section thirty-eight of the Interpretation Act, 1889 (which relates to the effect of repeals) any reference which is or is to be construed as a reference to a former enactment relating to agricultural holdings or an enactment repealed by the Agricultural Holdings (Scotland) Act, 1908 (other than a reference in such an enactment or this Act or such a reference as is mentioned in sub-section (1) of section *ninety-five* of this Act or adapted by the last foregoing sub-section) shall, so far as the operation of the enactment in which the reference occurs is material for the purposes of the enactments specified in the Eighth Schedule to this Act so far as continued in force by virtue of the last foregoing section, be construed in like manner as if this Act had not passed, and otherwise shall (save where the context otherwise requires) be construed as a reference to the corresponding provision of this Act.

(7) Any document referring to a former enactment relating to agricultural holdings or an enactment repealed by the Agricultural Holdings (Scotland) Act, 1908, shall, so far as it or its

operation is material for the purposes of the enactment specified in the Eighth Schedule to this Act so far as continued in force by virtue of the last foregoing section, be construed in like manner as if this Act had not passed, and otherwise shall be construed as referring to the corresponding provision of this Act.

(8) Nothing in this Act shall affect the provisions of the Allotments (Scotland) Act, 1922, or be construed as repealing—

    (*a*) section twenty-six of the Agriculture (Miscellaneous War Provisions) Act, 1940 (which excludes the operation of the Agricultural Holdings (Scotland) Acts, 1923 and 1931, in relation to certain tenancies granted during the war period);

    (*b*) section fifteen of the Agriculture (Miscellaneous Provisions) Act, 1943 (which relieves occupiers of agricultural land from liabilities and loss of compensation resulting from directions given under Defence Regulations); or

    (*c*) any enactment contained in Part II of the Agriculture (Scotland) Act, 1948.

(9) Any person holding office or acting or serving under or by virtue of a former enactment relating to agricultural holdings shall continue to hold his office or to act or serve as if he had been appointed by or by virtue of the corresponding provision of this Act.

(10) Notwithstanding sub-section (2) of section thirty-eight of the Interpretation Act, 1889, rights to compensation conferred by this Act shall be in lieu of rights to compensation conferred by any former enactment relating to agricultural holdings.

(11) Save to the extent to which it is otherwise provided by sub-sections (6) and (10) of this section, the mention of particular matters in this section shall not be taken to affect the general application of section thirty-eight of the Interpretation Act, 1889.

Savings for other rights, &c.

100.—Subject to the provisions of sub-section (2) of section *twelve* and sub-section (1) of section *sixty-eight* of this Act in particular, and to any other provision of this Act which otherwise expressly provides, nothing in this Act shall prejudicially affect any power, right or remedy of a landlord, tenant or other person, vested in or exercisable by him by virtue of any other Act or law, or under any custom of the country, or otherwise, in respect of a lease or other contract, or of any improvements deteriorations, away-going crops, fixtures, tax, rate, teind, rent or other thing.[1]

[1] There are here very important qualifications. There are several provisions of the Act which would be embraced within the qualifications. The main ones are, of course, expressly referred to in the sections. So far as the rights referred to are not inconsistent with the express terms of the Act, they remain intact. See *Strang* v. *Stuart*, 1887, 14 R. 637. Section 74, relating to arbitration, strikes deep into these rights. This section might easily be misinterpreted. For cases under the corresponding section of the English Act see *Goldsack* v. *Shore*, [1950], 1 K.B. 708; *Kent* v. *Conniff*, [1953], 1 Q.B. 361.

101.[1]—(1) This Act may be cited as the Agricultural Holdings (Scotland) Act, 1949. *Short title and extent.*

(2) This Act shall extend to Scotland only.

[1] A similar Act was enacted for England and Wales in 1948. It also was amended by the 1958 Act.

SCHEDULES

Sections 11, 47, 50, 51, 52, 53, 63, 65, 79, 81, 86.

FIRST SCHEDULE[1]

IMPROVEMENTS BEGUN ON OR AFTER 1ST NOVEMBER, 1948, FOR WHICH COMPENSATION MAY BE PAYABLE

PART I[2]

IMPROVEMENTS TO WHICH CONSENT OF LANDLORD IS REQUIRED

1. Laying down of permanent pasture.
2. Making of water-meadows or works of irrigation.[3]
3. Making of gardens.
4. Planting of orchards or fruit bushes.[4]
5. Warping or weiring of land.[5]
6. Making of embankments and sluices against floods.
7. Making or planting of osier beds.
8. Haulage or other work done by the tenant in aid of the carrying out of any improvement made by the landlord for which the tenant is liable to pay increased rent.

PART II[6]

IMPROVEMENTS IN RESPECT OF WHICH NOTICE TO LANDLORD IS REQUIRED

9. Land drainage.
10. Construction of silos.
11. Making or improvement of farm access or service roads, bridges and fords.
12. Making or improvement of watercourses, ponds or wells, or of works for the application of water power for agricultural or domestic purposes or for the supply of water for such purposes.
13. Making or removal of permanent fences, including hedges, stone dykes and gates.
14. Reclaiming of waste land.
15. Renewal of embankments and sluices against floods.
16. Provision of stells, fanks, folds, dippers, pens and bughts necessary for the proper conduct of the holding.
17. Provision or laying on of electric light or power, including the provision of generating plant, fixed motors, wiring systems, switches and plug sockets.[7]
18. Erection, alteration or enlargement of buildings, and making or improvement of permanent yards, loading banks and stocks.[8]
19. Erection of hay or sheaf sheds, sheaf or grain drying racks,[9] and implement sheds.
20. Provision of fixed threshing mills, barn machinery and fixed dairying plant.

21. Improvement of permanent pasture by cultivation and re-seeding.
22. Provision of means of sewage disposal.
23. Repairs to fixed equipment, being equipment reasonably required for the efficient farming of the holding, other than repairs which the tenant is under an obligation to carry out.

PART III[10]

IMPROVEMENTS IN RESPECT OF WHICH CONSENT OF, OR NOTICE TO, LANDLORD IS NOT REQUIRED

24. Protecting fruit trees against animals.
25. Chalking of land.
26. Clay burning.
27. Claying of land.
28. Liming of land.
29. Marling of land.
30. Eradication of bracken, whins or broom growing on the holding at the commencement of the tenancy and, in the case of arable land, removal of tree roots, boulders, stones or other like obstacles to cultivation.[11]
31. Application to land of purchased manure (including artificial manure).[12]
32. Consumption on the holding of corn (whether produced on the holding or not) or of cake or other feeding stuff not produced on the holding by
    (a) horses, cattle, sheep or pigs; or
    (b) poultry folded on the land as part of a system of farming practised on the holding.[12]
33. Laying down temporary pasture with clover, grass, lucerne sainfoin, or other seeds, sown more than two years prior to the termination of the tenancy, in so far as the value of the temporary pasture on the holding at the time of quitting exceeds the value of the temporary pasture on the holding at the commencement of the tenancy for which the tenant did not pay compensation.[13]

[1] See Introduction. This schedule sets forth the temporary improvements for which (with one exception) compensation may be claimed. The exception is 'the continuous adoption of a special standard or system of farming' (see Introduction and section 56).
Where a tenant was required under Defence of the Realm Regulations to plough up parts of his farm (which he was under covenant not to convert from permanent pasture to tillage), held he had a good claim for temporary pasture (*Ware* v. *Davies*, 1932, 146 L.T. 130).

[2] See Opencast Coal Act, 1958, section 24.

[3] **'Making of water meadows or works of irrigation'**. It is unlikely that tenants in Scotland would seek to make water meadows but spray irrigation installations may require to be dealt with under this paragraph.

[4] **'Planting of orchards or fruit bushes'**. For provisions relating to fruit trees and bushes in market gardens see section 67 and Fourth Schedule.

[5] **'Warping or weiring of land'**. This is a method of improving land by

flooding it with muddy water. It is not known to have been practised in Scotland: the paragraph derives from the English Acts.

[6] See Agriculture Act, 1958, section 6 (5).

[7] **'Provision of electric light or power'.** The valuation of improvements under this paragraph commonly causes difficulty. The basis of valuation being 'the value to an incoming tenant' (section 49) the original cost is not usually relevant but an estimate of the cost of installation at the date of termination of the tenancy would require not only to be discounted in respect of the time since installation but to be further reduced if the installation is of a type which is obsolescent or inferior to the newer kinds.

[8] In the case of new leases there is an obligation imposed on the landlord to provide buildings necessary for the proper equipment of the farm. See also Housing (Scotland) Act, 1964 section (38) 2 and Housing (Scotland) Act, 1966 section 80 (2).

[9] Grain handling plant and storage bins are not specifically mentioned but would come within paragraphs 20 and 10 respectively.

[10] See Opencast Coal Act, 1958, section 26.

[11] **'Arable land'** presumably means arable *before* the improvement but containing obstacles to cultivation. Otherwise the paragraph could overlap with paragraph 14 in Part II—'reclaiming waste ground'.

[12] See Introduction. Purchased straw made into manure not a 'manure' (*Brunskill* v. *Atkinson*, 1884, Sol. J. 29). If it were consumed by stock compensation could be claimed for it as a feeding stuff. For calculation of values see Report published annually by the Scottish Standing Committee for the calculation of residual values of Fertilizers and Feeding Stuffs.

[13] See Introduction under 'Temporary Pasture', and *Mackenzie* v. *McGillivray*, 1921, 58 S.L.R. 488. Second year's grass is generally sown more than two years before the termination of the tenancy and is 'temporary pasture' within this definition (*Earl of Galloway* v. *M'Clelland*, 1915 S.C. 1062, per Lord Johnston). (See notes to section 53).

Sections 36, 39, 40, 41, 42, 44, 63, 65, 81, 86.

SECOND SCHEDULE[1]

IMPROVEMENTS BEGUN BEFORE 31ST JULY, 1931, FOR WHICH COMPENSATION MAY BE PAYABLE

PART I

IMPROVEMENTS FOR WHICH COMPENSATION IS PAYABLE IF CONSENT OF LANDLORD WAS OBTAINED TO THEIR EXECUTION

1. Erection, alteration, or enlargement of buildings.
2. Formation of silos.
3. Laying down of permanent pasture.
4. Making and planting of osier beds.
5. Making of watermeadows or works of irrigation.
6. Making of gardens.
7. Making or improvement of roads or bridges.
8. Making or improvement of watercourses, ponds, wells, or reservoirs, or of works for the application of water power or for supply of water for agricultural or domestic purposes.
9. Making or removal of permanent fences.
10. Planting of hops.
11. Planting of orchards or fruit bushes.

SECOND SCHEDULE

12. Protecting young fruit trees.
13. Reclaiming of waste land.
14. Warping or weiring of land.
15. Embankments and sluices against floods.
16. Erection of wirework in hop gardens.
17. Provision of permanent sheep dipping accommodation.
18. In the case of arable land the removal of bracken, gorse, tree roots, boulders, or other like obstructions to cultivation.

PART II

IMPROVEMENT FOR WHICH COMPENSATION IS PAYABLE IF NOTICE WAS GIVEN TO LANDLORD BEFORE EXECUTION THEREOF

19. Drainage.

PART III

IMPROVEMENTS FOR WHICH COMPENSATION IS PAYABLE WITHOUT CONSENT OF, OR NOTICE TO, LANDLORD OF THEIR EXECUTION

20. Chalking of land.
21. Clay-burning.
22. Claying of land or spreading blaes upon land.
23. Liming of land.
24. Marling of land.
25. Application to land of purchased artificial or other purchased manure.
26. Consumption on the holding by cattle, sheep, or pigs, or by horses other than those regularly employed on the holding, of corn, cake, or other feeding stuff not produced on the holding.
27. Consumption on the holding by cattle, sheep, or pigs, or by horses other than those regularly employed on the holding, of corn, proved by satisfactory evidence to have been produced and consumed on the holding.
28. Laying down temporary pasture with clover, grass, lucerne, sainfoin, or other seeds, sown more than two years prior to the termination of the tenancy, in so far as the value of the temporary pasture on the holding at the time of quitting exceeds the value of the temporary pasture on the holding at the commencement of the tenancy for which the tenant did not pay compensation.
29. Repairs to buildings, being buildings necessary for the proper cultivation or working of the holding, other than repairs which the tenant is himself under an obligation to execute.

[1] See notes to First Schedule.

THIRD SCHEDULE[1]

Sections 36, 39, 40, 41, 42, 43, 44, 63, 65, 81, 86.

IMPROVEMENTS BEGUN ON OR AFTER 31ST JULY, 1931, AND BEFORE 1ST NOVEMBER, 1948, FOR WHICH COMPENSATION MAY BE PAYABLE

PART I

IMPROVEMENTS FOR WHICH COMPENSATION IS PAYABLE IF CONSENT OF LANDLORD WAS OBTAINED TO THEIR EXECUTION

1. Erection, alteration, or enlargement of buildings.
2. Laying down of permanent pasture.
3. Making and planting of osier beds.
4. Making of water-meadows or works of irrigation.
5. Making of gardens.
6. Planting of orchards or fruit bushes.
7. Protecting young fruit trees.
8. Warping or weiring of land.
9. Making of embankments and sluices against floods.

PART II

IMPROVEMENTS FOR WHICH COMPENSATION IS PAYABLE IF NOTICE WAS GIVEN TO LANDLORD BEFORE EXECUTION THEREOF

10. Drainage.
11. Formation of silos.
12. Making or improvement of roads or bridges.
13. Making or improvement of watercourses, ponds or wells, or of works for the application of water power or for the supply of water for agricultural or domestic purposes.
14. Making or removal of permanent fences.
15. Reclaiming of waste land.
16. Repairing or renewal of embankments and sluices against floods.
17. Provision of sheep dipping accommodation.
18. The provision of electrical equipment other than moveable fittings and appliances.

PART III

IMPROVEMENTS FOR WHICH COMPENSATION IS PAYABLE WITHOUT CONSENT OF, OR NOTICE TO, LANDLORD OF THEIR EXECUTION

19. Chalking of land.
20. Clay-burning.
21. Claying of land or spreading blaes upon land.
22. Liming of land.
23. Marling of land.
24. Eradication of bracken, whins, or gorse growing on the holding at the commencement of a tenancy and in the case of arable

land the removal of tree roots, boulders, stones or other like obstacles to cultivation.
25. Application to land of purchased artificial or other purchased manure.
26. Consumption on the holding by cattle, sheep, or pigs, or by horses other than those regularly employed on the holding, of corn, cake, or other feeding stuff not produced on the holding.
27. Consumption on the holding by cattle, sheep, or pigs, or by horses other than those regularly employed on the holding, of corn proved by satisfactory evidence to have been produced and consumed on the holding.
28. Laying down temporary pasture with clover, grass, lucerne sainfoin, or other seeds, sown more than two years prior to the termination of the tenancy, in so far as the value of the temporary pasture on the holding at the time of quitting exceeds the value of the temporary pasture on the holding at the commencement of the tenancy for which the tenant did not pay compensation.
29. Repairs to buildings, being buildings necessary for the proper cultivation or working of the holding, other than repairs which the tenant is himself under an obligation to execute.

[1] See notes to First Schedule.

FOURTH SCHEDULE[1]

Sections 65, 66, 79.

MARKET GARDEN IMPROVEMENTS FOR WHICH COMPENSATION MAY BE PAYABLE

1. Planting of standard or other fruit trees permanently set out.
2. Planting of fruit bushes permanently set out.
3. Planting of strawberry plants.
4. Planting of asparagus, rhubarb, and other vegetable crops which continue productive for two or more years.
5. Erection, alteration or enlargement of buildings for the purpose of the trade or business of a market gardener.

1. See Opencast Coal Act, 1958, sec. 28.

FIFTH SCHEDULE

Section 4.

MATTERS FOR WHICH PROVISION IS TO BE MADE IN WRITTEN LEASES

1. The names of the parties.
2. Particulars of the holding with sufficient description, by reference to a map or plan, of the fields and other parcels of land comprised therein to identify the extent of the holding.
3. The term or terms for which the holding or different parts thereof is or are agreed to be let.
4. The rent and the dates on which it is payable.

5. An undertaking by the landlord in the event of damage by fire to any building comprised in the holding to reinstate or replace the building if its reinstatement or replacement is required for the fulfilment of his responsibilities to manage the holding in accordance with the rules of good estate management, and (except where the interest of the landlord is held for the purposes of a government department or a person representing His Majesty under section *eighty-six* of this Act is deemed to be the landlord, or where the landlord has made provision approved by the Secretary of State for defraying the cost of any such reinstatement or replacement as aforesaid) an undertaking by the landlord to insure to their full value all such buildings against damage by fire.

6. An undertaking by the tenant, in the event of the destruction by fire of harvested crops grown on the holding for consumption thereon, to return to the holding the full equivalent manurial value of the crops destroyed, in so far as the return thereof is required for the fulfilment of his responsibilities to farm in accordance with the rules of good husbandry, and (except where the interest of the tenant is held for the purposes of a government department or where the tenant has made provision approved by the Secretary of State in lieu of such insurance) an undertaking by the tenant to insure to their full value all dead stock on the holding and all such harvested crops as aforesaid against damage by fire.

Sections 75, 76, 99.

SIXTH SCHEDULE

PROVISIONS AS TO ARBITRATIONS[1]

Appointment of Arbiter

1. A person agreed upon between the parties or, in default of agreement,[2] appointed on the application in writing of either of the parties by the Secretary of State from among the members of the panel constituted under this Act for the purpose, shall be appointed arbiter.[3]

2. If a person appointed arbiter dies, or is incapable of acting,[4] or for seven days after notice from either party requiring him to act fails to act, a new arbiter may be appointed as if no arbiter had been appointed.

3. Neither party shall have power to revoke the appointment of the arbiter without the consent of the other party.

4. Every appointment, notice, revocation and consent under the foregoing provisions of this Schedule must be in writing.[5]

Particulars of Claim[6]

5. Each of the parties to the arbitration shall within [twenty-eight][7] days from the appointment of the arbiter deliver to him a statement of that party's case with all necessary particulars; and
 (*a*) no amendment or addition to the statement or particulars

delivered shall be allowed after the expiration of the said [twenty-eight] days except with the consent of the arbiter;
(b) a party to the arbitration shall be confined at the hearing to the matters alleged in the statement and particulars so delivered and any amendment thereof or addition thereto duly made.[8]

Evidence

6. The parties to the arbitration, and all persons claiming through them respectively,[9] shall, subject to any legal objection, submit to be examined by the arbiter on oath or affirmation in relation to the matters in dispute, and shall, subject as aforesaid, produce before the arbiter all samples, books, deeds, papers, accounts, writings, and documents, within their possession or power respectively which may be required or called for, and do all other things which during the proceedings the arbiter may require.

7. The arbiter shall have power to administer oaths, and to take the affirmation of parties and witnesses appearing, and witnesses shall, if the arbiter thinks fit, be examined on oath or affirmation.

Award

8. The arbiter shall make and sign his award within two months of his appointment or within such longer period as may, either before or after the expiry of the aforesaid period be agreed to in writing by the parties, or be fixed by the Secretary of State.[10]

9. The arbiter may, if he thinks fit, make an interim award for the payment of any sum[11] on account of the sum to be finally awarded.

10. The award shall be in such form as may be specified by statutory instrument made by the Secretary of State.[12]

11. The arbiter shall—
   (a) state separately in his award the amounts awarded in respect of the several claims referred to him; and
   (b) on the application of either party,[13] specify the amount awarded in respect of any particular improvement or any particular matter the subject of the award.[14]

12. Where by virtue of this Act compensation under an agreement is to be substituted for compensation under this Act for improvements, the arbiter shall award compensation in accordance with the agreement instead of in accordance with this Act.

13. The award shall fix a day not later than one month after delivery of the award for the payment of the money awarded as compensation, expenses or otherwise.[15]

14. The award to be made by the arbiter shall be final and binding on the parties and the persons claiming under them respectively.[16]

15. The arbiter may correct in an award any clerical mistake or error arising from any accidental slip or omission.[17]

Expenses[18]

16. The expenses of and incidental to the arbitration and award shall be in the discretion of the arbiter, who may direct to and by whom and in what manner those expenses or any part thereof are to be paid, and the expenses shall be subject to taxation by the auditor of the sheriff court on the application of either party, but that taxation shall be subject to review by the sheriff.[19]

17. The arbiter shall, in awarding expenses, take into consideration the reasonableness or unreasonableness of the claim of either party whether in respect of amount or otherwise, and any unreasonable demand for particulars or refusal to supply particulars, and generally all the circumstances of the case, and may disallow the expenses of any witness whom he considers to have been called unnecessarily and any other expenses which he considers to have been incurred unnecessarily.[20]

18. It shall not be lawful to include in the expenses of and incidental to the arbitration and award, or to charge against any of the parties, any sum payable in respect of remuneration or expenses to any person appointed by the arbiter to act as clerk or otherwise to assist him in the arbitration unless such appointment was made after submission of the claim and answers to the arbiter and with either the consent of the parties to the arbitration or the sanction of the sheriff.[21]

Statement of Case[22]

19. The arbiter may at any stage of the proceedings,[23] and shall, if so directed by the sheriff (which direction may be given on the application of either party), state a case for the opinion of the sheriff on any question of law arising in the course of the arbitration.[24]

20. The opinion of the sheriff on any case stated under the last foregoing paragraph shall be final unless, within such time and in accordance with such conditions as may be specified by act of sederunt, either party appeals[25] to the Court of Session, from whose decision no appeal shall lie.

Removal of arbiter and setting aside of award

21. Where an arbiter has misconducted himself the sheriff may remove him.[26]

22. When an arbiter has misconducted himself, or an arbitration or award has been improperly procured, the sheriff may set the award aside.[27]

Forms

23. Any forms for proceedings in arbitrations under this Act which may be specified by statutory instrument made by the Secretary of State shall, if used, be sufficient.[28]

---

[1] See 'Arbitration under the Act'—Introduction.

[2] **'In default of agreement'**. These words in the corresponding paragraph 1

SIXTH SCHEDULE 217

of the Sixth Schedule to the English Act of 1948 were held to mean only that the parties have not agreed. It is not necessary that they should attempt to agree and fail to do so: (*Chalmers Property Investment Co. Ltd.* v. *MacColl*, 1951 S.C. 24 per Lord President Cooper, p. 30; *F. R. Evans (Leeds) Ltd.* v. *Webster*, (1962) L.J. 703).

3 Where an action of reduction of letter of appointment on the ground that material facts were withheld from the knowledge of the Department, and that the arbiter was an interested party, held that the Department in appointing arbiters acted in an administrative capacity and that its selection of an arbiter from the panel could not be challenged (*Ramsay* v. *McLaren and Another*, 1936, S.L.T. 35).

4 'Incapable of acting'. This is not limited to incapacity by reason of physical or mental disability but includes incapacity from any cause . (*Dundee Corporation* v. *Guthrie*, 1969 S.L.T. 93 in which an arbiter had been removed and his award reduced and it was held competent to appoint a second arbiter).

5 Not necessarily holograph or tested (section 85).

6 Each party must lodge not only particulars but a statement of his case. The period of 28 days cannot be extended by the arbiter. No provision is made for answers to such statements but possibly the arbiter has authority to order answers if they appear to him necessary. If he does so he should get consent of parties. If one party alone lodges claim and particulars the other may lead evidence in rebuttal but may not set up a case of his own. *Collet* v. *Deely*, (1949), 100 L.J. 108. *Jamieson* v. *Clark*, (1951), 67 Sh.Ct.Rep. 17; *Re Bennion and National Provincial Banks' Arbitration*, (1965), 115 L.J. 302. As to form of claim, see *Robertson's Trustees* v. *Cunningham*, 1951, S.L.T. (Sh.Ct.) 89; *Simpson* v. *Henderson*, 1944, S.C. 365; *Adam* v. *Smythe*, 1948, S.C. 445.

7 Substituted for 'fourteen' by the Agriculture (Miscellaneous Provisions) Act, 1963 section 20.

8 Matters not included in the statement are inadmissible even if raised in evidence by the other party (*Stewart* v. *Brims*, 1969 S.L.T. (Sh.Ct.) 2).

9 Where, for example, the landlord has assigned his rights to the incoming tenant, who, on the other hand, has undertaken to satisfy the landlord's obligations to the away-going tenant.

10 See form of application for extension in Appendix. An award made after the expiry of the permitted period may be set aside for misconduct (*Halliday* v. *Semple*, 1960 S.L.T. (Sh.Ct.) 11).

The Secretary of State is given powers to issue regulations for expediting arbitrations (section 75).

11 The provision here is apparently restricted to an award for a money payment, and does not provide for an interim award with declaratory or other findings.

12 See form of award (Appendix I). An arbiter may not make an award *ad factum praestandum*.

13 By landlord or tenant respectively. This application need not be in writing.

14 Here there must be a figure put down in respect of any particular 'improvement or matter' the subject of the award. Thus either party could request the arbiter to state how much he had awarded in respect of any of the items of claim under the different items in the Schedules.

15 This was formerly not sooner than one month and not later than two months. It is not practicable to fix the particular day, because it is rarely, if ever, known at the time the award is signed on what date it will be delivered. It is sufficient to state that payment shall be made on (say) the 15th day after delivery of the award. The arbiter or his clerk should always endorse on the back of the award the date of delivery.

AGRICULTURAL HOLDINGS (SCOTLAND) ACTS

'**Delivery**'.—It has been held that actual delivery to the parties is not essential to completion of the award, and that the award is validly issued if signed and put into the clerk's hands for the purpose of being delivered (*McQuaker* v. *Phoenix Assurance Co.*, 1859, 21 D. 794).

[16] This is, of course, subject to the right to have the award reduced at common law or set aside under rule 22 or because the arbiter has improperly dealt with an agreement purporting to provide fair and reasonable compensation (*Bell* v. *Graham*, 1908 S.C. 1060).

[17] This may be done even after delivery of the award.

[18] See Introduction.

[19] The expenses of a stated case fall to be dealt with by the Sheriff and not by the arbiter (*McQuater* v. *Fergusson*, 1911 S.C. 640).

The fees of arbiters appointed by the Secretary of State fall to be fixed by him.

[20] The arbiter should, in general, follow the same course as regards expenses as that adopted by the Court although what he awards is always in his discretion.

[21] This rule is copied from the earlier Acts and is inconsistent in that the present rule (No. 5) makes no provision for answers. The purpose of the rule is to prevent arbiters making appointments when the questions at issue are trivial, and to avoid unnecessary expense in the earlier stages of the proceedings.

[22] See Introduction.

[23] Not after the award has been issued.

[24] The Court will not deal with a case till the arbiter has found the facts (*Ferguson* v. *Norman*, 4 Bing. N.C. 52). There may be an appeal from the Sheriff to the Court of Session (paragraph 20 of Schedule VI). Arbiter held right in refusing to state a case (*a*) as to competency of arbitration, and (*b*) as to conduct of the arbitration procedure (*Broxburn Oil Co., Ltd.* v. *Earl of Buchan*, 1926, 42 Sh.Ct.Rep. 300). The terms in which the case is stated are for the arbiter to decide. (*Forsyth-Grant* v. *Salmon*, 1961 S.C. 54).

[25] For procedure on appeal see Rules of Court of Session enacted by Act of Sederunt (Rules of Court, Consolidation and Amendment) 1964 as amended by Act of Sederunt (Rules of Court Amendment No. 1) 1965 (in accordance with decision in *Macnab* v. *Willison*, 1960 S.C. 83).

[26] See Introduction.

[27] '**Misconduct**'. Where an arbiter visited the farm with an architect in the absence of the parties it was assumed that evidence was received and the award was set aside. (*Ellis* v. *Lewin*, (1963), 107 S.J. 851).

[28] See forms in Appendix.

Section 94.

SEVENTH SCHEDULE

AMENDMENTS OF OTHER ACTS

The Small Landholders and Agricultural Holdings (Scotland) Act, 1931

In section twenty-six, for subsection (2) there shall be substituted the following subsection:—

'(2) This Part of this Act may be cited as the Small Landholders (Scotland) Act, 1931, and shall be construed as one with the Small Landholders (Scotland) Acts, 1886 to 1919, and those Acts and this Part of this Act may be cited together as the Small Landholders (Scotland) Acts, 1886 to 1931.'

SEVENTH SCHEDULE

The Hill Farming Act, 1946

The Hill Farming Act, 1946, shall, in its application to Scotland, have effect with the substitution for section nine thereof of the following section:—

'Operation of The Agricultural Holdings (Scotland) Act, 1949, in relation to improvement schemes.

9.—(1) Subject to the provisions of this section, the Agricultural Holdings (Scotland) Act, 1949, shall apply to improvements for which provision is made by an approved hill farming land improvement scheme as it applies to other improvements.

(2) Where a tenant of an agricultural holding within the meaning of the said Act of 1949 has carried out thereon an improvement specified in Part I or Part II of the First Schedule to that Act in accordance with provision in such a scheme for the carrying out of the improvement and for the tenant's being responsible for doing the work, being provision included in the scheme at the instance or with the consent of the landlord, then—

(a) In the case of an improvement specified in the said Part I, the landlord shall be deemed to have consented as mentioned in section *fifty* of that Act in relation to the improvement; or

(b) in the case of an improvement specified in the said Part II, the tenant shall be deemed to have given notice to the landlord as mentioned in section *fifty-one* of that Act in relation to the improvement and the landlord shall be deemed to have received the notice and to have given no such notice to the tenant as is mentioned in section *fifty-two* of that Act objecting to the carrying out of the improvement or to the manner in which the tenant proposes to carry out the work;

and any agreement as to compensation or otherwise made between the landlord and the tenant in relation to the improvement shall have effect as if it had been such an agreement on terms as is mentioned in the said section *fifty* or the said section *fifty-one* as the case may be.

(3) If on the ground of work being badly done the appropriate Minister withholds or reduces the improvement grant in respect of an improvement, he may direct that any right conferred by section eight of the Agricultural Holdings (Scotland) Act, 1949, to have the rent of an agricultural holding increased shall not be exercisable in respect of the improvement, or shall be exercisable only to such extent as may be specified in the direction, and any such direction given after that right has been exercised shall be retrospective and any excess rent paid shall be repaid accordingly.

(4) In assessing the amount of any compensation payable whether under the said Act of 1949 or under custom or agreement to the tenant of an agricultural holding, if it is shown to the satisfaction of the person assessing the compensation that the improvement or cultivations in respect of which the compensation is claimed was or were wholly

or in part the result of or incidental to work in respect of the cost of which an improvement grant has been paid or will be payable, the amount of the grant shall be taken into account as if it had been a benefit allowed to the tenant in consideration of his executing the improvement or cultivations, and the compensation shall be reduced to such extent as that person considers appropriate.'

Sections, 97, 98, 99.

EIGHTH SCHEDULE

ENACTMENTS REPEALED

Session and Chapter	Short Title	Extent of Repeal
13 & 14 Geo. 5. c. 10.	The Agricultural Holdings (Scotland) Act, 1923.	The whole Act.
13 & 14 Geo. 5. c. 25.	The Agriculture (Amendment) Act, 1923.	The whole Act.
19 & 20 Geo. 5. c. 25.	The Local Government (Scotland) Act, 1929.	In section forty-eight, the words from 'or by an arbiter' to the end of the section.
21 & 22 Geo. 5. c. 42.	The Agricultural Marketing Act, 1931.	In section nineteen, in paragraph (6) the words 'or other occupier of an agricultural holding,' and the words from 'or by an arbiter' to the end of the paragraph.
21 & 22 Geo. 5. c. 44.	The Small Landholders and Agricultural Holdings (Scotland) Act, 1931.	Part II. In section forty-one, in subsection (1) the words from 'and the Small Landholders Acts' to the end of the subsection.
1 Edw. 8 and 1 Geo. 6. c. 70.	The Agriculture Act, 1937.	Section five, so far as it relates to agricultural holdings.
2 & 3 Geo. 6. c. 48.	The Agricultural Development Act, 1939.	In section thirty, sub-section (2) so far as it relates to agricultural holdings.

EIGHTH SCHEDULE

Session and Chapter	Short Title	Extent of Repeal
6 & 6 Geo. 6. c. 16.	The Agriculture (Miscellaneous Provisions) Act, 1943.	Section twenty-one.
9 & 10 Geo. 6. c. 73.	The Hill Farming Act, 1946.	In section thirty-nine, in subsection (1), paragraph (c).
11 & 12 Geo. 6. c. 45.	The Agriculture (Scotland) Act, 1948.	Part I, except section eight in relation to notices to quit given before the commencement of this Act and except section twenty-five so far as relating to the provisions therein mentioned so far as continued in force by this Act. In section eighty-four, the words 'the Agricultural Holdings (Scotland) Acts, 1923 and 1931, or.' The First and Second Schedules. In the Third Schedule, in paragraph 2, the words from 'or a direction' to 'permanent pasture' where those words first occur, and in paragraph 4, the words from the beginning to 'this Act'. The Fourth and Ninth Schedules.

LIST OF APPENDICES.

		PAGE
I.	Forms,	224
II.	The Agricultural Holdings (Specification of Forms) (Scotland) Instrument, 1960,	277
III.	Sheep Stocks Valuation (Scotland) Act, 1937,	285
IV.	Excerpts from Hill Farming Act, 1946, relating to Sheep Stock Valuations,	288
V.	The Agricultural Records (Scotland) Regulations, 1948,	296
VI.	Excerpts from the Agriculture Act, 1958, relating to Scotland,	302
VII.	Excerpts from the Succession (Scotland) Act, 1964,	316
VIII.	Excerpts from The Agriculture (Miscellaneous Provisions) Act, 1968,	322

APPENDIX I.

NOTICES TO QUIT AND MISCELLANEOUS

PAGE

A. NOTICES AND INTIMATIONS BY LANDLORD AND TENANT AND AGREEMENTS, &C. - - - - - - - 224

B. APPLICATIONS TO SECRETARY OF STATE, - - - - 225

C. APPLICATIONS TO SCOTTISH LAND COURT, - - - - 225

D. ARBITRATION PROCEDURE, - - - - - - 226

E. ADDITIONAL FORMS, - - - - 226

APPENDIX I.

FORMS FOR USE IN CONNECTION WITH THE ACT.

NOTICES TO QUIT AND MISCELLANEOUS.

A. NOTICES AND INTIMATIONS BY LANDLORD AND TENANT AND AGREEMENTS, &C.

	PAGE
1. Notice to Quit by Landlord to Tenant (secs. 24 and 25)	227
2. Notice to Quit by Landlord to Legatee or 'acquirer' of Lease but not 'near relative' of deceased tenant (1958 Act, sec. 6 (3)),	228
3. Notice to Quit by Landlord to exclude additional payment under sec. 9 of 1968 Act. (1968 Act secs. 11 and 18),	228
4. Counter-Notice under sec. 25 (1) by Tenant to Notice to Quit by Landlord,	230
5. Notice by Tenant to Landlord requiring questions arising out of Notice to Quit to be referred to Arbitration (sec. 25),	230
6. Notice to Tenant to pay rent within two months sec. 25 (2) (e)),	231
7. Notice by Landlord to Tenant to remedy Breach of Conditions of Tenancy capable of being remedied (sec. 25 (2) (e)),	231
8. Notice by the Landlord to Tenant to Remove from Part of Farm (sec. 32),	232
9. Counter-Notice by the Tenant accepting Notice to Remove from Part as Notice to Remove from the Entire Holding (sec. 33),	232
10. Notice by Landlord to Tenant of Contract for Sale of Farm (sec. 31 (2)),	233
11. Agreement between Landlord and Tenant as to Notice to Quit being valid on sale of holding (sec. 31 (2)),	233
12. Notification by Tenant to Landlord that he elects that Notice to Quit shall remain in force following on sale of holding (sec. 31 (2)),	234
13. Demand for execution of Tenancy Agreement (secs. 4, 5),	234
14. Reference to arbitration by Landlord as to amount of compensation payable by Tenant on transfer of liability for maintenance of fixed equipment (sec. 6),	235
15. Requirement by Tenant for Arbitration as to claim against Landlord on transfer of liability for maintenance to fixed equipment (sec. 6 (2)),	236
16. Demand for Arbitration as to Rent (sec. 7),	236
17. Notice by Landlord requiring Increase of Rent on completion of Improvements (sec. 8),	237

APPENDIX I 225

18. Agreement between Landlord and Incoming Tenant with regard to the latter paying Outgoing Tenant's Claims for Compensation (sec. 11), - - - - - 238
19. Notice to Landlord that Tenant intends to Remove Fixtures, &c. (sec. 14), - - - - - - - - 238
20. Notice to Tenant by Landlord that he Elects to Purchase Fixtures, &c. (sec. 14), - - - - - - 239
21. Notice of Damage by Game (sec. 15), - - - - 239
22. Claim for Damage by Game (sec. 15), - - - - 240
23. Notification by Landlord or Tenant requiring a Record to be made under sec. 17, - - - - - - 240
24. Intimation to Landlord of Bequest of Lease (sec. 20), - 241
25. Intimation by Landlord Declining to accept Legatee as Tenant (sec. 20 (3)), - - - - - - - 242
26. Intimation to Landlord by Acquirer of Lease (sec. 21 (1)), - 242
27. Intimation by Landlord Declining to accept Acquirer as Tenant (sec. 21 (2)), - - - - - - - 243
28. Application by Tenant to Landlord for written consent to Improvements under Schedule I, Part I (sec. 50), - 243
29. Landlord's Consent in Writing to Execution of Improvements (Schedule I, Part I, sec. 50), - - - - 244
30. Notice by Tenant of Intention to make Improvements (Schedule I, Part II, sec. 51), - - - - - 244
31. Notice by Landlord to Tenant of objections to carrying out of Improvements (sec. 52 (1)), - - - - - 245
32. Notice by Landlord that he intends to execute the Improvements himself (sec. 52 (3)), - - - - - 246
33. Notice of Intention by the Tenant to Claim Compensation for Continuous Special System of Farming (sec. 56), - 247
34. Notice by Landlord of Intention to Claim Compensation from the Tenant for Deterioration of Holding (secs. 57 and 58), - - - - - - - - - 247
35. Notice of Intention to Claim by Tenant (sec. 68 (2)), - 248
36. Notice by Landlord to Tenant of change of Landlord (sec. 90 (4)), - - - - - - - - - 249
37. Statement of Case and Particulars of Claim (Schedule VI, 5), 249

B. APPLICATIONS TO THE SECRETARY OF STATE FOR SCOTLAND.

See also Appendix II for applications to the Secretary of State for the appointment of arbiters and for Extension of Time for making Award. These must be in the forms specified,

Application by Landlord or Tenant to extend the time for adjusting Claims at termination of tenancy (sec. 68 (3)), 252

C. APPLICATIONS TO SCOTTISH LAND COURT.

List of Forms available either from Scottish Land Court or from Sheriff Courts. - - - - - - - 253

D. ARBITRATION PROCEDURE.

		PAGE
1.	Joint Appointment of Arbiter by Landlord and Outgoing Tenant,	253
2.	Joint Intimation to Arbiter of his Appointment where he has been appointed by the Parties,	254
3.	Consent to Revocation of Arbiter's Appointment (Schedule VI),	255
4.	Extension by Parties of Time for Award.	255
5.	Minute of Devolution,	255
6.	Incidental Orders by Arbiter—	
	(a) Order by Arbiter for Claims,	256
	(b) Order to Produce Documents,	256
	(c) Intimation to Parties of Inspection,	257
	(d) Extension of Time,	257
	(e) Revisal,	257
	(f) To Answer Statement of Facts,	258
	(g) For Closing Record,	258
	(h) Closing and Proof,	258
	(i) Avizandum,	258
	(j) With Notes of Proposed Findings,	258
	(k) Representations Repelled,	259
7.	Application by an Arbiter to the Sheriff for the appointment of a Clerk in the Arbitration,	260
8.	Stated Case for Opinion of Court,	261
9.	Application for Order on Arbiter to State a Case,	262
10.	Application to have Arbiter Removed,	262
11.	Common Law Submission beteeen Outgoing and Incoming Tenants,	263
12.	Award in Common Law Submission,	267
13.	Minute of Acceptance by Arbiters and Nomination of Oversman and Clerk,	268
14.	Minute of Acceptance by Oversman following thereon,	269
15.	Minute of Prorogation by the Parties,	269
16.	Minute of Prorogation when Submission has expired,	269
17.	Minute of Prorogation by Arbiters,	270

E. ADDITIONAL FORMS.

1.	Removing of Tenant for Non-Payment of Rent (sec. 19),	270
2.	Claim by outgoing tenant against incoming tenant,	271
3.	Abbreviated Form of Lease,	272
4.	Supplementary Agreement in terms of Section 5 (3))	275

A. NOTICES TO QUIT AND MISCELLANEOUS NOTICES AND INTIMATIONS BY LANDLORD AND TENANT AND AGREEMENTS, ETC.

The following styles are not statutory and are suggestions only capable of modification to suit particular cases.

1.   Notice to Quit by Landlord to Tenant (secs. 24 and 25).

Place................................

Date................................

To [name, designation and address of the party in possession].

You are required to remove from [describe subjects] at the term of [or, if different terms, state them and the subjects to which they apply], in terms of lease [describe it] [or in terms of your letter of removal of date   ] [or otherwise as the case may be].

(The reason [or reasons] for giving this notice is [are] [quote from sec. 25, so far as applicable].

(Signed) A. B. [landlord].

Notes.

(1) This is a statutory form and is taken from Form H in the First Schedule to the Sheriff Court (Scotland) Act, 1907. It should be sent by registered post. There are, however, unlikely to be many cases in which a notice to quit without reasons should be given. See below and Introduction.

(2) If the landlord does not agree that the tenant is entitled to compensation for disturbance, he must add to this notice the reason for which it is given, as above. To exclude payment under sec. 9 of 1968 Act reasons must be given as in Form 3.

(3) This form is obligatory.

(4) Where two concurrent notices are sent or two reasons given in one notice see *French* v. *Elliott* [1960] 1 W.L.R. 40.

(5) See also Forms 2 and 3.

2.   Notice to Quit by Landlord to legatee or 'acquirer of lease' but not 'near relative' of deceased Tenant. (1958 Act sec. 6 (3)).

Place.................................

Date..................................

To [name, designation and address of the party in possession].

You are required to remove from [describe subjects] at the term of [or, if different terms, state them and the subjects to which they apply], in terms of lease [describe it] [or in terms of your letter of removal of date   ] [or otherwise as the case may be].

This notice is given in pursuance of sub-section (3) of Section 6 of the Agriculture Act, 1958. [The tenancy is being terminated for the purpose of my using the land for agriculture only][1]

(Signed) A. B. [landlord].

(1) See Agriculture (Miscellaneous Provisions) Act, 1968, sec. 11 (7). If these words are not included compensation will be payable under sec. 9 of the 1968 Act. The question whether the tenancy is being terminated for the purpose stated may be referred by the Tenant to the Land Court.

If notice is given in this form and either the tenant does not contest the statement or the Land Court finds that the subjects are to be used for agriculture only compensation for disturbance will be payable but not the additional sum under sec. 9 of the 1968 Act.

For suggested form of notice to quit to 'near relative' successor see form 3.

3.   Notice to Quit by Landlord to exclude additional payment under sec. 9 of 1968 Act (1968 Act secs. 11 and 18).

Place.................................

Date..................................

To [name, designation and address of the party in possession].

You are required to remove from [describe subjects] at the term of [or, if different terms, state them and the subjects to which they apply], in terms of lease [describe it] [or in terms of your letter of removal of date   ] [or otherwise as the case may be].

The carrying out of the purpose for which I propose to terminate the tenancy is desirable on the following grounds:

[Quote from paragraphs (a) (b) or (c) of sec. 26 (1)][1] or
I hereby intimate that unless this notice has effect I shall suffer hardship[2]

(or if tenant is 'near relative' successor to deceased tenant)[3]

    (*a*) This notice is given by reason of the fact that you have neither sufficient training in agriculture nor sufficient experience of farming to enable you to farm the said holding with reasonable efficiency, or

    (*b*) This notice is given in order to enable me to use the said holding for the purpose of effecting an amalgamation with [specify land with which holding is to be amalgamated], or

    (*c*) This notice is given by reason of the fact that you are the occupier of [specify land occupied] which has been occupied by you since before the death of [name and designation of deceased tenant] and is an agricultural unit capable of providing full time employment for you and for at least one other man.

                                (Signed) A. B. [landlord].

(1) See Agriculture (Miscellaneous Provisions) Act, 1968, sec. 11 (1) (a).

(2) See 1968 Act sec. 11 (1) (b).

(3) See 1968 Act secs. 18 and 19 and sec. 11 (1) (c). Note that if Land Court consent to application under sec. 26 (1) (e) (use of land for non-agricultural purposes) or consent under sec. 26 (1) (b) (sound estate management) or 1968 Act sec. 18 (2) but certify that they would also have been satisfied under 26 (1) (e) the additional sum under sec. 9 of the 1968 Act will be payable.

If notice is given in this form and the tenant omits to serve counter notice under sec. 25 (1) compensation for disturbance will be payable but not the additional sum under sec. 9 of the 1968 Act.

If the tenant serves counter notice and the landlord succeeds in an application to the Land Court under sec. 26 (1) compensation for disturbance will be payable but the additional sum under sec. 9 of the 1968 Act will not be payable unless either, (a) the reasons for the court's decision include that they are satisfied as to the matter mentioned in sec. 26 (1) (e) [use of land for non-agricultural purposes] or (b) the reasons for the decision include that they are satisfied as to the matter mentioned in sec. 26 (1) (b) (sound management of estates) or in paragraph (a) [lack of training or experience] or (c) [tenant having other agricultural land] of sec. 18 of 1968 Act but would have been satisfied also as to the matter mentioned in sec. 26 (1) (e) if it had been specified in the application.

4.   Counter-Notice under sec. 25 (1)
                         by Tenant to Notice to Quit
                         by Landlord.

                                    Place................................
                                    Date................................
To A.B. [design landlord].
        AGRICULTURAL HOLDINGS (SCOTLAND) ACT, 1949.
             HOLDING OF..........................
  With reference to your Notice dated   , 19   ,
to quit the above holding at the term of   , I
hereby require that sub-sec. (1) of sec. 25 of the Agricultural
Holdings (Scotland) Act, 1949, shall apply to the said notice.
                             (Signed) C.D. [tenant].

                          Notes.
  (1) It is recommended that this notice be sent by registered post and a copy retained.
  (2) It may be advisable to reserve right to object to the validity of the notice to quit.
  (3) Where the Notice to Quit states that it is given for any of the reasons stated in sec. 25 (2), the tenant must within one month, serve a notice requiring the question to be determined by arbitration (sec. 27).

5.   Notice by Tenant to Landlord re-
                         quiring questions arising out
                         of Notice to Quit to be re-
                         ferred to Arbitration (sec. 25).

                                    Place................................
                                    Date................................
To A.B. [design landlord].
        AGRICULTURAL HOLDINGS (SCOTLAND) ACT, 1949.
             HOLDING OF..........................
  With reference to your Notice to Quit the above holding,
dated   , 19   , 1 hereby intimate to you that
I require all questions arising out of the reasons stated in the said
notice to quit be determined by arbitration under the above Act.
                             (Signed) C.D. [tenant].

                          Notes.
  (1) This notice must be given within one month of the date of the Notice to Quit (sec. 27).

(2) Where the notice to quit states one or more of the reasons in sec. 25 (2), the tenant should require arbitration and not serve a counter-notice requiring the operation of sec. 25 (1). On the award of the arbiter being issued he may then serve a notice under sec. 25 (1) within one month of the date of the award (sec. 27 (2)).

(3) See *French* v. *Elliott* [1960] 1 W.L.R. 40 where two reasons given or two notices sent.

6. Notice to Tenant to pay rent within two months (sec. 25 (2) (e)).

Place............................
Date............................

To C.D. [design tenant].

AGRICULTURAL HOLDINGS (SCOTLAND) ACT, 1949.

HOLDING OF............................

I hereby give you notice that you are required to pay the rent amounting to £   due at   , 19 , in respect of the above holding within two months of this intimation.

(Signed) A.B. [landlord].

Note.—The service of this notice, which should be sent by registered post, is essential if the landlord is to found on the tenant's failure to pay in a notice to quit.

7. Notice[1] by Landlord to Tenant to remedy breach of conditions[2] of tenancy capable of being remedied (sec. 25 (2) (e)).

Place............................
Date............................

To C.D. [design tenant].

AGRICULTURAL HOLDINGS (SCOTLAND) ACT, 1949.

HOLDING OF............................

I hereby intimate to you as tenant of the above holding that you are required to remedy within [state time considered to be reasonable or say 'a reasonable time'][3] from the date of this intimation the following breaches of the conditions of your tenancy, which are capable of being remedied and which are not inconsistent with your responsibilities to farm in accordance with the rules of good husbandry, viz.:—[here specify the breach of conditions in detail with reference to the conditions of the tenancy].

(Signed) A.B. [landlord].

Notes.

(1) This notice should be sent by registered port and is essential if the landlord is to found on the tenant's failure in a notice to quit.

(2) Where it is desired to give two notices or two reasons see *French* v. *Elliott* [1960] 1 W.L.R. 40.

(3) The time need not be stated (*Morrison-Low* v. *Howison* 1961 S.L.T. (Sh. Ct.) 53 *Stewart* v. *Brims* 1969 S.L.T. (Sh. Ct.) 2.

8. Notice by the Landlord to Tenant to remove from part of farm (sec. 32).

Place...........................
Date............................

To C.D. [tenant].

AGRICULTURAL HOLDINGS (SCOTLAND) ACT, 1949.

HOLDING OF..........................

In terms of the Agricultural Holdings (Scotland) Act, 1949, sec. 32, I hereby give notice that you are required to remove at [specify date] from that area of land, part of the above holding of which you are the tenant [describe the part or preferably refer to a plan drawn to scale, so as clearly to identify the part]. The said ground is required for the following purpose[s], namely [set out the purposes in terms of sec. 32 of the said Act].

(Signed) A.B. [landlord].

Note.—This notice should be sent by registered post.

9. Counter-Notice by the Tenant accepting notice to remove from part as notice to remove from the entire holding (sec. 33).

Place...........................
Date............................

To A.B. [design landlord].

AGRICULTURAL HOLDINGS (SCOTLAND) ACT, 1949.
HOLDING OF..........................

Having received your notice dated　　　　　　　to remove at [insert date] from part of the above holding, I hereby accept said notice as a notice to remove from the entire holding occupied by me in terms of sec. 33 of the Agricultural Holdings (Scotland) Act, 1949, to take effect at the same time as the original notice.

(Signed) C.D. [tenant].

Note.—This notice must be served within 28 days of the service of the landlord's notice to quit or within 28 days of it being determined by arbitration that the notice is effective.

10.　　　　　　　　Notice by Landlord to Tenant of Contract for Sale of Farm (sec. 31 (2)).

Place..............................
Date..............................

To C.D. [design tenant].

AGRICULTURAL HOLDINGS (SCOTLAND) ACT, 1949.
HOLDING OF..........................

I hereby give you notice that on　　　　　　　, 19　　, I entered into a contract for the sale to E.F. [design purchaser] of the above-mentioned holding of which you are tenant.

This notice is given to you in terms of sec. 31 (2) of the Agricultural Holdings (Scotland) Act, 1949.

(Signed) A.B. [landlord].

Note.—This notice must be given within 14 days of the making of the Contract for Sale.

11.　　　　　　　　Agreement between Landlord and Tenant as to Notice to Quit being valid on sale of holding (sec. 31 (2)).

Place..............................
Date..............................

AGRICULTURAL HOLDINGS (SCOTLAND) ACT, 1949.
HOLDING OF..........................

We, A.B. [design], the landlord, and C.D. [design], the tenant, of the holding of   in the Parish of   and County of   , hereby agree that the Notice to Quit dated   , 19   , served by the landlord on the tenant shall [continue in force] [be of no effect] if a Contract for the Sale of the said holding is entered into by the landlord within three months of the date of this agreement: In witness whereof

12. Notification by Tenant to Landlord that he elects that Notice to Quit shall remain in force following on sale of holding (sec. 31 (2)).

Place................................
Date.................................

To A.B. [design landlord].

AGRICULTURAL HOLDINGS (SCOTLAND) ACT, 1949.
HOLDING OF..........................

With reference to your intimation to me dated 19   , of the sale of the above holding of which I am tenant, I hereby notify you that I elect that the Notice to Quit served on me and dated   , 19   , shall remain in force.

(Signed) C.D. [tenant].

Note.—This notice must be given before the expiry of one month from the receipt by the tenant of the notice of the making of the Contract.

13. Demand for execution of Tenancy Agreement (secs. 4, 5).

Place................................
Date.................................

To C.D. [tenant].

AGRICULTURAL HOLDINGS (SCOTLAND) ACT, 1949.
HOLDING OF..........................

I hereby give you notice to enter into a written lease containing provisions as to the terms of your tenancy of the above holding, in terms of sec. 4 of the Agricultural Holdings (Scotland) Act, 1949 [or agree to additions or revised terms of your existing lease in accordance with secs. 4, 5, &c.].

I enclose draft of the proposed agreement [or note of proposed provisions] and shall be obliged by your returning same approved or intimating any adjustments you propose. In the event of agreement not being reached within six months from this date the terms of the lease will be referred to arbitration under the Act.

(Signed) A.B. [landlord].

Notes.
(1) At the expiry of six months, if no agreement is reached, the landlord or tenant should send notice requiring the other party to arbitrate under the section, and, if this is not agreed, application should be made to the Secretary of State to appoint an arbiter.
(2) The form should be adjusted when it is to be served by the tenant.

14. Reference to arbitration by Landlord as to amount of compensation payable by Tenant on transfer of liability for maintenance of fixed equipment (sec. 6).

Place................................
Date................................

To C.D. [design tenant].

AGRICULTURAL HOLDINGS (SCOTLAND) ACT, 1949.
HOLDING OF............................

Whereas in virtue of sec. 4 of the above Act liability for the maintenance or repair of certain fixed equipment, as specified in the list annexed hereto, has been transferred by the Arbiter from you to me as from   , 19   , in terms of Award dated   , 19   , I hereby require that the compensation to be paid by you to me in respect of said transfer of liability up to the said date of transfer shall be settled by arbitration under the Act in terms of sec. 6.

(Signed A.B. [landlord].

List of Equipment.

Note.—The reference must be made within one month from the date on which the transfer of liability takes effect.

15.   Requirement by Tenant for Arbitration as to claim against Landlord on transfer of liability for maintenance of fixed equipment (sec. 6 (2)).

Place............................
Date.............................

To A.B. [design landlord].

AGRICULTURAL HOLDINGS (SCOTLAND) ACT, 1949.
HOLDING OF...........................

Whereas in virtue of sec. 4 of the above Act the liability for the maintenance or repair of certain fixed equipment ,asspecified in the list annexed hereto, has been transferred as from   , 19   , from you to me in terms of agreement between us dated   , 19   [or award by E.F., the arbiter, dated   , 19   ], I hereby require that my claim in respect of your previous failure to discharge your liability for such maintenance or repair shall be settled by arbitration under sec. 75.

(Signed) C.D. [tenant].

List of Equipment.

Note.—This notice must be given within one month from the date on which the transfer liability takes effect.

16.   Demand by the Landlord for Arbitration as to Rent (secs. 7 and 8).

Place............................
Date.............................

To C.D. [design tenant].

AGRICULTURAL HOLDINGS (SCOTLAND) ACT, 1949.
HOLDING OF...........................

I hereby, in terms of the Agricultural Holdings (Scotland) Act, 1949, sec. 7, demand arbitration as to the rent to be paid for the above holding from and after the term of being the next ensuing term at which I could terminate the tenancy by notice to quit, given at this date [or, as to the additional rent to be paid in respect of the following improvements (specify, and refer to sec. 8)].

Please acknowledge receipt of this demand and suggest the names of two or three arbiters whom you would agree to appoint. Failing agreement, I shall apply to the Secretary of State to make an appointment.[4]

(Signed) A.B. [landlord].

Notes.
(1) Although not essential, it is well to send this notice by registered letter.
(2) This form may be adapted to the case of a tenant demanding arbitration as to the rent under the same sections. The part in square brackets is appropriate to sec. 8.
(3) When there is more than one ish the notice should refer to the first in time.
(4) It is often better to have the appointment made by the Secretary of State, since arbiters so appointed are subject to the provisions of The Tribunals and Inquiries Act, 1958.

17. Notice by Landlord requiring increase of rent on completion of improvements (sec. 8).

Place..................................
Date..................................

To C.D. [design tenant].

AGRICULTURAL HOLDINGS (SCOTLAND) ACT, 1949.

HOLDING OF............................

I hereby intimate to you as tenant of the above holding that I require the rent thereof to be increased by an amount equal to the increase in the rental value of the improvements carried out by me as from   , 19   , the date of their completion.

I annex a note of the improvements and consider that the rent should be increased by £   per annum. Failing agreement, I require the increase in rent to be determined by arbitration.

This notice is given in terms of sec. 8 of the above Act.

(Signed) A.B. [landlord].

Statement of Improvements.

(Detail the improvements and their cost).

Note.—This notice requires to be served within six months of the completion of the improvements.

18.   Agreement between Landlord and Incoming Tenant with regard to the latter paying Outgoing Tenant's Claims for Compensation (sec. 11).

AGRICULTURAL HOLDINGS (SCOTLAND) ACT, 1949.
HOLDING OF..........................

Whereas I, A.B. [design], the landlord of the holding of   , have let the said holding to C.D. [design] as incoming tenant at the term of   , it is hereby agreed between us that I the said C.D. will settle [or refund to the said A.B. the amount of] the Claim of E.F., the outgoing tenant, in respect of compensation for improvements on the holding in respect of [specify the particular improvements in terms of Part III of the First Schedule of the Act] and that up to a maximum of £   : In witness whereof

                     (Signed) A.B. [landlord].
                     (Signed) C.D. [incoming tenant].

19.   Notice to Landlord that Tenant intends to remove fixtures, etc. (sec. 14).

                              Place...............................
                              Date................................

To A.B. [design landlord].

AGRICULTURAL HOLDINGS (SCOTLAND) ACT, 1949.
HOLDING OF..........................

In terms of the above Act (sec. 14), I hereby give you notice that I intend at Whitsunday [or Martinmas, as the case may be] next, or within six months thereafter, to remove the following fixtures [and/or buildings, as the case may be] erected by me [or acquired by me from   on   ] on the above holding, namely [specify the fixtures and /or buildings in such manner that they can be identified, and, if necessary, where and when and from whom any of the buildings or fixtures were acquired].

                                (Signed) C.D. [tenant].

Note.—This notice must be served at least one month before both the exercise of the right to remove the fixtures or buildings and the termination of the tenancy. The landlord may give counter-notice, see form 20.

20.   Notice to Tenant by Landlord
                             that he elects to purchase
                             fixtures, &c. (sec. 14).

                             Place..........................
                             Date...........................

To C.D. [design tenant].

AGRICULTURAL HOLDINGS (SCOTLAND) ACT, 1949.
HOLDING OF...........................

With reference to your notice dated intimating that you intend to remove certain fixtures [and/or buildings] from the above holding, of which you are [were] my tenant, at Whitsunday [or Martinmas] next, or within six months thereafter, I hereby give you notice in terms of the Agricultural Holdings (Scotland) Act, 1949, sec. 14 (3) that I elect to purchase the same [or the following (specify)] at a price, failing agreement, which will be determined by arbitration. Please acknowledge receipt of this notice.

(Signed) A.B. [landlord].

Note.—See note to form 19. This counter-notice must be given before the expiration of the tenant's notice.

21.   Notice of damage by game (sec. 15).

                             Place..........................
                             Date...........................

To A.B. [design landlord].

AGRICULTURAL HOLDINGS (SCOTLAND) ACT, 1949.
HOLDING OF...........................

TAKE NOTICE that the field [describe by name or reference to Ordnance Survey] now in oats [or as the case may be] on the above holding, of which I am tenant, has been and is being damaged by game other than ground game, and that it is my intention to claim compensation therefor under the Agricultural Holdings (Scotland) Act, 1949, sec. 15. You may inspect the damage within [state a reasonable time] before the crop is removed. Please acknowledge receipt of this notice.

(Signed) C.D. [tenant].

22. Claim for damage by game (sec. 15).

Place...............................
Date...............................

To A.B. [design landlord].

AGRICULTURAL HOLDINGS (SCOTLAND) ACT, 1949.
HOLDING OF..........................
Referring to my notice to you dated
[or as the case may be] in terms of the Agricultural Holdings (Scotland) Act, 1949, sec. 15, I hereby claim the sum of £
for damage to oats [or as the case may be] on the above holding of which I am tenant, by game in or about [specify date]. I am prepared to refer the matter to arbitration under the Acts if you do not agree to the amount claimed.

(Signed) C.D. [tenant].

Note.—This notice requires to be given within one month after the end of the calendar year.

23. Notification by Landlord or Tenant requiring a Record to be made under sec. 17.

Place...............................
Date...............................

To C.D. [design tenant].

AGRICULTURAL HOLDINGS (SCOTLAND) ACT, 1949.
HOLDING OF..........................
In terms of Section 17 of the Agricultural Holdings (Scotland) Act, 1949, I hereby require a Record to be made of the fixed equipment on, and of the cultivation of, the above holding, of

which you are tenant. I accordingly suggest the following names of persons whom I am prepared to accept for the purpose of making the Record, viz. [names]. Failing agreement on a person to make the Record I shall apply to the Secretary of State to make an appointment.

(Signed) A.B. [landlord].

Note.—Adapt this form when it is to be sent by the tenant, who may also require the Record to refer to improvements he has carried out or to fixtures and buildings he is entitled to remove (sec. 17 (1) (a) and (b).

24.   Intimation to Landlord of bequest of lease, 1949 Act (sec. 20); 1958 Act (sec. 6).

Place............................

Date............................

To A.B. [design landlord].

AGRICULTURAL HOLDINGS (SCOTLAND) ACT, 1949,
AS AMENDED BY AGRICULTURE ACT, 1958.

HOLDING OF...........................

In terms of sec. 20 of the Agricultural Holdings (Scotland) Act, 1949, I hereby intimate to you as landlord of the above holding that C.D., tenant of said holding, who died on by his last Will and Testament, dated   [of which a copy is herewith enclosed], bequeathed to me the current lease of said holding, and I hereby intimate that it is my intention to take up the lease.

I shall be glad to supply you with any information which you may reasonably require in connection with my resources or capacity.

Please acknowledge receipt and let me know if you agree to accept me as tenant.

(Signed) J.H. [legatee].

Stamp. Lease duty on acceptance.

Note.—This intimation must, if possible, be made within twenty-one days of the tenant's death.

25.   Intimation by Landlord declining to accept legatee as Tenant (sec. 20 (3)).

                                               Place............................
                                               Date..............................

To J.H. [design legatee].

                AGRICULTURAL HOLDINGS (SCOTLAND) ACT, 1949,
                   AS AMENDED BY AGRICULTURE ACT, 1958.
                        HOLDING OF...........................

With reference to your letter of   intimating that the late C.D., tenant of the above holding, by his last Will and Testament bequeathed to you the current lease of said holding, I hereby intimate in terms of the Agricultural Holdings (Scotland) Act, 1949, sec. 20 (3), that I object to receive you as tenant under the said lease. [The reasons for my objection are   ].

                                      (Signed) A.B. [landlord].

Note.—This notice must be given within one month of receipt of the notice from the legatee.

26.   Intimation to Landlord of acquisition of lease from executor 1949 Act (sec. 21); 1958 Act (sec. 6).

                                             Place............................
                                               Date..............................

To A.B. [design landlord].

               AGRICULTURAL HOLDINGS (SCOTLAND) ACT, 1949,
                AS AMENDED BY SUCCESSION (SCOTLAND) ACT, 1964
                        HOLDING OF...........................

In terms of sec. 21 of the Agricultural Holdings (Scotland) Act, 1949, I hereby intimate to you as landlord of the above holding that X.Y. (design) as executor of C.D. tenant of said holding, who died on   transferred to me on [insert date of transfer] the current lease of said holding, and I hereby intimate that it is my intention to take up the lease.

I shall be glad to supply you with any information which you may reasonably require in connection with my resources or capacity.

Please acknowledge receipt and let me know if you agree to accept me as tenant.

(Signed) J.H. [legatee].

Note.—This intimation must, if possible, be made within twenty-one days of the date of transfer.

27. Intimation by Landlord declining to accept 'acquirer' as Tenant (sec. 21 (2)).

Place...............................

Date...............................

To J.H. [design acquirer].

AGRICULTURAL HOLDINGS (SCOTLAND) ACT, 1949, AS AMENDED BY SUCCESSION (SCOTLAND) ACT, 1964.

HOLDING OF............................

With reference to your letter of   intimating that the current lease of the above holding has been transferred to you I hereby intimate in terms of the Agricultural Holdings (Scotland) Act, 1949, sec. 21 (2), that I object to receive you as tenant under the said lease. [The reasons for my objection are   ].

(Signed) A.B. [landlord].

Note.—This notice must be given within one month of receipt of the notice from acquirer.

28. Application by Tenant to Landlord for written consent to Improvements under Schedule I, Part I (sec. 50).

Place...............................

Date...............................

To A.B. [design landlord].

AGRICULTURAL HOLDINGS (SCOTLAND) ACT, 1949.
HOLDING OF............................

I hereby intimate that I propose to carry out on the above holding the following improvements referred to in Part I of the First Schedule of the above Act and I request you to give your written consent thereto [unconditionally] [or on such terms as to compensation or otherwise as may be agreed on between us in writing], viz.:—[specify the proposed improvements in detail, using the appropriate words in the Schedule].

(Signed) C.D. [tenant].

Note.—If certain conditions are proposed they should be mentioned.

29.   Landlord's Consent in Writing to Execution of Improvements (Schedule I, Part I, sec. 50).

Place..................................
Date..................................

To C.D. [design tenant].

AGRICULTURAL HOLDINGS (SCOTLAND) ACT, 1949.
HOLDING OF............................

I hereby consent to your executing on the above holding the following improvements embraced in the First Schedule, Part I of the above Act, namely [here specify the improvements mentioned in the Schedule so far as applicable and the conditions as to compensation, &c., on which the consent is given]. Please notify me when they have been completed.

(Signed) A.B. [landlord].

Note.—If the landlord proposes conditions additional to those proposed by the tenant the whole conditions acceptable to the landlord should be detailed. See *Turnbull* v. *Millar*, 1942, S.C. 521.

30.   Notice by Tenant of intention to make improvements (Schedule I, Part II, sec. 51).

Place..................................
Date..................................

To A.B. [design landlord].

AGRICULTURAL HOLDINGS (SCOTLAND) ACT, 1949.
HOLDING OF............................

In terms of Section 51 of the Agricultural Holdings (Scotland) Act, 1949, I hereby intimate that it is my intention, on the expiry of three months from this date, to execute the improvements embraced under Part II, First Schedule to the Act, on the above holding as specified in the statement annexed hereto. Please acknowledge receipt of this notice.

(Signed) C.D. [tenant].

Annexed Statement.

Specify the several improvements particularly, under the heads, detailed in Nos. 9—23 of the Schedule and amplify where necessary.

Notes.

(1) In order to entitle the tenant to compensation for improvements embraced in Part II of the First Schedule of the Act, it is necessary for him to give such notice not less than three months before beginning to execute the improvements.

(2) It is necessary to follow the different heads in the schedule and to state before beginning the work with some particularity the improvement proposed, its situation, and the manner in which the work is to be carried out. The landlord must not be left in doubt as to what is intended. It is generally desirable to supply an estimate of cost, with specification made up by a practical man, and a plan where necessary.

(3) Landlords are referred to the provisions of sec. 52.

31. Notice by Landlord to Tenant of objection to carrying out of Improvements (Schedule I, Part II, sec. 52 (1)).

Place...............................
Date...............................

To C.D. [tenant].

AGRICULTURAL HOLDINGS (SCOTLAND) ACT, 1949.
HOLDING OF............................

With reference to your letter dated intimating your intention to make (in accordance with the Agricultural Holdings (Scotland) Act, 1949, sec. 51) on the above holding the improvements therein specified, I hereby intimate in terms of sec. 52 (1) that I object to the carrying out of those improvements [or to the manner in which you propose to carry out those improvements].

                                                        (Signed) A.B. [landlord].

Notes.

(1) This notice must be given within one month after receiving notice from the tenant of his intention to make improvements.

(2) On receipt of this notice the tenant may, without further notice, apply to the Land Court for approval of the carrying out of the proposed improvements.

32.   Notice by Landlord that he intends to execute the Improvements himself (sec. 52 (3)).

                                            Place............................

                                            Date.............................

To C.D. [tenant].

                 AGRICULTURAL HOLDINGS (SCOTLAND) ACT, 1949.

                 HOLDING OF........................

With reference to your letter dated   , intimating your intention to make (in accordance with the Agricultural Holdings (Scotland) Act, 1949, sec. 51) on the above holding the improvements therein specified and the approval of the Scottish Land Court having now been obtained thereto, such approval being dated   , I hereby intimate, in terms of sec. 52 (3), that I undertake to execute those improvements myself. Please acknowledge receipt of this intimation. The amount of increased rent payable in respect of these improvements will be settled, in the absence of agreement, by arbitration under the Act.

                                        (Signed) A.B. [landlord].

Notes.

(1) After giving this intimation the landlord may, unless the tenant's notice is previously withdrawn, proceed to do the work himself in any reasonable and proper manner.

(2) This notice must be given within one month of the date of the decision of the Scottish Land Court.

FORMS

33. Notice of intention by the Tenant to claim Compensation for Continuous Special System of Farming (sec. 56).

Place............................
Date............................

To A.B. [design landlord].

AGRICULTURAL HOLDINGS (SCOTLAND) ACT, 1949.

HOLDING OF............................

I hereby give you notice (in accordance with the Agricultural Holdings (Scotland) Act, 1949, sec. 56) that I intend to claim from you compensation under that Act in respect of the continuous adoption by me (in the years   ) of a special standard or system of farming on the above holding.

(Signed) C.D. [tenant].

Note.—This notice must be given one month before termination of the tenancy; the existence of a record is a prerequisite.

34. Notice by Landlord of intention to claim Compensation from the Tenant for Deterioration of Holding (secs. 57 and 58).

Place............................
Date............................

To C.D. [design tenant].

AGRICULTURAL HOLDINGS (SCOTLAND) ACT, 1949.

HOLDING OF............................

I hereby give you notice that (in accordance with the Agricultural Holdings (Scotland) Act, 1949 (sec. 57) I intend to claim compensation from you (under that section) at the termination of your tenancy of the above holding, in respect of the undernoted dilapidation [or deterioration or damage], the value of the farm having been deteriorated during the said tenancy by your failure.

(Signed) A.B. [landlord].

Notes.

(1) This notice must be given not later than three months before the termination of the tenancy.

(2) In the case of a lease entered into after 31st July, 1931 (such claim is competent only where the failure has occurred after the date of a 'record'.

(3) The above form may be adjusted to meet the case of a claim for general deterioration under sec. 58, or at common law.

(4) In the case of a lease entered into after 1st November, 1948, all claims including one under the lease, are competent only if there is a record (sec. 59 (2)).

35.   Notice of intention to claim by Waygoing Tenant (sec. 68 (2)).

Place...................................
Date...................................

To A.B. [design landlord].

AGRICULTURAL HOLDINGS (SCOTLAND) ACT, 1949.

HOLDING OF..........................

I hereby intimate to you as landlord of the above holding my intention to make the following claims, under and in accordance with the above Act, on the termination of my tenancy of the said holding at [state date or dates of termination of tenancy], viz.:—

(1) Claims for compensation under the said Acts for improvements embraced in the Schedules to the Act [or as the case may be], and (2) additional claims, all as set forth in the annexed statement.

Please acknowledge receipt.

(Signed) C.D. [tenant].

Annexed Statement.

1. Claims under the Agricultural Holdings (Scotland) Act, 1949 [specify using the words of the Schedules to the Act].

2. Additional claims as follows:—

I. CLAIMS UNDER PART I, FIRST SCHEDULE.
    State the nature of each improvement for which prior consent in writing was given by the landlord.

II. CLAIMS UNDER PART II, FIRST SCHEDULE.
    State each improvement under the heads contained in the Statute so far as applicable.

III. CLAIMS UNDER PART III, FIRST SCHEDULE.
    State the particulars under the heads in the Statute.

IV. ADDITIONAL CLAIMS BY THE TENANT.
  (a) IN RESPECT OF INCREASED VALUE OF HOLDING (SEC. 56).
  (b) CLAIM FOR BREACH OF CONTRACT.
      These particulars need not be given very fully, but should be sufficient to indicate what is complained of and the date or dates and refer to the lease if there is a breach in one or other of the clauses or conditions thereof.
  (c) CLAIM FOR VALUE OF BUILDINGS OR FIXTURES TAKEN OVER BY LANDLORD.
      [Specify the buildings or fixtures]
  (d) CLAIM FOR COMPENSATION FOR DISTURBANCE.
  (e) CLAIM FOR ADDITIONAL SUM FOR REORGANISATION OF TENANT'S AFFAIRS (1968 ACT, SEC. 9).

Note.—When the claim for disturbance is for more than one year's rent (maximum two years' rent), one month's notice must be given to the landlord of the sale of implements, &c., which form the basis of the claim, and an opportunity of making a valuation must also be afforded (sec. 35.)

36. Notice by Landlord to Tenant of change of Landlord (sec. 90 (4)).

Place.............................
Date.............................

To C.D. [design tenant].

AGRICULTURAL HOLDINGS (SCOTLAND) ACT, 1949.
HOLDING OF...........................

I hereby give you notice in terms of sec. 90 (4) of the above Act that as from   , 19   , I cease to be entitled to receive the rents and profits of the above holding and E.F. [design] is now entitled to receive the same. All notices or other documents requiring to be served on the landlord of the holding should be served on the said E.F. as from the said date.

(Signed) A.B. [landlord].

37. Statement of Case and Particulars of Claim (Schedule VI, 5).

AGRICULTURAL HOLDINGS (SCOTLAND) ACT, 1949.
HOLDING OF...........................
A.B. [design], Landlord.
C.D. [design], Tenant.

AGRICULTURAL HOLDINGS (SCOTLAND) ACTS

Statement of Case—Landlord's Claim.

1. The claimant is landlord of the holding of..................... in the Parish of......................and the County of..................... and C.D. is [was] tenant of the said holding in terms of lease between the parties dated........................, a copy of which is produced herewith and referred to.

2. The lease expired at Whitsunday, and separation of crop, 19...... [or otherwise as the case may be], since when the tenant has been sitting on tacit relocation. The tenant quitted the farm at Whitsunday, and separation of crop, 19......, in consequence of notice to quit by the landlord, dated....................., copy of which is produced herewith.

3. In terms of the said lease the tenant was bound to maintain the buildings, fences, &c., in good and tenantable repair during the period of his tenancy, his obligations being defined as follows: [take in from lease].

4. No repairs have been carried out by the tenant during his occupancy of the holding. In addition, the tenant has allowed the farm generally to deteriorate, has failed to keep the ground clean and in a good state of fertility and he has not left it in the rotation prescribed in the lease [give details].

5. The landlord has frequently called on the tenant to implement his obligations under the lease [or at common law], which he has failed to do. In particular, reference is made to letters addressed to the tenant, dated....................., which are produced.

6. The tenant being bound under his lease to maintain the said subjects and at common law cultivate the farm in accordance with the rules of good husbandry and having failed to do so is liable to the landlord in damages, which are reasonably estimated at a sum of £   . Particulars are annexed hereto.

7. On..................., 19......, the landlord gave notice to the tenant of his intention to claim for general deterioration of the holding in terms of sec. 58 of the Agricultural Holdings (Scotland) Act, 1949.

8. A record of the farm was made by...........................on ...., a copy of which is produced herewith, the date of the record being earlier than the date of the tenant's failure to perform his obligations as condescended on.

9. In the course of negotiations to settle the claim above referred to, the tenant by letter dated.................................... admitted liability in respect of the repairs to buildings but disputed the amount as excessive and denied liability for the other claim. This arbitration is therefore necessary.

PARTICULARS OF CLAIMS.

Amount Claimed.
1. Dilapidations to buildings [specify these in detail with reference to particular buildings] -   -   -   -   - £
2. Dilapidation of fences [specify with reference to fields or number in record] -   -   -   -   -   -   -
3. Dilapidation of drains [do.] -   -   -   -   -   -
4. Cost of restoration of fertility in respect of the tenant's failure to restore fertility in respect of crops sold off the farm in exercise of his rights under sec. 12 -   -   -
5. Failure to leave the farm in the rotation prescribed in the lease in the following respects:—[detail] -   -   -
6. General deterioration in respect of failure to leave the land clean and in good condition of fertility (sec. 58) -   -

Statement of Case re Tenant's Claims.

1. Reference is made to the foregoing statement of the landlord's case.

2. With regard to the tenant's claim for the unexhausted value of manures and feeding stuffs, the claim is excessive and should be reduced.

3. Reference is made to the tenant's claim for compensation for disturbance. The landlord by letter dated........................ (copy herewith produced) called on the tenant to remedy within two months breaches of the condition of his tenancy with regard to repairs as above condescended on. The tenant at the date of the notice to quit had failed to remedy these breaches and that fact was stated in the notice to quit.

4. The breaches in question being capable of being remedied and being within the meaning of sec. 25 (2) (*e*) of the Agricultural Holdings (Scotland) Act, 1949, the tenant is not entitled to compensation for disturbance and the landlord is not liable accordingly.

Notes.
(1) Adapt this form in the case of a tenant's claim.
(2) The statement of case and particulars must be lodged with the Arbiter within twenty-eight days of his appointment. There is no provision that a copy must be sent to the other party to the arbitration, but in practice this should be done. See note to Arbitration form 6 (a).

(3) The lodgment must be timeous to entitle the arbiter to proceed. Where a party fails to lodge timeously he cannot set up an affirmative case. (*Collett* v. *Deely*, 1949, 100 L.J. 108; *Jamieson* v. *Clark*, 1951, 67 *Sh. Ct. Rep.* 17).

(4) There is no provision in the Act for Answers being lodged and this does not appear to be competent, as parties are confined at the hearing to what is alleged in the statement of case and particulars lodged. If they are allowed, the formal consent of parties should be obtained.

(5) Where at the end of a tenancy parties have been in negotiation for four months or more (sec. 68 (3)), it is assumed with justification that each knows the other's case and can answer it in the original statement of case.

(6) It is envisaged by paragraph 5 (a) of the Sixth Schedule that amendments or additions may be made to the original statement of case compared with the adjustment of record in civil actions in the Courts.

B. APPLICATIONS TO THE SECRETARY OF STATE FOR SCOTLAND

See also Appendix II for applications for the appointment of arbiters.

Application by Landlord or Tenant to extend the time for settling claims at termination of tenancy (sec. 68 (3)).

To the Secretary of State for Scotland.

AGRICULTURAL HOLDINGS (SCOTLAND) ACT, 1949.
HOLDING OF..........................

I, A.B. [landlord]/C.D. [tenant] of the above holding in the Parish of   and County of
the tenancy of which terminated at [specify term or terms], hereby apply under sec. 68 (3) of the above Act for an extension of two months [or a further extension of two months] of the period within which claims between the parties may be settled by agreement, without resort to arbitration.

(Signed) A.B. [landlord]
or
(Signed) C.D. [tenant].

Notes.
(1) The application may be signed by one or both parties.
(2) It must be made before the lapse of the initial period of four months, or, in the case of a second application, before the lapse of six months, from the termination of the tenancy.

C. APPLICATIONS TO SCOTTISH LAND COURT.

Forms for most of the applications which require to be made to the Scottish Land Court have been prepared by the Court officials and those are available free of charge either from The Principal Clerk, The Scottish Land Court, 1 Grosvenor Crescent, Edinburgh, 12, or from the offices of the Sheriff Clerk in each Sheriffdom. The following Forms, with the number given to each by the Land Court, apply to the 1949 and 1958 Acts.

34.	Application by Legatee under Section 20, 1949 Act.
35.	Application to terminate interest of heir at law made by landlord under Section 21, 1949 Act.
36.	Application for consent to the operation of a notice to quit made by landlord under Sections 25 & 26, 1949 Act, and Section 3 of 1958 Act.
37.	Application by Landlord under Section 28, 1949 Act, and Section 3 of 1958 Act.
69.	Joint Application under Section 78, 1949 Act.
73.	Application by Tenant under Section 6 (5), 1958 Act.

D. ARBITRATION PROCEDURE.

1. Minute of Appointment of Arbiter by Landlord and outgoing Tenant (Sch. VI).

AGRICULTURAL HOLDINGS (SCOTLAND) ACT, 1949.

HOLDING OF..........................

We, A.B. [designation and address] landlord of the farm of in the parish of and county of , and C.D. [designation and address], outgoing tenant of the said farm at the term of , 19 , hereby appoint E.F. [designation and address] arbiter under the Agricultural Holdings (Scotland) Act, 1949, to determine what sum or sums, if any, are payable by either of us to the other in respect of the questions and claims detailed in the annexed Schedule[s] or the compensation (if any) payable under the Agricultural Holdings (Scotland) Act, 1949, by me, the said A.B., to me, the said C.D., in respect of damage by game to the crops of the latter on said farm during the year ending the day of , 19 .

In witness whereof
[Date]
Adopted as holograph or witnessed.

(Signed) A.B. [landlord].

(Signed) C.D. [tenant].

Notes.

(1) The Tribunals and Inquiries Act, 1958, does not apply to arbiters appointed by agreement. If either party desires the arbiter to be bound to give reasons for his decision, his better course is to apply to the Secretary of State for Scotland for the appointment of an arbiter. It is not necessary for parties to try to agree upon the person to be appointed. *Chalmers Property Investment Co. Ltd.* v. *MacColl,* 1951 S.C. 24, per Lord President Cooper, p. 30; *F. R. Evans (Leeds), Ltd.* v. *Webster* (1962) 112 L.J. 703.

(2) It is important to state the questions and claims as clearly and fully as possible.

(3) An appointment by agreement of parties need not be made from the panel of arbiters appointed inder the Act.

(4) An appointment may be made by the Secretary of State on the application of either party.

(5) An appointment by the parties is not complete until it is delivered. See *Chalmers Property Investment Co. Ltd.* v. *MacColl, supra.*

2. Joint intimation to Arbiter of completion of his minute of appointment where he has been appointed by the Parties.

Place............................
Date............................

To R.W. [design arbiter].

AGRICULTURAL HOLDINGS (SCOTLAND) ACT, 1949.

HOLDING OF..........................

We enclose minute of your appointment by us as arbiter under the above Act to deal with the claims therein referred to, and we have to request that you will commence to act thereunder within seven days from this date.

Copies of our respective statements of case and particulars of claims are enclosed [or will follow].

We consent to the appointment of a clerk in the reference.

Please acknowledge receipt and confirm your acceptance of the appointment.

(Signed) A.B. [landlord].

(Signed) C.D. [tenant].

Notes.
(1) Statements of case and full particulars must be lodged with the arbiter within 28 days of his appointment.
(2) The award must be made within two months after the appointment unless the time is extended by the parties jointly or by the Secretary of State for Scotland (Schedule VI (8)).

3. Consent to revocation of Arbiter's appointment (Sch. VI.).

To the Secretary of State for Scotland.

AGRICULTURAL HOLDINGS (SCOTLAND) ACT, 1949.
HOLDING OF..............................

We hereby consent to the revocation of the appointment of E.F. as arbiter, in the arbitration between us under the Agricultural Holdings (Scotland) Act, 1949, with reference to the above holding.

Dated at , the day of , 19 .
(Signed) A.B. [landlord.
(Signed) C.D. [tenant].

4. Extension by Parties of time for award.

Place..............................
Date..............................

To R.W. [design arbiter].

ARBITRATION—HOLDING OF..............................

We extend the time for the issue of your award in the arbitration between us until the day of , 19 .

(Signed) A.B. [landlord].
(Signed) C.D. [tenant].

5. Minute of Devolution by Arbiters on Oversman.

We, A.B., and C.D., the arbiters under the foregoing submission having differed in opinion regarding the determination of the matters submitted to us, hereby devolve the said reference and submission and whole matters therein contained upon G.H., the oversman: IN WITNESS WHEREOF

Notes.
(1) It is competent to devolve part only of the question or claim.
(2) This style has no application to the single arbiter procedure laid down in the 1949 Act.

6. Incidental orders by Arbiter.

(Some of these orders may not be appropriate in statutory arbitrations)

(*a*) Order by Arbiter for Claims.

Place...........................
Date............................

AGRICULTURAL HOLDINGS (SCOTLAND) ACT, 1949.
ARBITRATION—HOLDING OF.........................

Having, on   , 19   , been appointed by the Secretary of State for Scotland [or by the parties] as arbiter in the arbitration between A.B. and C.D. relating to the above holding, I hereby appoint E.F. [design] to be clerk and legal adviser in the arbitration; further I require the parties to lodge with me [or the clerk] their claims and also allow them to see and answer the claim of the other party within   days from this date.

(Signed) R.W. [arbiter].

Note.—If his expenses are to be recovered, a clerk can only be appointed after the claims and answers are lodged and with the consent of the parties or the Sheriff (Sch. VI, 18). Rule 5 makes no reference to the lodging of answers and it is doubtful if the arbiter can make an order for answers unless they come within the scope of the words 'amendment or addition' in that rule. It is thought that 'amendment or addition' should be by way of adjustment on the original document for each party. If parties consent to the lodging of answers they would be personally barred from objecting to the procedure. The rule of course only applies to arbitrations under the Act and not at common law.

(*b*) Order to produce documents.

Place...........................
Date............................

ARBITRATION—HOLDING OF.........................

I require you to attend at [place] on the day of   , 19   , at   o'clock, and to bring with you and produce the documents, receipts, &c., mentioned in the annexed list, so far as the same may be in your possession or within your power.

(Signed) R.W. [arbiter].

To C.D. [design].

[Subjoin List of Documents referred to].

Note.—Where parties refuse to produce documents, &c., in response to an order as above, or where a necessary witness declines to attend and give evidence, application may be made to the Sheriff for an order to compel him to attend, give evidence, and produce documents.

(c) Intimation to Parties of inspection.

Place..................................
Date..................................

To A.B. [design].

ARBITRATION—HOLDING OF..........................

I hereby notify you that I require parties to meet me at the above holding on the   day of   19   at   o'clock, when I propose to inspect the farm buildings, fences, &c., [and hear parties on their respective claims] [and objections] [or as the case may be].

(Signed) R.W. [arbiter].

(d) Extension of time.

The arbiter on cause shown extends the time for the first parties lodging their claim to   and for the second parties lodging answers thereto to

Note.—See note on page 256 as to the competency of answers in a statutory arbitration.

(e) Revisal.

The arbiter allows parties to adjust their statements of case and claims [and answers] respectively and to intimate their adjustments to the other party within 14 days of the date hereof.

(f) To answer statement of facts.

The arbiter allows the first parties to answer the statement of facts for the second parties by [date].

(g) For closing record.

The arbiter having considered the revised statement of case and claim No.   of process and the revised answers No.   of process with the productions and whole process closes the record, and appoints parties to be heard on the preliminary pleas [or otherwise] at   on the   day of   .

(h) Closing and proof.

The arbiter closes the record on the revised statement of case and answers Nos.   and   of process; allows the parties a proof of their respective averments [or allows the first party a proof of his averments and the second party a conjunct probation] the claimant A.B. to lead in the proof. Appoints the proof to commence in   on [date] at [hour]. Further the arbiter respectfully recommends to the Lords of Council and Session [or the Sheriff of   ] to grant warrant for citing witnesses and havers on the application of either party.

(i) Avizandum.

The arbiter having heard the proof adduced for both parties and their agents thereon and on the whole cause makes avizandum.

(j) Proposed findings.

.... Arbitration.

The arbiter, having inspected the lands of   and buildings let therewith in presence of the parties and their agents, and having thereafter heard the evidence adduced and the statements of parties' agents thereon, and having carefully considered the respective claims, now proposes to find and determine as follows:—

1. *Tenant's* Claim.
   To allow the following tenant's claims:—

   Amount claimed.   *Amount allowed.*

   (a) For disturbance
   (b) For additional sum to assist in reorganisation of Tenant's affairs.
   (c) For unexhausted manures
   (d) For feeding stuffs [or otherwise according to claim]

2. *Proprietor's Claims.*
   To allow the following claims [as above].

   Amount claimed.   *Amount allowed.*

The arbiter further proposes that the sum of £   , being the difference between the said respective claims, shall be payable by the said   to the said   on or before the twenty-first day after the issue of the award following on these proposed findings.

The arbiter further proposes that his and the clerk's fees and expenses shall be payable by the parties equally and that, otherwise, each party shall pay his own expenses [or otherwise as the case may be].

The arbiter allows the parties or either of them until the   day of   , 19   , to lodge written representations against these proposed findings, if so advised.

(Signed) R.W. [arbiter].

Note.—It is usual to add a note on any points of law or fact which appear to require explanation. An arbiter appointed by the Secretary of State must, if required to do so, give reasons for his decision. See the Tribunals and Inquiries Act, 1958.

(k) Representations repelled.

The arbiter having considered the representations lodged for the [landlord or tenant] adheres to his proposed findings and will issue his award accordingly.

Note.— It is not desirable to make proposed findings final without issuing an award.

7.   Application by an Arbiter to the Sheriff for the appointment of a Clerk in the arbitration (Sch. VI).

SHERIFFDOM OF   AT

A.B. [design arbiter], Pursuer,
against
C.D. [design tenant] and E.F. [design landlord], Defenders.

The pursuer craves the Court—
To grant authority to him to appoint a clerk in the arbitration under the Agricultural Holdings (Scotland) Act, 1949, between the said defenders, and to find the defenders jointly and severally [or the defender C.D. or as the case may be] liable in expenses, and to decern.

Condescendence.

1. The pursuer was appointed by the Secretary of State for Scotland [or by the defenders jointly] as arbiter between them to deal with certain claims or questions under the Agricultural Holdings (Scotland) Act, 1949.

2. Statements of case and particulars of the claims have been duly lodged with the pursuer and are produced herewith and he is of opinion that, on account of the magnitude and importance of the arbitration [and/or on account of the fact that questions of law have arisen in the arbitration] it is expedient that he should have the advice and assistance of a properly qualified clerk.

3. The pursuer having requested the defenders and they [or the defender C.D. or as the case may be] having refused to consent to the appointment of a clerk he finds it necessary to make this application to the Court for sanction to make the appointment.

Plea-in-Law.

In the circumstances condescended on, the Court should grant authority to appoint a clerk as craved and find the defenders jointly and severally [or the defender C.D. or otherwise] liable in expenses.

8.   Stated case for the opinion of the Sheriff.

SHERIFFDOM OF   AT

Case stated by

R.S. [design], arbiter in the arbitration under the Agricultural Holdings (Scotland) Act, 1949,

between

A.B. [design], landlord of the holding of M, in the parish of T and County of X,

and

C.D. [design], outgoing [or present] tenant of said farm.

1. This is an arbitration under the Agricultural Holdings (Scotland) Act, 1949 [or as the case may be] brought before me as arbiter acting under joint appointment by the said A.B. and C.D. [or appointment by the Secretary of State for Scotland] to determine the following claims and questions in terms of the said statute [set forth]. The appointment was dated   and the period for issuing my award was extended by the parties [or by the Secretary of State] to

2. After certain procedure, in the course of which I inspected the farm, proof was led and the parties were heard on their respective claims and objections, I found, *inter alia*, the following facts proved or admitted:—

[Narrate the facts bearing on the question of law on which the opinion of the Court is required].

3. The said A.B. on those facts contends that [landlord's contentions].

4. The said C.D., on the other hand, contends that [tenant's contentions].

5. After I issued notes of my proposed findings [or as the case may be] the said C.D. [or the said A.B.] requested me to state a case for the opinion of the Court on the following

Questions of Law.

[State the question or questions in such form that the Court may answer them by a simple affirmative or negative].

This case is stated by me,

R.S., Arbiter.

9. Application to the Sheriff for order on Arbiter to state a case.

SHERIFFDOM OF   AT

A.B. [design], recently tenant of the farm of   in the county of   , Pursuer,
against
C.D. [design], the landlord of said farm, and E.F. [design arbiter], Defenders.

The pursuer craves the Court—
To ordain the defender, the said E.F., as arbiter in the arbitration under the Agricultural Holdings (Scotland) Act, 1949, between the pursuer and the said C.D., to state a case for the opinion of the Sheriff upon the following question [or questions] of law which has [have] arisen in the course of the said arbitration, namely [specify the question(s) of law referred to]; and in the meantime to interdict, prohibit and discharge the said E.F. from pronouncing or issuing any award, and to find the defenders jointly and severally [or the defender E.F. or as the case may be] liable in expenses and to decern.

Condescendence.

[Set out the facts of the case, in so far as relevant to the legal question, and particularly aver that the said question [or questions] of law have arisen in the course of the arbitration.]

Pleas-in-Law.

1. The said question(s) of law having arisen in the course of the arbitration, and the same being proper question(s) for the determination of the Court, the defender, the said E.F., ought to be ordained to state a case for the opinion of the Sheriff thereon in terms of Rule 19 of the Sixth Schedule to the Agricultural Holdings (Scotland) Act, 1949.

2. In the circumstances condescended on, the defenders [or the defender E.F. or otherwise] should be held liable [jointly and severally] in expenses.

10. Application to the Sheriff to have Arbiter removed.

SHERIFFDOM OF   AT

A.B. [design], Pursuer,
against
C.D. [design] and E.F. [design], Defenders.

The pursuer craves the Court—
To remove the defender E.F. from the office of arbiter in the arbitration between the pursuer and the defender C.D.; to find the pursuer entitled to expenses against the said E.F.; and to decern.

Condescendence.

[Here set out the facts in numbered paragraphs narrating the course of the arbitration and setting forth in detail the misconduct alleged].

Plea-in-Law.

The arbiter, having been guilty of misconduct as condescended on, should be removed as craved.

11.   Common Law submission between outgoing and incoming Tenants.

We, A.B. [design], the outgoing tenant of the farm of   in the parish of   and county of   (hereinafter called the 'outgoing tenant'), and C.D. [design], the incoming tenant of the said farm (hereinafter called the 'incoming tenant'). Considering that by lease of the said farm entered into between the outgoing tenant and E.F. [design], the landlord of the said farm, dated   , 19   , provision is made whereby the landlord undertook that he, or the incoming tenant of the said farm, at the termination of the said lease, would take over from the outgoing tenant, on the terms therein set forth, various claims under the said lease [which is referred to and herein held as repeated *brevitatis causa*]; Further considering that the outgoing tenant's away-going takes place at [insert term or terms and date of the month and year], and that the incoming tenant will then enter as tenant of the said farm; Further considering that the outgoing tenant has agreed with the incoming tenant that the latter, in place of the landlord [where that is the case] should take over from the former at the termination of the said lease by arbitration as herein provided the various things embraced in Part I of the First Schedule annexed hereto; Further considering that the incoming tenant has also agreed to take over at the said [term] from the outgoing tenant on the same basis the fixtures, buildings, &c. [as the case may be], specified in Part II of the First Schedule hereto annexed; Further considering that the incoming tenant has also agreed to settle the claims of the outgoing tenant against the landlord for the improvements alleged to have been made by the outgoing tenant under [Part III of] the First Schedule to the Agricultural Holdings (Scotland) Act, 1949, in so far as specified in Part III of the First Schedule hereto annexed, it being agreed that the landlord has received timeous and sufficient particulars of the claims to meet the requirements under the said Act, and that the compensation for the said improvements shall be determined by arbitration, as herein provided, and shall be on the same basis as could have been claimed against the landlord under the said Act; Further considering that the outgoing tenant has undertaken to pay to the incoming tenant, under agreement with the landlord, all sums [if any] due in respect of the claims competent to the landlord under the said Act, and/or the said lease for failure of the outgoing tenant to implement the conditions of his tenancy in so far as specified in the Second Schedule hereto annexed; Therefore, the said parties hereby submit and refer to the amicable decision, final sentence and decree-arbitral to be pronounced by J.K. [design] and G.H. [design], arbiters mutually chosen, or in the case of their differing in opinion, then, in so far as they may differ, by an oversman to be named by them before entering on the business of the reference, to ascertain, fix and determine the sums payable by the outgoing and incoming tenants to each other, in respect of the several claims or matters embraced in the schedules hereto annexed, with power to the arbiters and/or the oversman to call for and receive the claims of parties, to decide any question of ownership or other incidental question which it may be necessary to decide in order to enable the submission to be carried out, to hear parties, to take such probation, order such measurements and take such advice or assistance from solicitors, engineers, wrights, men of skill, and others [without, however, being bound to do so], as they or he shall think proper, fix the time of payment of the sum found to be due under the submission, to prorogate the submission from time to time; and whatever the arbiters or oversman shall determine in the premises by any award or decree-arbitral, interim or part or final, to be pronounced by them or him, both parties bind and oblige themselves, their heirs and successors respectively to implement and fulfil to each other [under penalty of £   to be paid by the party failing, to the party observing or willing to observe the same over and above performance]; Declaring (1) that the arbiters and/or oversman shall proceed with the reference and issue their awards in the manner and at the times usual in the district or otherwise as may be directed hereby; (2) that the death of either or both of the parties shall not be allowed to interrupt or terminate the reference, the heirs or representatives of any deceased party being bound to proceed and to implement the award or awards; (3) in the event of any arbiter dying during the subsistence of the reference the party who nominated him shall forthwith appoint a successor in the reference, and the oversman shall continue in office, and, in the event of the death of the oversman, a successor to him shall be appointed by the arbiters; (4) any interim or part award shall remain operative notwithstanding any change of arbiters or oversman, and any new arbiter or oversman shall give effect to the same in any subsequent interim or part final award in the same way as if such new arbiter or oversman had been in office from the outset; (5) the arbiters shall, if called on by either party, furnish along with any award, interim or final, full details showing in detail the manner in which each of the sums which may be found payable to or by either party to the other is arrived at including separate valuations of the various items in the schedules; (6) in the event of the arbiters differing on certain matters but not on others, they may devolve on the oversman only the matter of difference; Finally we direct that the fees and expenses of and incidental to the arbitration shall be borne by the parties equally [or as may be directed by the arbiters or oversman]; And the parties consent to the registration hereof and of any prorogations or devolutions and interim or part or final decrees-arbitral to follow hereon for preservation and execution: IN WITNESS WHEREOF

Stamp 10s.

FIRST SCHEDULE.
Part I.
Claims by the Outgoing Tenant.

(1) Grass seeds sown with the awaygoing white crop.
(2) Dung made after sowing the last or awaygoing green crop.
(3) The turnip crop [if at consuming value, state the fact].
(4) The awaygoing white crop [with straw unless it is steelbow].
(5) Grain drying plant and bins.
(6) Sheep stock.

Part II.
Further Claims by the Outgoing Tenant.

The following buildings, fixtures, fences, &c. [as the case may be], agreed to be taken over by the incoming tenant from the outgoing tenant [identify buildings], &c.

Part III.
Further Claims by the Outgoing Tenant.

Improvements to be valued on the same basis as under the Agricultural Holdings (Scotland) Act, 1949. Specify the different improvements in terms of the First Schedule to the Act of 1949, so far as applicable.

SECOND SCHEDULE.
Claims as by the Landlord agreed by him to be credited to the Incoming Tenant.

Claims for failure by the outgoing tenant to implement the conditions of the lease in respect of

[Dilapidations to farmhouse, steading, farm servants' cottages].

[Failure to leave the ditches and drains in tenantable order and not properly scoured and clear].

[Failure to leave the fences in reasonable repair].

[Failure to cultivate the farm and/or to leave the same conform to the rules of good husbandry, and, in particular, to leave the farm in the prescribed rotation and/or in a clean and in a fertile condition, the following fields particularly being left in a dirty condition with wrack, knot-grass, thistles, and other noxious weeds].

[Refer to clauses of the lease so far as relevant].

THIRD SCHEDULE.

Any other question or claim connected with the outgoing or ingoing of the parties respectively that may with the consent of the parties in writing be remitted to the arbiters in the course of the arbitration.

12.   Award in Common Law submission.

We, E.F. and G.H. [design], the arbiters appointed by deed of submission dated   , entered into by and between A.B. [design], the outgoing tenant of the farm of
in the parish of   and county of   at the term of   , and C.D. [design], the incoming tenant of the said farm, at the said term whereby they submitted and referred to the decree-arbitral to be pronounced by us as arbiters mutually chosen or, in the event of difference, by an oversman to be named by us before entering on the submission, which is held as incorporated herein and repeated *brevitatis causa*, the claims therein referred to and specified in the schedules thereto and hereto annexed; Having accepted the said submission (conform to acceptance dated   endorsed on the deed of submission), and having appointed an oversman (and having prorogated the submission conform to minute of prorogation dated
also endorsed on the submission); And having received and considered the respective claims of parties, together with the lease and all other documents produced, and inspected the farm, allowed and taken proof, and heard parties or their solicitors (issued proposed findings and considered the representations thereon) and now being well advised on the matters submitted to us, we do hereby give forth and pronounce our final sentence and decree-arbitral as follows, namely:—

(In the First Place) We award and determine that the said C.D. shall pay to the said A.B. the following sums: (*a*) the sum of £   in respect of the matters specified in Part I of the First Schedule hereto annexed, (*b*) the sum of £   in respect of the claims specified in Part II of the First Schedule hereto annexed, and (*c*) the sum of £   in respect of the claims specified in Part III of the First Schedule hereto annexed;

(In the Second Place) We award and determine that the said A.B. shall pay to the said C.D. the sum of £   in respect of the matters specified in the Second Schedule hereto, with interest on the said sums due respectively to the said A.B. and C.D. at the rate of 5 per cent per annum from the date of payment hereinafter mentioned until paid; And we hereby fix the [date] as the date for payment of the foregoing sums; and

(In the Third Place) We direct that the said A.B. and C.D. shall be jointly and severally liable for our own and the oversman's [and the clerk's] fees and expenses, with equal liability and mutual relief *inter se*, and that otherwise the parties shall pay their own expenses [or as may be directed in terms of the submission]; [And we decern and ordain both parties to implement and fulfil their respective parts of this decree-arbitral under the penalty of £   following any direction in the deed of submission]; [And we appoint the said deed of submission and this decree-arbitral to be recorded in the books of Council and Session, all in terms and to the effect of the consent to registration contained in the deed of submission]: IN WITNESS WHEREOF

Notes.

(1) The Schedules will repeat the Schedules to the submission, the amount awarded in respect of the claims in each Schedule, or part of a Schedule, being stated, or 'nil' as the case may be, and the total being shown as due by each party.

(2) Where, as is usual in the case of crops, any sums are awarded and paid on account, these will be stated and deducted from the sum finally awarded.

(3) For particulars which require to be stated in sheep stock valuations see Appendix IV for Part III of the Hill Farming Act, 1946.

13.   Minute of acceptance by Arbiters and nomination of Oversman and Clerk.

We, A.B. and C.D., both designed in the foregoing minute of reference, hereby accept office as arbiters and we appoint E.F. [design] to be oversman and G.H. [design] to be clerk [and legal assessor]: IN WITNESS WHEREOF

Note.—This Minute is usually endorsed on the deed of submission. No stamp is required.

14. Minute of acceptance by Oversman following thereon.

I, E.F., designed in the foregoing minute, hereby accept office as oversman: IN WITNESS WHEREOF

15. Minute of prorogation by the Parties.

We, the parties to the foregoing deed of submission hereby prorogate the same to the   day of
next to come and of new appoint the arbiters therein named and confer on them the whole powers therein mentioned, all in terms of the said deed of submission: And we agree and declare that the orders already pronounced by the said arbiters and the nomination of an oversman by them and the whole procedure which has already taken place under the submission shall remain effectual: IN WITNESS WHEREOF
Stamp 6d.

Note.—This Minute is usually endorsed on the deed of submission.

16. Minute of prorogation when submission has expired.

We, A.B. and C.D., the parties to the foregoing submission, considering that the arbiters therein named by minute dated   appointed E.F. [design] to be oversman and that the time for making an award has expired without any award having been made and that we have agreed to adopt and homologate the whole proceedings in the submission and to renew the submission to the said arbiters and to the said E.F. as oversman. Therefore we hereby adopt and homologate the whole proceedings in the submission and renew the submission in terms of the foregoing deed of submission to the said arbiters and to the said E.F. as oversman appointed by them whom failing before giving forth a final award to any other oversman whom the said arbiters may appoint and we hereby prorogate and extend the time within which they and he may determine the question thereby and hereby referred till the lapse of   from the last date of this minute: IN WITNESS WHEREOF
Stamp 10s.

17.   Minute of prorogation by Arbiters.

We, the arbiters appointed by the foregoing deed of submission, hereby prorogate the submission to [date]: IN WITNESS WHEREOF

Stamp nil.

Note.—Power of prorogation is generally conferred in deeds of submission. It is exercised by means of a minute of prorogation, which is usually endorsed on the submission. It does not require to be tested., but it is advisable that it should be.

E. ADDITIONAL FORMS.

1.   Removing of Tenant for non-
                            payment of rent (sec. 19).

SHERIFFDOM OF   AT

A.B. [design], Pursuer,
against
C.D. [design], Defender.

The pursuer craves the Court—
To ordain the defender to flit and remove himself, his family, servants, sub-tenants, cottars, and dependants, with their cattle, goods and gear, furth of and from the farm and lands and pertinents of   in the parish of   and county of   at the term of Martinmas [or Whitsunday] next, 19   , and to leave the said farm and others void and redd to the end that the pursuer or others in his name may enter thereto and peaceably possess and enjoy the same, and that under the pain of ejection; And further to ordain the defender to make payments to the pursuer of the sum of £   , with the interest thereof at the rate of 5 per cent. per annum from the   day of   , 19   , until payment, and with expenses.

Condescendence.

1. The pursuer is proprietor of the farm of
The defender is tenant thereof under lease by the pursuer in his favour, dated   , 19   , for the period of

years from and after the term of Martinmas, 19   , at the annual rent of £   , payable half-yearly by equal portions at the terms of Whitsunday and Martinmas.

2. The rent due by the defender for said farm is in arrear to the extent of £   , conform to statement herewith produced and referred to. The defender has been repeatedly applied to for payment of the said arrears, but he refuses or delays to pay the same, and the present action has therefore been rendered necessary.

3. This application, so far as it relates to the crave for a decree of removing, is brought under sec. 19 of the Agricultural Holdings (Scotland) Act, 1949.

Pleas-in-Law.

1. Six months' rent of the said farm of
being due and unpaid, the pursuer is entitled to decree of removing as craved.

2. The defender being due and resting-owing to the pursuer the sum sued for, the pursuer is entitled to decree therefor as craved, with expenses.

Note.—Under sec. 19 of the Act a tenant whose rent is six months in arrear, and who cannot find caution for the arrears and one year's further rent, may be removed at the next ensuing term of Whitsunday or Martinmas. No notice of the intention to raise the action is necessary. The tenant has the rights of an ordinary waygoing tenant as at the term of removal.

2.   Claim by outgoing Tenant against incoming Tenant.

Place..............................
Date..............................

To E.F. [design, incoming tenant].

AGRICULTURAL HOLDINGS (SCOTLAND) ACT, 1949.
HOLDING OF..........................

I, C.D., outgoing tenant of the above holding, having seen a letter dated   addressed to you by A.B. [design] the landlord of the said farm consenting to the payment by you of compensation due to me for improvements under the Agricultural Holdings (Scotland) Act as the same shall be ascertained, failing agreement, by arbitration, but limited to the sum of £   , and having accepted as I hereby accept your obligation to make payment of the said compensation, now claim to be paid on quitting said farm at the term of Whitsunday, 19   , the following, viz.:—

The sum of £   being the value of the improvements effected by me on said farm by the application thereto of purchased artificial or other purchased manures all as detailed in the Schedule annexed [or otherwise as the case may be].

(Signed) A.B. [outgoing tenant].

3. Abbreviated Form of Lease incorporating provisions required by the Agricultural Holdings (Scotland) Act, 1949.

LEASE
BETWEEN

A.B. [design], Landlord of the Farm of
in the Parish of   and County of
(hereinafter called 'the landlord')

AND

C.D. [design], Tenant of the said farm (hereinafter called 'the tenant').

1. The landlord lets to the tenant excluding all assignees and sub-tenants, legal or voluntary, the farm of   in the Parish of   and County of   , extending to   acres or thereby, which measurement is not guaranteed, all as previously occupied by E.F. [design former tenant] and all as delineated and coloured red on the plan annexed and signed as relative hereto.

2. The lease will be for a period of fifteen years from and after Whitsunday (28th May, 19   ,) and separation of crop in that year, with mutual breaks at Whitsunday and Separation of Crop in the years 19   and 19   .

3. The rent will be £   *per annum*, payable at Whitsunday (15th May) and Martinmas (11th November) in each year, commencing the payment of the first half-year's rent at the term of Martinmas, 19   , and thereafter at Whitsunday and Martinmas

each year, with 5 per cent. interest on each term's payment until paid.

4. There are reserved to the landlord (*a*) all shootings and fishings on the subjects let, with the exclusive right of taking and killing game, subject to the tenant's right under The Ground Game (Scotland) Acts, (*b*) all woods, timber and plantations, (*c*) all minerals, sand, gravel and clay with right to work and remove the same, subject to payment of surface damages and an appropriate adjustment of rent as the same may be fixed by arbitration, (*d*) power to alter marches and excamb land with neighbouring proprietors, (*e*) power to resume at any time on giving one month's notice any part or parts of the farm [not exceeding   acres in any one year] for any purpose (other than agricultural or pastoral), including, without prejudice to the foresaid generality, fencing, planting, the erection of houses or other buildings, working mineral quarries or sandpits, making ditches or drains, &c., (*f*) all existing wayleaves with power to grant further wayleaves, subject to payment for surface damages, (*g*) all common roads and means of access to other parts of the estate, (*h*) right to terminate the lease forthwith if the tenant becomes bankrupt or is sequestrated, grants a trust deed for behoof of his creditors or allows one-half year's rent to remain unpaid when the next half-year's rent falls due [or fails to implement other specific obligations].

5. The tenant shall be bound always to reside on the farm and to keep the same fully stocked and equipped with his own *bona fide* property. He shall cultivate and manage the farm according to the rules of good husbandry, shall not break up any permanent pasture (being the fields Nos.   of the Ordnance Survey) and shall not add to or alter any buildings or the fixed equipment without the written authority of the landlord. At the termination of the lease, whether at its natural expiry or at a break, he shall leave the arable land in the following rotation, viz.:—[specify].

6. The landlord undertakes that, at the commencement of the tenancy or as soon as is reasonably possible thereafter, he will put the fixed equipment on the holding, as defined in sec. 93 of the Agricultural Holdings (Scotland) Act, 1949, into a thorough state of repair and will provide such buildings and other fixed equipment as will enable the tenant (assuming he is reasonably skilled in husbandry) to maintain efficient production as respects both the

kind of produce in use to be produced on the holding and the quantity thereof, and the landlord will further, during the tenancy, effect such replacement or renewal of the buildings or other fixed equipment as may be rendered necessary by natural decay or by fair wear and tear; [the tenant agrees that the landlord's undertaking to put the fixed equipment on the holding in a thorough state of repair at the commencement of the tenancy has been duly implemented;] the liability of the tenant in relation to the maintenance of fixed equipment on the holding shall extend only to a liability to maintain the same on the holding in as good a state of repair (natural decay and fair wear and tear excepted) as it was in immediately after it was put in repair as aforesaid or, in the case of equipment provided, improved, replaced or renewed during the tenancy, as it was in immediately after it was so provided, improved, replaced or renewed. The march fences shall be the sole responsibility of the landlord [or otherwise as the case may be]. The tenant shall keep all hedges properly trimmed and cut and in good order, and shall paint all iron work at least every four year, during the currency of the lease. The tenant shall, free of charges carry out all cartages required in connection with the repair, replacement or renewal of fixed equipment.

7. The landlord and tenant agree that a record of the condition of the fixed equipment [or a record of the fixed equipment on and of the cultivation of the holding] shall be made forthwith in compliance with the provisions of sec. 5 of the Agricultural Holdings (Scotland) Act, 1949.

8. The landlord undertakes and binds himself and his executors and representatives (1) in the event of damage by fire to any building comprised in the holding to reinstate or replace the building if its reinstatement or replacement is required for the fulfilment of his responsibilities to manage the holding in accordance with the rules of good estate management and (2) to insure to their full value all such buildings against damage by fire.

9. The tenant undertakes and binds himself and his heirs, executors and representatives (1) in the event of the destruction by fire of harvested crops grown on the holding for consumption thereon, to return to the holding the full equivalent manurial value of the crops destroyed, in so far as the return thereof is required for the fulfilment of his responsibilities to farm in accordance with the rules of good husbandry; and (2) to insure to their value all

dead stock on the holding and all such harvested crops as aforesaid against damage by fire.

10. The tenant shall take over the waygoing tenant's white crop at mutual valuation and shall pay to the outgoing tenant the amount of his claim for improvements under Part III of the First Schedule of the Agricultural Holdings (Scotland) Act, 1949, limited to a maximum sum of £   . At his waygoing the tenant shall make over to the landlord or the incoming tenant the whole of the waygoing white crop at mutual valuation [or specify the basis of valuation].

11. Both parties consent to the registration of this lease for preservation and execution: IN WITNESS WHEREOF

4. Supplementary Agreement in terms of Section 5 (3).

AGREEMENT
BETWEEN
A.B. [design], Landlord of the holding of   in the Parish of   and County of
(hereinafter called 'the Landlord')
AND
C.D. [design], Tenant of the said holding (hereinafter called 'the Tenant').

WHEREAS by Lease dated   the Landlord let to the Tenant the holding of   in the Parish of   and County of   and the parties have agreed that their obligations under said Lease be varied in manner underwritten THEREFORE they hereby agree as follows, viz.:—

1. Notwithstanding the terms of said Lease the Tenant hereby accepts the fixed equipment on the holding as being in a thorough state of repair and sufficient in all respects to enable him to maintain efficient production and further the Tenant undertakes that he will during the tenancy effect at his own expense on behalf of the Landlord such replacement or renewal of the buildings or other fixed equipment on the holding as may be rendered necessary by natural decay or by fair wear and tear.

2. Except in so far as varied hereby, the parties confirm the terms of the said Lease: IN WITNESS WHEREOF
Stamp 6d.

Note.—An agreement of this kind may be entered into immediately after the Lease has been signed. See *Secretary of State for Scotland* v. *Sinclair*, 1960 S.L.C.R., 10.

APPENDIX II

STATUTORY INSTRUMENTS
1960 No. 1337 (S. 69.)

LANDLORD AND TENANT

AGRICULTURAL HOLDINGS, SCOTLAND

THE AGRICULTURAL HOLDINGS (SPECIFICATION OF FORMS) (SCOTLAND) INSTRUMENT, 1960

Made - - - - *27th July,* 1960
Coming into Operation - 1*st August,* 1960

The Secretary of State, in exercise of the powers conferred on him by paragraphs 10 and 23 of the Sixth Schedule to the Agricultural Holdings (Scotland) Act, 1949(a), and after consultation with the Council on Tribunals as provided in section 8 of the Tribunals and Inquiries Act, 1958(b), hereby makes the following instrument:—

1.—(1) This instrument may be cited as the Agricultural Holdings (Specification of Forms) (Scotland) Instrument, 1960, and shall come into operation on the   day of   , 1960.

(2) In this instrument 'the Act' means the Agricultural Holdings (Scotland) Act, 1949, and other expressions have the same meaning as in the Act.

(3) The Interpretation Act, 1889(c), applies to the interpretation of this instrument as it applies to an Act of Parliament.

2.—The form specified in the First Schedule to this instrument shall, modified as circumstances may require, be the form of an award in an arbitration under the Act.

3. The form specified in the Second Schedule to this instrument may be used for proceedings in arbitrations under the Act as follows:—

(a) for the making of an application for appointment by the Secretary of State of an arbiter to determine claims, questions or differences (except as to determination of rent) arising between the landlord and tenant of an agricultural holding—Form A;

(b) for the making of an application for appointment by the Secretary of State of an arbiter to determine the rent of an agricultural holding—Form B;

(c) for the making of an application to the Secretary of State by an arbiter for extension of time for making his award in an arbitration—Form C.

4.—The Agricultural Holdings (Scotland) Rules of 1909(d), the Agricultural Holdings (Scotland) Forms of 1912, the Agricultural Holdings (Scotland) Forms of 1926 and the Agricultural Holdings (Scotland) Forms of 1934 are, to the extent to which all or any of them are in operation immediately before the date at which this instrument comes into operation, hereby revoked.

Given under the seal of the Secretary of State for Scotland this Twenty-seventh day of July, 1960.

M. CAMPBELL,
Secretary.

Department of Agriculture and Fisheries for Scotland,
St. Andrew's House,
Edinburgh, 1.

(a) 12 & 13 Geo. 6. c. 75.   (b) 6 & 7 Eliz. 2. c. 66.   (c) 52 & 53 Vict. c. 63.
(d) S.R. & O. 1912/581 (Rev. I, p. 115: 1912, p. 262).

FIRST SCHEDULE
FORM OF AWARD
AGRICULTURAL HOLDINGS (SCOTLAND) ACT, 1949

Award in Arbitration between A. B. (name and address), the [outgoing] tenant, and C. D. (name and address), the landlord, with regard to the holding known as (insert name of holding, parish and county), [lately] in the occupation of the said tenant.

Whereas under the Agricultural Holdings (Scotland) Act, 1949, the claims, questions or differences set forth in the Schedule to this Award are referred to arbitration in accordance with the provisions set out in the Sixth Schedule to the said Act:

And whereas by appointment dated the   day of
19 , signed by (on behalf of) the said tenant and landlord [or, as the case may be—given under the seal of the Secretary of State], I (insert name and address), was duly appointed under the said Act to be the arbiter for the purpose of

1 ⎰ settling the said claims
  ⎨ settling the said questions or differences
  ⎩ determining the rent to be paid in respect of the said holding as from[2]

in accordance with the provisions set out in the Sixth Schedule to the said Act:

[And whereas the time for making my Award has been extended by
1 ⎰ the written agreement of the said tenant and landlord, dated
  ⎨ the   day of   19 ,
  ⎩ order of the Secretary of State, dated the   day of   19 ,
to the   day of   19 .]

And whereas I, the said (insert name)   , having accepted the appointment as arbiter, and having heard the parties (agent for the parties) and examined the documents and other productions lodged and the evidence led and having fully considered the whole matters referred to me, do hereby make my final Award as follows:—[3]

I award and determine that the said landlord shall pay to the said tenant the sum of   pounds   shillings and   pence, as compensation in respect of the claims set forth in the [first part of the] Schedule to this Award, the amount awarded in respect of each claim being as there stated.

I award and determine that the said tenant shall pay to the said landlord the sum of   pounds   shillings and   pence, in respect of the claims set forth in the [second part of the] Schedule to this Award, the amount awarded in respect of each claim being as there stated.

I determine the questions or differences set forth in the [third part of the] Schedule to this Award, as follows, namely:—

I fix and determine the rent to be paid by the said tenant to the said landlord, as from[2]   to be the sum of   per annum.

I award and direct that each party shall bear his own expenses and one half of the other expenses of and incidental to the arbitration and Award, including my remuneration [and that of the clerk]
(or otherwise as the arbiter may see fit to direct in light of the provisions of section 76(4) of, and paragraphs 16-18 of the Sixth Schedule to, the 1949 Act)
and that, subject to the provisions of the Agricultural Holdings (Scotland) Act, 1949, all sums, including any expenses, payable under or by virtue of this Award shall be so paid not later than[4]

In witness whereof I have signed this Award this day of   19   , in the presence of the following witnesses.

Signature..............................

Designation...........................

Address.................................... (Arbiter)

Signature..............................

Designation...........................

Address...............................

Schedule to the above Award

Claims,[5] questions or differences to be determined.

Part I—
  Claims made by the tenant.

Part II—
  Claims made by the landlord.

Part III—
  Questions or differences (including questions of rent).

In the case of appointment under the seal of the Secretary of State, the arbiter must, if either party so requests, state the reasons for any determination arrived at.

[1] Adapt to meet the circumstances.

[2] Insert date from which revised rent is to run. (Where variation of rent under section 7 of the 1949 Act is concerned, the date will be the next ensuing day on which the tenancy could have been terminated by notice to quit given at the date of demanding the reference of the rent question to arbitration—usually a term of Whitsunday or Martinmas.)

[3] Such of the following four paragraphs as may be appropriate should be incorporated in the award, adaptations to meet the particular circumstances being made as necessary.

[4] The date of payment specified must not be later than one calendar month after the delivery of the Award.

[5] Where claims are made under the First, Second, Third or Fourth Schedules to the Act, the amounts awarded must, if either party so requires, be shown separately against each numbered item as set out in those Schedules. Where claims are made by either party under agreement or custom and not under statute, the amounts awarded must be separately stated.

SECOND SCHEDULE

FORM A

(Application for appointment by the Secretary of State of an arbiter to determine claims, questions or differences (except as to determination of rent) arising between the landlord and tenant of an agricultural holding.)

AGRICULTURAL HOLDINGS (SCOTLAND) ACT, 1949

To the Secretary of State,

In default of agreement between the landlord and the tenant of the holding specified in the Schedule to this application as to the person to act as arbiter and in the absence of any provision in any lease or agreement between them relating to the appointment of an arbiter, I/we hereby apply to the Secretary of State to appoint an arbiter for the purpose of settling the claims, questions or differences set out in the Schedule to this application.

Signature....................
[1]....................
Date....................

SCHEDULE

Particulars required	Replies
1. Name and address of holding.	Holding: Parish: County:
2. Name and address of landlord.	
3. Name and address of landlord's agent.[2]	
4. Name and address of tenant.	
5. Name and address of tenant's agent.[2]	
6. If the tenancy has terminated state date of termination.	
7. If an extension of time has been granted under section 68(3) of the Agricultural Holdings (Scotland) Act, 1949, for the settlement of claims, state date on which extension expires.	
8. Approximate acreage of holding.	
9. Description of holding.[3]	
10. Nature of claim to be referred to arbitration. (a) Claims for compensation for improvements by the tenant, and short particulars of any further claims by the tenant.	

(b) Short particulars of any claims by the landlord.

11. Questions or differences to be referred to arbitration.

[1] State whether landlord or tenant. If an agent signs, state on whose behalf he is signing. The appointment will be expedited if the application is made by *both* parties.
[2] If no agent, insert 'None' in second column.
[3] Describe holding briefly, e.g., mixed, arable, dairying, market garden.

Copies of this form may be obtained from the Department of Agriculture and Fisheries for Scotland, St. Andrew's House, Edinburgh, 1.

Form B

(Application for appointment by the Secretary of State of an arbiter to determine the rent of an agricultural holding.)

Agricultural Holdings (Scotland) Act, 1949

To the Secretary of State,

In default of agreement between the landlord and the tenant of the holding specified in the Schedule to this application as to the person to act as arbiter and in the absence of any provision in any lease or agreement between them relating to the appointment of an arbiter, I/we hereby apply to the Secretary of State to appoint an arbiter to determine the rent to be paid for the said holding as from[1]

Signature....................
[2]....................
Date....................

Schedule

Particulars required	Replies
1. Name and address of holding.	Holding: Parish: County:
2. Name and address of landlord.	
3. Name and address of landlord's agent.[3]	
4. Name and address of tenant.	
5. Name and address of tenant's agent.[3]	

6. Approximate acreage of holding.

7. Description of holding.[4]

8. Date of demand in writing for reference to arbitration.

9. Date at which tenancy of holding could be terminated by notice to quit.

10. (a) Date of commencement of tenancy.
    (b) Effective date of any previous increase or reduction of rent.
    (c) Effective date of any previous direction of an arbiter that the rent continue unchanged.

[1] Where variation of rent under section 7 of the 1949 Act is concerned, the date will be the next ensuing day on which the tenancy could have been terminated by notice to quit given at the date of demanding the reference of the rent question to arbitration—usually a term of Whitsunday or Martinmas.

[2] State whether landlord or tenant. If an agent signs, state on whose behalf he is signing. The appointment will be expedited if the application is made by *both* parties.

[3] If no agent, insert 'None' in second column.

[4] Describe holding briefly, e.g., mixed, arable, dairying, market garden.

Copies of this form may be obtained from the Department of Agriculture and Fisheries for Scotland, St. Andrew's House, Edinburgh, 1.

Form C

(Application to the Secretary of State by an arbiter for extension of time for making his award in an arbitration.)

Agricultural Holdings (Scotland) Act, 1949

To the Secretary of State,

As the time for making the Award in the arbitration detailed below will expire/expired on the   day of   19   , I hereby apply for an extension of the time for making the said Award to the   day of   19   .

(Signature of arbiter
or arbiter's clerk).....................
Date.....................

Details to be supplied:—

1. Name of holding and parish and county in which situated.

2. Name and address of landlord (and agent, if any).
3. Name and address of tenant (and agent, if any).
4. Name and address of arbiter (and clerk, if any).
5. (*a*) Date on which arbiter appointed.
   (*b*) Whether appointed by agreement of parties or by the Secretary of State.

EXPLANATORY NOTE

(This Note is not part of the instrument but is intended to indicate its general purport.)

The forms specified in this instrument take the place of the forms specified in the Agricultural Holdings (Scotland) Forms of 1934 (AG.100) made by the Department of Agriculture for Scotland on 17th December, 1934, in exercise of the powers vested in them by the Agricultural Holdings (Scotland) Acts, 1923 and 1931.

APPENDIX III

SHEEP STOCKS VALUATION (SCOTLAND) ACT, 1937

AN ACT TO AMEND THE LAW WITH RESPECT TO VALUATIONS OF SHEEP STOCK IN SCOTLAND

10th June, 1937

Be it enacted by the King's most Excellent Majesty, by and with the advice and consent of the Lords Spiritual and Temporal, and Commons, in this present Parliament assembled, and by the authority of the same, as follows:—

1.—(1) Where in pursuance of any lease of an agricultural holding whether entered into before or after the passing of this Act the tenant is required at the termination of the tenancy to leave the stock of sheep on the holding to be taken over by the landlord or the incoming tenant at a price or valuation to be fixed by arbitration, the arbiter shall, in his award, show the basis of valuation[1] of each class of stock and state separately any amounts included in respect of acclimatisation or hefting or of any other consideration or factor for which he has made special allowance.

(2) Where an arbiter fails to comply with any requirements of the foregoing sub-section, his award may be set aside by the sheriff.[1]

2.—(1) In any arbitration in pursuance of a lease entered into after the passing of this Act[2] as to the price or value of sheep stock to be taken over at the termination of the tenancy by the landlord or incoming tenant, the arbiter may, at any stage of the proceedings, and shall, if so directed by the Sheriff (which direction may be given on the application of either party), submit, in the form of a stated case for the decision of the sheriff, any question of law arising in the course of the arbitration.

(2) The decision of the sheriff on any question submitted in pursuance of the foregoing subsection shall be final unless within such time, and in accordance with such conditions, as may be

prescribed by Act of Sederunt, either party appeals to the Court of Session, from whose decision no appeal shall lie.

(3) Where any question is submitted in pursuance of subsection (1) of this section for the decision of the sheriff, and the arbiter is satisfied that, whatever the decision on the question may be, the sum ultimately to be found due will be not less than a particular amount, it shall be lawful for the arbiter, pending the decision of such question, to make an order directing payment to the outgoing tenant of such sum, not exceeding that amount, as the arbiter may think fit, to account of the sum that may ultimately be awarded.

3.—(1) Any question or difference as to the price or value of sheep stock required in terms of any lease (whether entered into before or after the passing of this Act) to be taken over at the termination of the tenancy by the landlord or incoming tenant may, if both parties agree, in lieu of being determined in the manner provided in the lease, be determined by the Land Court, and the Land Court shall, on the joint application of the parties, determine such question or difference accordingly.

(2) The provisions of the Small Landholders (Scotland) Acts, 1886 to 1931, with regard to the Land Court shall, with any necessary modifications, apply for the purpose of the determination of any such question or difference as aforesaid in like manner as those provisions apply for the purpose of the determination by the Land Court of matters referred to them under those Acts.

4.—In this Act, unless the context otherwise requires—
    the expression 'agricultural holding' means a piece of land held by a tenant which is wholly or in part pastoral, and which is not let to the tenant during his continuance in any office, appointment, or employment held under the landlord;

    the expression 'arbiter' includes an oversman and any person required to determine the value or price of sheep stock in pursuance of any provision in the lease of an agricultural holding, and the expression 'arbitration' shall be construed accordingly;

the expressions 'lease', 'landlord', and 'tenant' have the like meaning as in the Agricultural Holdings (Scotland) Act, 1923.

5.—(1) This Act may be cited as the Sheep Stocks Valuation (Scotland) Act, 1937.

(2) This Act shall extend to Scotland only.

[1] Failure to show basis of valuation is a good ground for reduction of award (*Dunlop* v. *Mundell,* 1943 S.L.T. 286). See also *Paynter* v. *Rutherford and ors.,* 1940 S.L.T. (Sh.Ct.) 18.

[2] Held that sec. 2(1) did not apply to arbitration in pursuance of leases entered into before the passing of the Act (*Pott's Jud. Factor* v. *Glendinning and ors.,* 1949 S.L.T. 190).

APPENDIX IV

EXCERPTS FROM HILL FARMING ACT, 1946

RELATING TO SHEEP STOCK VALUATIONS

Valuation of Sheep Stocks (Scotland)

Rules as to valuation of sheep stocks.

28.—(1) In any arbitration in pursuance of any lease of an agricultural holding in Scotland entered into after the commencement of this Act[1] as to the value of sheep stock to be taken over at the termination of the tenancy by the landlord or the incoming tenant, the arbiter shall fix the value of the sheep stock in accordance, in the case of a valuation made in respect of a tenancy terminating at Whitsunday in any year, with the provisions of Part I of the Second Schedule to this Act, or, in the case of a valuation made in respect of a tenancy terminating at Martinmas in any year, with the provisions of Part II of the said Schedule.

1 Edw. 8 & 1 Geo. 6. c. 34.

(2) Sub-section (1) of section one of the Sheep Stocks Valuation (Scotland) Act, 1937 (which requires certain particulars to be given in an arbiter's award) shall, in relation to an arbitration to which sub-section (1) of this section applies, have effect as if for the words from 'show the basis' to the end of the sub-section there were substituted the words 'state separately the particulars set forth in Part III of the Second Schedule to the Hill Farming Act, 1946'.

Valuation by Land Court of sheep stocks.
1 Edw. 8 & 1 Geo. 6. c. 34.

29.—(1) Section three of the Sheep Stocks Valuation (Scotland) Act, 1937 (which relates to the determination by the Land Court of questions as to the value of sheep stocks) shall, in relation to any question or difference as to the value of sheep stock required in terms of a lease entered into after the commencement of this Act to be taken over at the termination of the tenancy by the landlord or the incoming tenant, have effect as if for the words 'may, if both parties agree' and the words 'on the joint application of the parties' there were **substituted** respectively the words 'shall, if either party so desires' and the words 'on the application of that party'.

[1] *Chapman* v. *Lockhart*, 1950, C.L.Y. 4507/8.

(2) The Land Court shall determine any question or difference which they are required to determine under the said section three as amended by the last foregoing sub-section in accordance with the appropriate provisions of the Second Schedule to this Act.

30. Where any question as to the value of any sheep stock has been submitted for determination to the Land Court or to an arbiter, the outgoing tenant shall, not less than twenty-eight days before the determination of the question, submit to the Court or to the arbiter, as the case may be, a statement of the sales of sheep from such stock during the preceding three years in the case of a valuation made in respect of a tenancy terminating at Whitsunday, or during the current year and in each of the two preceding years in the case of a valuation made in respect of a tenancy terminating at Martinmas. The outgoing tenant shall also submit such sale-notes and other evidence as may be required by the Court or the arbiter to vouch the accuracy of such statement. *Production of documents for purposes of valuation of sheep stocks.*

Any document submitted by the outgoing tenant in pursuance of this section shall be open to inspection by the other party to the valuation proceedings.

31. The three last preceding sections and the Second Schedule to this Act shall be construed as one with the Sheep Stocks Valuation (Scotland) Act, 1937, and may be cited with that Act as the Sheep Stocks Valuation (Scotland) Acts, 1937 and 1946. *Construction and citation of ss. 27 to 29*

39.—(1) This Act shall, in its application to Scotland, have effect subject to the following modifications:— *Provisions as to Scotland.*

   (a) in sub-section (2) of section three there shall be inserted after the words 'having an interest' the words 'as proprietor or as tenant';
   (b) any question which under section five is to be determined by an official arbitrator shall be determined by the Land Court;
   (c) in section nine for a reference to any provision of the Agricultural Holdings Act, 1923, there shall be substituted a reference to the corresponding provision of the Agricultural Holdings (Scotland) Act, 1923; *13 & 14 Geo. 5. c. 10.*

K

(d) in section fourteen for sub-sections (1) and (2) there shall be substituted the following sub-section—

'(1) Subject to the provisions of the next succeeding section, subsidy payments falling to be made, in accordance with a hill sheep scheme, in respect of sheep comprised in a flock of any of the relevant days shall be made to the person maintaining the flock on that day; and subsidy payments falling to be made, in accordance with a hill cattle scheme, in respect of cattle grazed on any land shall be paid to the person who, at the beginning of such day as may be specified in the scheme, is the occupier of the land:

Provided that, in the case of sheep or cattle belonging to a landholder within the meaning of the Small Landholders (Scotland) Acts, 1886 to 1931, and grazed on land which is a common pasture or grazing, provision may be made by a hill sheep scheme or a hill cattle scheme for the making of the subsidy payments to the clerk of the committee appointed under those Acts for the management of such common pasture or grazing.'

(e) in section thirty-five the word 'summarily' and the words from 'and a complaint' to the end of the section shall be omitted;

(f) unless the context otherwise requires, the following expressions shall have the meanings hereby assigned to them respectively, that is to say—

'lease' in relation to a common pasture or grazing includes regulations made or approved by the Land Court under the Small Landholders (Scotland) Acts, 1886 to 1931;

'making muirburn' includes setting fire to or burning any heath or muir; and

'tenant' means a tenant for agricultural or pastoral purposes, and, in the case of a common pasture or grazing, includes the committee appointed under the Small Landholders (Scotland) Acts, 1886 to 1931.

HILL FARMING ACT, 1946

(2) The provisions of the Small Landholders (Scotland) Acts, 1886 to 1931, with regard to the Land Court shall, with any necessary modifications, apply for the purpose of the determination of any matter which they are required by or under this Act to determine, in like manner as those provisions apply for the purpose of the determination by the Land Court of matters referred to them under those Acts.

SECOND SCHEDULE

Section 28.

PROVISIONS AS TO VALUATION OF SHEEP STOCKS IN SCOTLAND

PART I

Provisions as to a valuation made in respect of a tenancy terminating at Whitsunday

1. The Land Court or the arbiter (in Part I and Part II of this Schedule referred to as 'the valuer') shall ascertain the number of, and the prices realised for, the ewes and the lambs sold off the hill from the stock under valuation at the autumn sales in each of the three preceding years, and shall determine by inspection the number of shotts present in the stock at the time of the valuation.

2. The valuer shall calculate an average price per ewe, and an average price per lamb, for the ewes and lambs sold as aforesaid for each of the three preceding years. In calculating the average price for any year the valuer shall disregard such number of ewes or lambs so sold in that year, being the ewes or lambs sold at the lowest prices, as bears the same proportion to the total number of ewes or lambs so sold in that year as the number of shotts as determined bears to the total number of ewes or lambs in the stock under valuation.

3. The valuer shall then ascertain the mean of the average prices so calculated for the three preceding years for ewes and for lambs respectively. The figures so ascertained or ascertained, in a case to which the next succeeding paragraph applies, in accordance with that paragraph, are in this Part of this Schedule referred to as the 'three year average price for ewes' and the 'three year average price for lambs'.

4. In the case of any sheep stock in which the number of ewes or the number of lambs sold off the hill at the autumn sales during the preceding three years has been less than half the total number of ewes or of lambs sold, the three-year average price for ewes or the three-year average price for lambs, as the case may be shall, in lieu of being ascertained by the valuer as aforesaid, be determined by the Land Court on the application of the parties; and the Land Court shall determine such prices by reference to the prices realised at such sales for ewes and for lambs respectively from similar stocks kept in the same district and under similar conditions.

5. The three-year average price for ewes shall be subject to adjustment by the valuer within the limits of ten shillings [twenty per cent][1] upwards or downwards as he may think proper having regard to the general condition of the stock under valuation and to the profit which the purchaser may reasonably expect it to earn. The resultant figure shall be the basis of the valuation of the ewes, and is in this Part of this Schedule referred to as the 'basic ewe value'.

The valuer shall similarly adjust the three-year average price for lambs, and the resultant figure shall be the basis for the valuation of the lambs and is in this Part of this Schedule referred to as the 'basic lamb value'.

6. In making his award the valuer shall value the respective classes of stock in accordance with the following rules, that is to say—

    (a) ewes of all ages (including gimmers) shall be valued at the basic ewe value with the addition of fifteen shillings per head [thirty per cent of such value];[1]

    (b) lambs shall be valued at the basic lamb value; so, however, that twin lambs shall be valued at such price as the valuer thinks proper;

    (c) ewe hoggs shall be valued at two-thirds of the combined basic values of a ewe and a lamb subject to adjustment by the valuer within the limits of five shillings [ten per cent][1] per head upwards or downwards as he may think proper, having regard to their quality and condition;

    (d) tups shall be valued at such price as in the opinion of the valuer represents their value on the farm having regard to acclimatisation or any other factor for which he thinks it proper to make allowance;

(e) eild sheep shall be valued at the value put upon the ewes subject to such adjustment as the valuer may think proper having regard to their quality and condition; and

(f) shotts shall be valued at such value not exceeding two-thirds of the value put upon good sheep of the like age and class on the farm as the valuer may think proper.

Part II

Provisions as to a valuation made in respect of a tenancy terminating at Martinmas

1. The valuer shall ascertain the number of, and the prices realised for, the ewes sold off the hill from the stock under valuation at the autumn sales in the current year and in each of the two preceding years, and shall calculate an average price per ewe so sold for each of the said years. In calculating the average price for any year the valuer shall disregard one-tenth of the total number of ewes so sold in that year, being the ewes sold at the lowest prices.

2. The mean of the average prices so calculated shall be subject to adjustment by the valuer within the limits of five shillings [ten per cent][1] upwards or downwards as he may think proper having regard to the general condition of the stock under valuation and to the profit which the purchaser may reasonably expect it to earn. The resultant figure shall be the basis of the valuation of the ewes and is in this Part of this Schedule referred to as the 'basic ewe value'.

3. In making his award the valuer shall assess the respective classes of stock in accordance with the following rules, that is to say—

(a) ewes of all ages (including gimmers) shall be valued at the basic ewe value with the addition of fifteen shillings per head [thirty per cent of such value];[1]

(b) ewe lambs shall be valued at the basic ewe value subject to adjustment by the valuer within the limits of five shillings [ten per cent][1] per head upwards or downwards as he may think proper having regard to their quality and condition; and

(c) tups shall be valued at such price as in the opinion of the valuer represents their value on the farm having regard to acclimatisation or any other factor for which he thinks it proper to make allowance.

Part III

Particulars required to be shown in an arbiter's award

1. The three-year average price for ewes and the three-year average price for lambs ascertained under Part I, or the mean of the average prices calculated under Part II, of this Schedule, as the case may be.
2. Any amount added or taken away by way of adjustment for the purpose of fixing the basic ewe value or the basic lamb value, and the grounds on which such adjustment was made.
3. The number of each class of stock valued (ewes and gimmers of all ages with lambs being taken as one class, and eild ewes and eild gimmers being taken as separate classes at a Whitsunday valuation, and ewes and gimmers of all ages being taken as one class at a Martinmas valuation) and the value placed on each class.
4. Any amount added or taken away by way of adjustment in fixing the value of ewe hoggs at a Whitsunday valuation, or the value of ewe lambs at a Martinmas valuation, and the grounds on which such adjustment was made.

Part IV

Interpretation

In this Schedule the expressions 'ewe', 'gimmer', 'eild ewe', 'eild gimmer', 'lamb', 'ewe hogg', 'shott', 'eild sheep' and 'tup' shall be construed as meaning respectively sheep of the classes customarily known by those designations in the locality in which the flock under valuation is maintained.

[1] In the case of Leases entered into on or after 15th May, 1963, the percentage adjustments shown in brackets are substituted for the adjustments of fixed amount: Agriculture (Miscellaneous Provisions) Act, 1963, Sec. 21.
[2] There is no equivalent in Part II to paragraph 4 of Part I and accordingly it is not competent even when most of the ewes are sold privately to make a

valuation by reference to similar stocks in the district (*Macpherson* v. *Secretary of State for Scotland*, 1967, S.L.T. (Land Ct.) 9). It has been observed by the Land Court that it is probable that in course of time the second Schedule will become inoperable because stock is being sold privately and not by auction (*Secretary of State for Scotland* v. *Anderson*, 1967, *S.L.C.R. App.* 117).

APPENDIX V

THE AGRICULTURAL RECORDS (SCOTLAND) REGULATIONS, 1948

Made - - - *22nd December,* 1948
Laid before Parliament *29th December,* 1948
Coming into Operation *5th January,* 1949

In exercise of the powers conferred on me by section 37 of the Agricultural Holdings (Scotland) Act, 1923 (a), as amended by section 24 of the Agriculture (Scotland) Act, 1948 (b), and of all other powers enabling me in that behalf, I hereby make the following regulations:—

1.—(1) These regulations may be cited as the Agricultural Records (Scotland) Regulations, 1948, and shall come into operation on the fifth day of January, 1949.

(2) The Interpretation Act, 1889(c), shall apply to the interpretation of these regulations as it applies to the interpretation of an Act of Parliament.

2. A record of a holding or part thereof, or of the fixed equipment on a holding, made under the said section 37, amended as aforesaid, or under sub-section (1) of section 13 of the Agriculture (Scotland) Act, 1948, shall be in the form set out in the First Schedule hereto so far as applicable to such record, or in a form to the like effect, and shall be made in accordance with the Rules relative to the said form contained in the Second Schedule hereto.

Dated this 22nd day of December, 1948.

Arthur Woodburn,
One of His Majesty's Principal
Secretaries of State.

St. Andrew's House,
Edinburgh, 1.

(a) 13 & 14 GEO. 5. c. 10.   (b) 11 & 12 GEO. 6. c. 45.
(c) 52 & 53 VICT. c. 63.

FIRST SCHEDULE

Form of Record

NOTE.—Where the record is a record of

(a) the whole holding (which includes fixed equipment and cultivation), the full form of record so far as applicable will be used;

(b) part of the holding, the full form of record so far as applicable will be used. Particulars of the whole holding require to be furnished under Head I; the remaining Heads and the relative map and plan will refer only to the part of the holding specified;

(c) the fixed equipment only, Heads II, VIII and IX of the form do not apply, and Heads III and IV apply only in so far as these relate to fixed equipment.

In all cases where, in terms of section 37 of the Agricultural Holdings (Scotland) Act, 1923, the tenant requires that a record be made of any existing improvements executed by the tenant or for which the tenant with the consent in writing of his landlord has paid compensation to an outgoing tenant, or of any fixtures or buildings which under section 29 of the said Act the tenant is entitled to remove, such a record shall be made under the appropriate Head or Heads of this form.

Record

made under the Agricultural Holdings (Scotland) Acts,
1923 to 1948,
of

the holding (*or* part of the holding, *or* the fixed equipment on the holding, *as the case may be*) known as..........................
in the Parish of...............and County of.................
being a record made on..................19...., by...........
..........................nominated for that purpose under the said Acts on......................19..... A map of the ground and a plan of the buildings thereon are attached and signed as relative to the record.

I   *Particulars of tenancy*
    Landlord...
    Tenant...
    Rent...
    Entry..
    Duration...
    Expiry...
    Breaks...
    Whether continued on tacit relocation...................

II. *Area*   *Acres*   *Acres*
    1. Arable ... ...
    2. Permanent Pasture, as shown in
        lease ... ...
    3. Hill Land ... ...
    4. Rough Grazing (not included
        above) ... ...

   *5. Other areas:—
         Gardens ... ... ⎫
         Buildings ... ... ⎪
         Roads ... ... ⎪
         Water... ... ... ⎬ ....
         Shelter belts and woodlands ⎪
         Stackyards ... ... ⎪
         Folds ... ... ⎭

                        Total ... ...

III. *Compensation*
      Nature and amount of compensation paid by tenant to former tenant.

IV. *Consideration or allowances*
      Any consideration or allowances which have been made
          (i) by the landlord to the tenant;
          (ii) by the tenant to the landlord.

V. *Buildings*
          (i) General description.
          (ii) Detailed description.

VI. *Fences, Hedges, Dykes and Walls, Gates, Roads and Paths, Water Supplies, Water Courses, Ditches and Drains*
          (*a*) Fences
          (*b*) Hedges
          (*c*) Dykes and Walls
          (*d*) Gates
          (*e*) Roads and Paths
          (*f*) Water Supplies
          (*g*) Water Courses, Ditches and Drains

VII. *Other items of fixed equipment*

VIII. *Cultivation*

IX. *Any other items*

*Separate figures, if readily available, should be given for these items.

SECOND SCHEDULE
Rules

1. *Map and plan*
    (a) *Holding.*—The map of the arable ground and the land occupied by buildings shall be on a scale of not less than 25 inches to the mile. The map of the remainder of the holding shall, wherever practicable, be on the same scale but in the case of hill farms and extensive holdings a map on a smaller scale may be used. In all cases the relative Ordnance Survey sheet number and edition, parcel or field numbers, areas and orientation shall be shown.
    (b) *Buildings.*—The plan or plans of permanent buildings (farmhouse, steading buildings, outbuildings and cottages) shall be a single-line drawing or drawings on a scale of not less than 1/32nd inch to one foot, showing the outline and any internal main subdivisions of the buildings, and shall be suitably lettered in relation to the record.

2. *Buildings*
    (a) A general descriptive note of the whole range and location of buildings shall be given, followed by a detailed description of each building.
    (b) (i) Buildings shall be listed and described in the following order, adapted as necessary to suit the case:—
    　　(a) Farm dwelling-house and any outbuildings
    　　(b) Cottar houses and any outbuildings
    　　(c) Other houses and any outbuildings
    　　(d) Farm steading buildings and any outbuildings
    　　(e) Stells, fanks, folds, dippers, pens and bughts.
    　(ii) External and internal condition of buildings, equipment and fixtures, including electrical equipment and troughs, and condition of roofs and all roofing materials, including slates, nails and sarking, shall be detailed.
    　(iii) Details of all defects and deficiencies shall be shown.
    　(iv) Sewage disposal and drainage systems of houses and steading shall, so far as reasonably practicable, be detailed.

3. *Fences, Hedges, Dykes and Walls, Gates, Roads and Paths, Water Supplies, Water-Courses, Ditches and Drains*

    Detailed description of each item, related by number to the map, shall be given, under sub-headings, as follows:—
    (a) *Fences.*—Construction, material and condition and whether boundary or internal.
    (b) *Hedges.*—Species, height and condition of hedges and condition of any supplementary fencing.
    (c) *Dykes and Walls.*—Type, height and condition, including wire, if any.

(d) *Gates.*—Type, location and condition of all gates and gateposts.

(e) *Roads and Paths.*—Location, type of construction and condition, including condition of side drains, offlets, culverts and bridges.

(f) *Water Supplies.*—Supplies to houses, steading buildings, farm cottages and fields, detailing, with description of condition, sources of supplies, storage tanks, filters, piping, fittings and pumping installations.

(g) *Water-Courses, Ditches and Drains.*—Length, width, depth, running condition, condition of verges, sides and bottoms of ditches, and open drains; drain outlets; watering places, ponds, mill-lades, sluices, dams and, in the case of water-courses, condition of banks and field drain outlets.

All available information regarding the drainage system or systems of the holding shall be shown on the map.

4. *Other items of fixed equipment*

All items of 'fixed equipment', as defined in section 86(3) of the Agriculture (Scotland) Act, 1948, shall be included.

5. *Cultivation*

Cropping Schedule shall be detailed and shall show sequence of cropping for each field or part of field for the past four years. Condition of grass (including permanent pasture) and other crops shall be detailed with reference to appropriate numbering, field by field, on map. The nature and extent of infestation of crops and land by weeds shall be specified. Any evidence of lime and other mineral deficiency shall be noted. A note shall be added as to whether dung, straw, hay or turnips are sold off the holding. In the case of arable land, the presence of tree roots, boulders, stones, or other like obstacles to cultivation shall be noted.

6. *Any other items*

This shall include items not falling under the foregoing Heads such as:—

> hill land (heather burning, condition of grazing, etc., and approximate area and location of bracken, whin, etc.);
> gardens, including cottar house gardens (condition, size and location);
> game and
> injurious animals and birds, as defined in section 39(3) of the Agriculture (Scotland) Act, 1948.

EXPLANATORY NOTE

(*This Note is not part of the Regulations, but is intended to indicate their general purport*)

Under section 37 of the Agricultural Holdings (Scotland) Act,

1923, as amended by the Agriculture (Scotland) Act, 1948, records of holdings are to be made in the circumstances set out in the Acts. The regulations prescribe the form of records to be used.

In view of the consolidation of the provisions of the Agricultural Holdings (Scotland) Acts, 1923 to 1948, in the Agricultural Holdings (Scotland) Act, 1949, the references in the attached Regulations to sections of the 1923 and 1948 Acts should be read as references to the corresponding sections of the 1949 Act, as follows:—

Agricultural Holdings (Scotland Act, 1923		*Agricultural Holdings (Scotland) Act*, 1949
For Section 37	*read*	Section 17
,, Section 29		Section 14
Agriculture (Scotland) Act, 1948		
For Section 13	*read*	Section 5
,, Section 86(3)		Section 93

APPENDIX VI

EXCERPTS FROM THE AGRICULTURE ACT, 1958.
(6 & 7 ELIZ. 2. CH. 71.)

An Act to amend the Agriculture Act, 1947, the Agricultural Holdings Act, 1948, the Agriculture (Scotland) Act, 1948, and the Agricultural Holdings (Scotland) Act, 1949; to require the landlord of an agricultural holding in certain cases to provide, repair or alter fixed equipment on the holding; to amend Part II of the Landlord and Tenant Act, 1954, as to tenancies of agricultural land excluded therefrom; to amend the Schedule to the Corn Production Acts (Repeal) Act, 1921, and section twenty-one of the Hill Farming Act, 1946; and for purposes connected with the matters aforesaid.

[1st August, 1958]

Repeal of powers of supervision, direction and dispossession under Part II of Agriculture Act 1947, and Part II of Agriculture (Scotland) Act, 1948.

1.—(1) Not applicable to Scotland.

(2) So much of Part II of the Agriculture (Scotland) Act, 1948 (in this Act referred to as 'the Scottish Act of 1948') as provides for warning notices, and for the giving of directions to and the dispossession of owners or occupiers on grounds of bad estate management or bad husbandry, that is to say sections twenty-seven to thirty-four of that Act, shall cease to have effect.

Amendments as to fixing of rents of agricultural holdings.

2. . . . and in section seven of the Agricultural Holdings (Scotland) Act, 1949 (in this Act referred to as 'the Scottish Act of 1949'), the following paragraph shall be inserted at the end of sub-section (1) (which enables the landlord or tenant of an agricultural holding to demand a reference to arbitration of the question what rent should be payable in respect of the holding)—

'For the purposes of this sub-section the rent properly payable in respect of a holding shall be the rent at which, having regard to the terms of the tenancy (other than those relating to rent), the holding might reasonably be expected to be let in the open market by a willing landlord to a willing tenant, there being disregarded (in addition to the matters referred to in the next following sub-section) any effect on rent of the fact that the tenant who is a party to the arbitration is in occupation of the holding.'

3.¹—(1) There shall be transferred to the Agricultural Land Tribunal the functions conferred on the Minister of Agriculture Fisheries and Food (in this Act referred to as 'the Minister') by sections twenty-four and twenty-five of the Act of 1948 (which provide for the giving or withholding by the Minister of consent to the operation of notices to quit agricultural holdings) and by section twenty-seven thereof (which relates to the grant by the Minister of certificates of bad husbandry for the purposes of notices to quit). Amendments as to notices to quit agricultural holdings.

(2) The following sub-section shall be substituted for sub-section (1) of section twenty-five of the Act of 1948 (which requires the Minister to withhold his consent to the operation of a notice to quit an agricultural holding unless he is satisfied as to certain matters)—

'(1) The Agricultural Land Tribunal shall consent under the last foregoing section to the operation of a notice to quit an agricultural holding or part of an agricultural holding if, but only if, they are satisfied as to one or more of the following matters, being a matter or matters specified by the landlord in his application for their consent, that is to say—

    (a) that the carrying out of the purpose for which the landlord proposes to terminate the tenancy is desirable in the interests of good husbandry as respects the land to which the notice relates, treated as a separate unit; or

    (b) that the carrying out thereof is desirable in the interests of sound management of the estate of which the land to which the notice relates forms part or which that land constitutes; or

    (c) that the carrying out thereof is desirable for the purposes of agricultural research, education, experiment or demonstration, or for the purposes of the enactments relating to small-holdings or allotments; or

    (d) that greater hardship would be caused by withholding than by giving consent to the operation of the notice; or

(e) that the landlord proposes to terminate the tenancy for the purpose of the land's being used for a use, other than for agriculture, not falling within paragraph (b) of sub-section (2) of the last foregoing section:

Provided that, notwithstanding that they are satisfied as aforesaid, the Tribunal shall withhold consent to the operation of the notice to quit if in all the circumstances it appears to them that a fair and reasonable landlord would not insist on possession.'

(3) The foregoing provisions of this section shall apply to Scotland subject to the following modifications, that is to say—

- (a) for references to the Minister and to the Agricultural Land Tribunal there shall be substituted respectively references to the Secretary of State and to the Land Court;
- (b) for references to sections twenty-four, twenty-five and twenty-seven of the Act of 1948 there shall be substituted respectively references to sections twenty-five, twenty-six and twenty-eight of the Scottish Act of 1949; and
- (c) in the sub-section substituted for sub-section (1) of section twenty-six of the Scottish Act of 1949, in paragraph (c) after the word 'smallholdings' there shall be inserted the words 'or such holdings as are mentioned in section sixty-four of the Agriculture (Scotland) Act, 1948,' and in paragraph (e) for the words 'paragraph (b)' there shall be substituted the words 'paragraph (c)'.

(4) This section shall come into operation on the appointed day.

[1] See also Sec. 18(2) of the Agriculture (Miscellaneous Provisions) Act, 1968.

Provisions as to succession to holdings in Scotland.

6.—(1) Sub-section (1) of section twenty of the Scottish Act of 1949 (which empowers the tenant of an agricultural holding by will or other testamentary writing, to bequeath his lease of the holding to any person) shall have effect, in relation to any tenant of an agricultural holding whose death occurs after the expiration of a period of one month beginning with the date of the passing of this Act, with the substitution for the words 'to any person' of the words 'to any member of his family'.[1]

(2) For the purposes of the said sub-section (1), as amended as aforesaid, the expression 'member of his family' means the wife or husband of the tenant or his son-in-law or daughter-in-law or a person adopted by the tenant or any person who, failing nearer heirs, would be entitled to succeed in case of intestacy to the lease of the holding.

In this sub-section 'adopted' means adopted in pursuance of an adoption order made under the Adoption Act, 1950, or any enactment repealed by that Act, or under any corresponding enactment of the Parliament of Northern Ireland.

(3) Subject to the provisions of the next following sub-section, where notice to quit is given to the tenant of an agricultural holding, being a tenant who after the passing of this Act has acquired right to the lease of the holding [by virtue of section 16 of the Succession (Scotland) Act, 1964][2] or as a legatee by virtue of section twenty of the Scottish Act of 1949, then if—

(a) the notice to quit is given in accordance with sub-section (1) of section twenty-four of that Act so as to terminate the tenancy not earlier than the term (being the term stipulated in the lease as the term of outgo or the corresponding term in any succeeding year) next occurring after the first anniversary of the date on which the tenant acquired right to the lease as aforesaid and not later than the term next occurring after the second anniversary of that date, or

(b) in a case where at that date the unexpired term of the lease exceeded two years, the landlord gives notice to quit when it becomes legally competent for him to give such notice; and

(c) it is stated in the notice to quit that it is given in pursuance of this sub-section;

the provisions of sub-section (1) of section twenty-five of that Act shall not apply to the notice to quit.[3]

(4) In relation to such a tenant as is mentioned in the last foregoing sub-section who has acquired right to the lease within the period of seven years commencing with the passing of this Act paragraph (a) of the said sub-section shall have effect with the substitution for the words 'the first anniversary' of the words 'the second anniversary' and for the words 'the second anniversary' of the words 'the third anniversary'.

(5) Such a tenant as is mentioned in sub-section (3) of this section not be deprived on quitting the holding in pursuance of a notice to quit given in accordance with that sub-section of any right to compensation for an improvement specified in Part II of the First Schedule to the Scottish Act of 1949 which has been carried out on the holding between the first day of November, nineteen hundred and forty-eight, and the passing of this Act by reason only of the failure of the person by whom the improvement was carried out to give notice to the landlord in accordance with section fifty-one of that Act of his intention to carry out the improvement:

Provided that—

(a) a claim for compensation shall not be made by virtue of this sub-section unless the Land Court, on an application made to them in that behalf by the tenant, are satisfied that if notice of intention to carry out the improvement had been duly given as aforesaid the landlord would not have given notice in accordance with section fifty-two of that Act objecting to the carrying out of the improvement or if he had given such notice the Secretary of State, in pursuance of that section, would not have withheld his approval to the carrying out of the improvement, and authorise the making of the claim; and

(b) the compensation payable by virtue of this sub-section shall not exceed either such sum as fairly represents the value of the improvement to an incoming tenant or such sum as is equal to the capital cost of the improvement less one-tenth of such cost for each complete year which has elapsed between the time at which the carrying out of the improvement was completed and the time at which the tenant quitted the holding, whichever is the less.

[1] See amendment effected by Succession (Scotland) Act, 1964.
[2] Words in square brackets introduced by Succession (Scotland) Act, 1964.
[3] See Sec. 18 of Agriculture (Miscellaneous Provisions) Act, 1968, which provides that this sub-section shall not apply to a 'near relative' successor. For notice to quit see forms 2 and 3 and notes thereto.

. . .

Minor and consequential amendments.

8.—(1) The enactments specified in Part I of the First Schedule to this Act (being enactments applying to England and Wales) and the enactments specified in Part II of that Schedule (being enactments applying to Scotland) shall have effect subject to the amendments in that Schedule, being minor amendments or amendments consequential on the foregoing provisions of this Act.

. . . .

9.—(1) In this Act the following expressions have the meanings hereby assigned to them respectively, that is to say:— *Interpretation.*
- 'Act of 1947' means the Agriculture Act, 1947;
- 'Act of 1948' means the Agricultural Holdings Act, 1948;
- 'agricultural holding', as respects England and Wales has the meaning assigned to it by section one of the Act of 1948, and as respects Scotland has the meaning assigned to it by section one of the Scottish Act of 1949;
- 'the appointed day' means such day as the Minister or, in relation to Scotland, the Secretary of State, may by order made by statutory instrument appoint, and different days may be appointed by such orders in relation to different provisions of this Act;
- 'contract of tenancy' and 'fixed equipment' have the meanings assigned to them by section ninety-four of the Act of 1948;
- 'Land Court' means the Scottish Land Court;
- 'landlord' and 'tenant', as respects England and Wales, have the meanings assigned to them by section ninety-four of the Act of 1948, and as respects Scotland have the meanings assigned to them by section ninety-three of the Scottish Act of 1949;
- 'lease', as respects Scotland, means a letting of land for a term of years, or for lives, or for lives and years, or from year to year;
- 'the Minister' means the Minister of Agriculture, Fisheries and Food;
- 'Scottish Act of 1948' means the Agriculture (Scotland) Act, 1948;
- 'Scottish Act of 1949' means the Agricultural Holdings (Scotland) Act, 1949.

(2) References in this Act to any enactment shall be construed, except where the context otherwise requires, as references to that enactment as amended by or under any other enactment, including this Act.

308 AGRICULTURAL HOLDINGS (SCOTLAND) ACTS

Repeals, savings and transitional provisions.

10.—(1) The enactments specified in the Second Schedule to this Act are, in consequence of the foregoing provisions of this Act, hereby repealed to the extent specified in the third column of that Schedule—

(*a*) in the case of the enactments specified in Part I of that Schedule, on the passing of this Act, and

(*b*) in the case of the enactments specified in Part II of that Schedule, on the appointed day.

(2) The repeal by virtue of this Act of provisions contained in Part II of the Act of 1947 shall not affect the operation of section ninety-five of that Act (which applies certain of those provisions for the purposes of special directions to secure production).

(3) The repeal by virtue of this Act of section twenty-one of the Act of 1947 shall not affect the operation of sub-section (5) of section twenty of the Mineral Workings Act, 1951 (which applies the definition of 'owner' in the said section twenty-one for the purposes of the said section twenty).

(4) The repeal by virtue of this Act of provisions contained in Part II of the Scottish Act of 1948 shall not affect the operation of section thirty-five of that Act (in relation to which certain of those provisions have effect for the purposes of special directions to secure production).

(5) The repeal by virtue of this Act of section seventy-one of the Scottish Act of 1948 shall not affect the operation of sub-section (6) of section twenty-one of the Crofters (Scotland) Act, 1955 (which applies the provisions of the said section seventy-one to the reference to the Land Court of certain proposals of the Crofters Commission) or the power of the Secretary of State to make regulations under the said section as so applied.

(6) The enactments specified in the Third Schedule to this Act are hereby repealed to the extent specified in the third column of that Schedule, being to that extent spent.

(7) The transitional provisions set out in the Fourth Schedule to this Act shall have effect.

Short title and extent.

11.—(1) This Act may be cited as the Agriculture Act, 1958.

(2) This Act, except sub-section (2) of section eight thereof, shall not extend to Northern Ireland.

SCHEDULES

FIRST SCHEDULE
MINOR AND CONSEQUENTIAL AMENDMENTS

. . . .

PART II[1]
SCOTLAND
Agriculture (Scotland) Act, 1948

30. Section thirty-six (which relates to special directions as to stocking of deer forests and grouse moors) shall cease to have effect.

31. In the Third Schedule, in paragraph 1, the words 'section twenty nine or' and in paragraph 5 the words 'sub-section (4) of section thirty and' shall be omitted.

Agricultural Holdings (Scotland) Act, 1949

32. For section nine there shall be substituted the following section—

'9.—(1) Where under the lease of an agricultural holding, whether entered into before or after the commencement of this Act, provision is made for the maintenance of specified land, or a specified proportion of the holding, as permanent pasture, the landlord or the tenant may, by notice in writing served on his tenant or landlord, demand a reference to arbitration under this Act of the question whether it is expedient in order to secure the full and efficient farming of the holding that the amount of land required to be maintained as permanent pasture should be reduced.

(2) On a reference under the foregoing sub-section the arbiter may by his award—
    (*a*) direct that the lease shall have effect subject to such modifications of the provisions thereof as to land which is to be maintained as permanent pasture or is to be treated as arable land, and as to cropping, as may be specified in the direction; and
    (*b*) if he gives a direction reducing the area of land which under the lease is to be maintained as permanent pasture, order that the lease shall have effect

---
[1] Part I is applicable to England and Wales.

as if it provided that on quitting the holding on the termination of the tenancy the tenant should leave as permanent pasture, or should leave as temporary pasture sown with seeds mixture of such kind as may be specified in the order, such area of land (in addition to the area of land required by the lease, as modified by the direction, to be maintained as permanent pasture) as may be so specified, so however that the area required to be left as aforesaid shall not exceed the area by which the land required by the lease to be maintained as permanent pasture has been reduced by virtue of the direction.'

33. In section twelve, in sub-section (3) (which provides for a question whether a tenant has so exercised his rights under sub-section (1) of that section as to injure or deteriorate his holding to be determined for certain purposes by the Secretary of State) for the words from 'determined by the Secretary of State' to 'a certificate of the Secretary of State' there shall be substituted the words 'determined by arbitration; and a certificate of the arbiter'; and in sub-section (5) after the words 'section nine of this Act' there shall be inserted the words 'or an arbiter has directed under the said section nine'.

34. In section twenty-four (which relates to the giving of notices to quit), in paragraph (*a*) of sub-section (6) after the words 'other purposes' there shall be inserted the words '(not being agricultural purposes)'.

35. On the appointed day, in section twenty-five—
   (*a*) for references to the Secretary of State there shall be substituted references to the Land Court;
   (*b*) paragraph (*a*) of sub-section (2) shall cease to have effect; and
   (*c*) at the end of paragraph (*b*) of sub-section (2) there shall be added the words 'and it is stated in the notice that it is given by reason of the matter aforesaid'.

36. On the appointed day, in section twenty-six—
   (*a*) sub-sections (2) to (4) shall cease to have effect;
   (*b*) in sub-section (5) the words 'the Secretary of State or' in each place where they occur shall be omitted; and
   (*c*) in sub-section (6) for the reference to the Secretary of

State there shall be substituted a reference to the Land Court.

37. On the appointed day, for section twenty-seven there shall be substituted the following section—

'27.—(1) An application by a landlord for the consent of the Land Court under section twenty-five of this Act to the operation of a notice to quit shall be made within one month after service on the landlord by the tenant of a counter-notice requiring that sub-section (1) of that section shall apply to the notice to quit.

(2) A tenant to whom has been given a notice to quit in connection with which any question arises under sub-section (2) of section twenty-five of this Act shall, if he requires such question to be determined by arbitration under this Act, give notice to the landlord to that effect within one month after the notice to quit has been served on him; and where the award of the arbiter in an arbitration so required is such that the provisions of sub-section (1) of section twenty-five of this Act would have applied to the notice to quit if a counter-notice had been served within the period limited by that sub-section the period within which a counter-notice may be served under that sub-section shall be extended up to the expiration of one month from the issue of the arbiter's award.

(3) Where such an arbitration as is referred to in the last foregoing sub-section has been required by the tenant, or where an application has been made to the Land Court for their consent to the operation of a notice to quit, the operation of the notice to quit shall be suspended until the issue of the arbiter's award or of the decision of the Land Court, as the case may be.

(4) Where the decision of the Land Court giving their consent to the operation of a notice to quit, or the award of the arbiter in such an arbitration as is referred to in sub-section (2) of this section, is issued at a date later than six months before the date on which the notice to quit is expressed to take effect, the Land Court, on application made to them in that behalf at any time not later than one month after the issue of the decision or award aforesaid, may postpone the

operation of the notice to quit for a period not exceeding twelve months.

(5) If the tenant of an agricultural holding receives from the landlord notice to quit the holding or a part thereof and in consequence thereof gives to a sub-tenant notice to quit that holding or part, the provisions of sub-section (1) of section twenty-five of this Act, shall not apply to the notice given to the sub-tenant; but if the notice to quit given to the tenant by the landlord does not have effect, the notice to quit given as aforesaid by the tenant to the sub-tenant shall not have effect.

For the purposes of this sub-section a notice to quit part of the holding which under the provisions of section thirty three of this Act is accepted by the tenant as notice to quit the entire holding shall be treated as a notice to quit the holding.

(6) Where notice is served on the tenant of an agricultural holding to quit the holding or a part thereof, being a holding or part which is subject to a sub-tenancy, and the tenant serves on the landlord a counter-notice in accordance with the provisions of sub-section (1) of section twenty-five of this Act, the tenant shall also serve on the sub-tenant notice in writing that he has served such counter-notice on the landlord, and the sub-tenant shall be entitled to be a party to any proceedings before the Land Court for their consent to the notice to quit.'

38. On the appointed day, for section twenty-eight there shall be substituted the following section—

'28. For the purposes of paragraph (*d*) of sub-section (2) of section twenty-five of this Act, the landlord of an agricultural holding may apply to the Land Court for a certificate that the tenant is not fulfilling his responsibilities to farm in accordance with the rules of good husbandry, and the Land Court, if satisfied that the tenant is not fulfilling his said responsibilities, shall grant such a certificate.'

39. Section twenty-nine (which empowers the Secretary of State to give to the tenant of an agricultural holding, being a holding in respect of which a certificate of bad husbandry under section twenty-five of the Scottish Act of 1949 is in force, directions for securing that the holding does not further deteriorate before the termination of the tenancy) shall cease to have effect.

40. For section thirty there shall be substituted the following section—

'Penalty for breach of condition accompanying consent to notice to quit.

30.—(1) Where, on giving consent under section twenty-five of this Act to the operation of a notice to quit an agricultural holding or part of an agricultural holding, the Land Court imposed a condition under section twenty-six of this Act for securing that the land to which the notice to quit related would be used for the purpose for which the landlord proposed to terminate the tenancy, and it is proved, on an application to the Land Court on behalf of the Crown—

(*a*) that the landlord has failed to comply with the condition within the period allowed thereby, or

(*b*) that the landlord has acted in contravention of the condition,

the Land Court may by order impose on the landlord a penalty of an amount not exceeding two years' rent of the holding at the rate at which rent was payable immediately before the termination of the tenancy, or, where the notice to quit related to a part only of the holding, of an amount not exceeding the proportion of the said two years' rent which it appears to the Land Court is attributable to that part.

(2) A penalty imposed under this section shall be a debt due to the Crown and shall, when recovered, be paid into the Exchequer.'

41. On the appointed day, in section fifty-two (which empowers the Secretary of State to approve the carrying out by the tenant of certain long-term improvements)—

(*a*) for references to the Secretary of State there shall be substituted references to the Land Court;

(*b*) in sub-section (2) the words from 'after giving notice' to 'so to do', the words from 'after affording' to 'appointed by the Secretary of State' and the words from 'and in either case' to the end of the sub-section, shall be omitted; and

(c) in sub-section (4) the words from 'after affording' to 'appointed by the Secretary of State' shall be omitted.

42. In section sixty-three, in sub-section (1), for the words 'paragraph (ii) of section nine' there shall be substituted the words 'paragraph (b) of sub-section (2) of section nine'.

43. On the appointed day, in section sixty-six (which empowers the Secretary of State to direct that, for the purposes of certain improvements to be carried out by the tenant, an agricultural holding shall be treated as a market garden)—

(a) for references to the Secretary of State there shall be substituted references to the Land Court; and

(b) in sub-section (1) the words from 'and after affording' to 'appointed by the Secretary of State' shall be omitted.

44. On the appointed day, sections seventy-one and seventy-two (which respectively provide for representations to the Secretary of State as to the taking of action by him, and for proposals as to such action to be referred to the Land Court) shall cease to have effect.

Reserve and Auxiliary Forces (*Protection of Civil Interests*) *Act*, 1951

45. On the appointed day in section twenty-one (as read with section twenty-four)—

(a) in sub-sections (2) and (3) for references to the Secretary of State there shall be substituted references to the Land Court;

(b) sub-sections (4) and (7) shall cease to have effect;

(c) in sub-section (5) for paragraph (c) there shall be substituted the following paragraph—

'(c) the Scottish Land Court has not before the beginning of his period of residence protection decided whether to give or withhold consent to the operation of the notice to quit; and

(d) in sub-section (6) the reference to section twenty-seven of the Scottish Act of 1949 shall be construed as a reference to that section as originally enacted and not as amended by this Act.

46. On the appointed day in section twenty-four, in paragraph (*b*), the words from 'for references to the Agricultural Land Tribunal' to 'appeals thereto' and paragraph (*c*) shall be omitted.

The Second and Third Schedules deal with Repealed Enactments. The Fourth Schedule deals with Transitional Provisions. These have not been printed, but readers may have occasion to refer to the Transitional Provisions which are of temporary effect.

EXCERPTS FROM THE SUCCESSION (SCOTLAND) ACT, 1964 (1964, c.41)

Part III
Administration and Winding Up of Estates

Assimilation for purposes of administration etc. of heritage to moveables.

14.—(1) Subject to sub-section (3) of this section the enactments and rules of law in force immediately before the commencement of this Act with respect to the administration and winding up of the estate of a deceased person so far as consisting of moveable property shall have effect (as modified by the provisions of this Act) in relation to the whole of the estate without distinction between moveable property and heritable property; and accordingly on the death of any person (whether testate or intestate) every part of his estate (whether consisting of moveable property or heritable property) falling to be administered under the law of Scotland shall, by virtue of confirmation thereto, vest for the purposes of administration in the executor thereby confirmed and shall be administered and disposed of according to law by such executor.

(2) Provision shall be made by the Court of Session by act of sederunt made under the enactments mentioned in section 22 of this Act (as extended by that section) for the inclusion in the confirmation of an executor, by reference to an appended inventory or otherwise, of a description, in such form as may be so provided, of any heritable property forming part of the estate.

(3) Nothing in this section shall be taken to alter any rule of law whereby any particular debt of a deceased person falls to be paid out of any particular part of his estate.

Provisions as to transfer of heritage.

15.—(1) Section 5(2) of the Conveyancing (Scotland) Act 1924 (which provides that a confirmation which includes a heritable security shall be valid title to the debt thereby secured) shall have effect as if any reference therein to a heritable security, or to a debt secured by a heritable security, included a reference to any interest in heritable property which has vested in an executor in pursuance of the last foregoing section by virtue of a confirmation:

Provided that a confirmation shall not be deemed for the purposes of the said section 5(2) to include any such interest unless a description of the property, in accordance with any act of sederunt such as is mentioned in subsection (2) of the last foregoing section, is included or referred to in the confirmation.

(2) Where in pursuance of the last foregoing section any heritable property has vested in an executor by virtue of a confirmation, and it is necessary for him in distributing the estate to transfer that property—

- (a) to any person in satisfaction of a claim to legal rights or the prior rights of a surviving spouse out of the estate, or
- (b) to any person entitled to share in the estate by virtue of this Act, or
- (c) to any person entitled to take the said property under any testamentary disposition of the deceased,

the executor may effect such transfer by endorsing on the confirmation (or where a certificate of confirmation relating to the property has been issued in pursuance of any act of sederunt, on the certificate) a docket in favour of that person in the form set out in Schedule 1 to this Act, or in a form as nearly as may be to the like effect, and any such docket may be specified as a midcouple or link in title in any deduction of title; but this section shall not be construed as prejudicing the competence of any other mode of transfer.

(2) This section shall not prejudice the operation of section 16 of the Crofters Holdings (Scotland) Act 1886 or section 20 of the Agricultural Holdings (Scotland) Act 1949 (which relate to bequests in the case of agricultural leases).

Provisions relating to leases

16.[1]—(1) This section applies to any interest, being the interest of a tenant under a lease, which is comprised in the estate of a deceased person and has accordingly vested in the deceased's executor by virtue of section 14 of this Act; and in the following provisions of this section 'interest' means an interest to which this section applies.

(2) Where an interest—

- (a) is not the subject of a valid bequest by the deceased, or

(b) is the subject of such a bequest, but the bequest is not accepted by the legatee, or

(c) being an interest under an agricultural lease, is the subject of such a bequest, but the bequest is declared null and void in pursuance of section 16 of the Act of 1886 or section 20 of the Act of 1949 or becomes null and void under section 10 of the Act of 1955,

and there is among the conditions of the lease (whether expressly or by implication) a condition prohibiting assignation of the interest, the executor shall be entitled, notwithstanding that condition to transfer the interest to any one of the persons entitled to succeed to the deceased's intestate estate, or to claim legal rights or the prior rights of a surviving spouse out of the estate, in or towards satisfaction of that person's entitlement or claim: but shall not be entitled to transfer the interest, to any other person without the consent—

(i) in the case of an interest under an agricultural lease, being a lease of a croft within the meaning of section 3(1) of the Act of 1955, of the Crofters Commission;

(ii) in any other case, of the landlord.

(3) If in the case of any interest—

(a) at any time the executor is satisfied that the interest cannot be disposed of according to law and so informs the landlord, or

(b) the interest is not so disposed of within a period of one year or such longer period as may be fixed by agreement between the landlord and the executor or, failing agreement, by the sheriff on summary application by the executor—

(i) in the case of an interest under an agricultural lease which is the subject of a petition to the Land Court under section 16 of the Act of 1886 or an application to that court under section 20 of the Act of 1949, from the date of the determination or withdrawal of the petition or, as the case may be, the application,

(ia) in the case of an interest under an agricultural lease which is the subject of an application by the legatee to the Crofters Commission under section 10(1) of the Act of 1955, from the date of any refusal by the Commission to determine that the bequest shall not be null and void,

(i*b*) in the case of an interest under an agricultural lease which is the subject of an intimation of objection by the landlord to the legatee and the Crofters Commission under section 10 (3 of the Act of 1955, from the date of any decision of the Commission upholding the objection,

(ii) in any other case from the date of death of the deceased,

either the landlord of the executor may, on giving notice in accordance with the next following sub-section to the other, terminate the lease (in so far as it relates to the interest) notwithstanding any provision therein, or any enactment or rule of law, to the contrary effect.

(4) The period of notice given under the last foregoing sub-section shall be—

(*a*) in the case of an agricultural lease, such period as may be agreed, or, failing agreement, a period of not less than one year and not more than two years ending with such term of Whitsunday or Martinmas as may be specified in the notice; and

(*b*) in the case of any other lease, a period of six months:

Provided that paragraph (*b*) of this sub-section shall be without prejudice to any enactment prescribing a shorter period of notice in relation to the lease in question.

(5) Sub-section (3) of this section shall not prejudice any claim by any party to the lease for compensation or damages in respect of the termination of the lease (or any rights under it) in pursuance of that sub-section; but any award of compensation or damages in respect of such termination at the instance of the executor shall be enforceable only against the estate of the deceased and not against the executor personally.

(6) Where an interest is an interest under an agricultural lease, and—

(*a*) an application is made under section 3 of the Act of 1931 or section 13 of the Act of 1955 to the Land Court for an order for removal, or

(b) a reference is made under section 27(2) of the Act of 1949 to an arbiter to determine any question which has arisen under section 25(2)(*f*) of that Act in connection with a notice to quit,

the Land Court shall not make the order, or, as the case may be, the arbiter shall not make an award in favour of the landlord, unless the court or the arbiter is satisfied that it is reasonable, having regard to the fact that the interest is vested in the executor in his capacity as executor, that it should be made.

(7) Where an interest is not an interest under an agricultural lease, and the landlord brings an action of removing against the executor in respect of a breach of a condition of the lease, the court shall not grant decree in the action unless it is satisfied that the condition alleged to have been breached is one which it is reasonable to expect the executor to have observed, having regard to the fact that the interest is vested in him in his capacity as an executor.

(8) Where an interest is an interest under an agricultural lease and is the subject of a valid bequest by the deceased, the fact that the interest is vested in the executor under the said section 14 shall not prevent the operation, in relation to the legatee, of paragraphs (*a*) to (*h*) of section 16 of the Act of 1886, or, as the case may be, sub-sections (2) to (7) of section 20 of the Act of 1949, or as the case may be, sub-sections (2) to (7) of section 10 of the Act of 1955.

(9) In this section—
'agricultural lease' means a lease of a holding within the meaning of the Small Landholders (Scotland) Acts 1886 to 1931 or of the Act of 1949, or a lease of a croft within the meaning of section 3(1) of the Act of 1955;
'the Act of 1886' means the Crofters Holdings (Scotland) Act 1886;
'the Act of 1931' means the Small Landholders and Agricultural Holdings (Scotland) Act 1931;
'the Act of 1949' means the Agricultural Holdings (Scotland) Act 1949;
'the Act of 1955' means the Crofters (Scotland) Act 1955;
'lease' includes tenancy.

[1] As amended by the *Law Reform* (*Miscellaneous Provisions*) (*Scotland*) *Act*, 1968.

Part VI

Miscellaneous and Supplementary

. . . .

29.—(1) A bequest by a tenant of his interest under a tenancy or lease to any one of the persons who, if the tenant had died intestate, would be, or would in any circumstances have been, entitled to succeed to his intestate estate by virtue of this Act shall not be treated as invalid by reason only that there is among the conditions of the tenancy or lease an implied condition prohibiting assignation.

_{Right of tenant to bequeath interest under lease.}

(2) This section shall not prejudice the operation of section 16 of the Crofters Holdings (Scotland) Act 1886 or section 20 of the Agricultural Holdings (Scotland) Act 1949 (which relate to bequests in the case of agricultural leases).

Excerpts from the
AGRICULTURE (MISCELLANEOUS) PROVISIONS) ACT 1968

. . . .

Part II
Additional Payments to Tenant Farmers

Additional payments to tenants quitting agricultural holdings.
1948 c. 63.

9.—(1) Where under the Agricultural Holdings Act 1948 (hereafter in this Part of this Act referred to as ('the principal Act') compensation for disturbance in respect of an agricultural holding or part of such a holding becomes payable by the landlord to the tenant of the holding, then, subject to the provisions of this Part of this Act, there shall be payable by the landlord to the tenant, in addition to any compensation payable by the landlord to the tenant, a sum to assist in the reorganisation of the tenant's affairs of the amount prescribed by sub-section (2) of this section.

(2) Subject to the provisions of this Part of this Act, the sum payable in pursuance of sub-section (1) of this section shall be equal to four times the annual rent of the holding or, in the case of part of a holding, four times the appropriate portion of that rent, at the rate at which the rent was payable immediately before the termination of the tenancy of the holding or part to which the said compensation relates.

1948 c. 63.

1949 c. 75.

(3) In the application of this section to Scotland, in sub-section (1) for the references to the Agricultural Holdings Act 1948 and the principal Act there shall be substituted respectively references to the Agricultural Holdings (Scotland) Act 1949 and the principal Scottish Act.

Provisions supplementary to s. 9 in Scotland.

11.—(1) Subject to the provisions of this section, no sum shall be payable in pursuance of section 9 of this Act in consequence of the termination of the tenancy of an agricultural holding or part of such a holding by virtue of a notice to quit[1] in a case where—

    (*a*) the relevant notice contains a statement that the carrying out of the purpose for which the landlord proposes to terminate the tenancy is desirable on any of the grounds mentioned in paragraphs (*a*) to (*c*) of section 26(1) of the principal Scottish Act and, if an application for consent in respect of the notice is made to the Scottish Land Court (hereafter in this section referred to as 'the court') in pursuance of section 25(1) of the principal Scottish Act, the court consent to its operation and state in the reasons for their decision that they are satisfied as to any of the matters so mentioned; or

(*b*) the relevant notice contains a statement that the landlord will suffer hardship unless the notice has effect and, if such an application as aforesaid is made in respect of the notice, the court consent to its operation and state in the reasons for their decision that they are satisfied that greater hardship would be caused by withholding consent than by giving it; or

(*c*) the relevant notice is a notice to which, apart from the provisions of section 18 or section 19 of this Act, section 6(3) of the Agriculture Act 1958 would apply and, if an application for consent in respect of the notice is made to the court in pursuance of the said section 25(1), the court consent to its operation and state in the reasons for their decision that they are satisfied with regard to the matter mentioned in paragraph (*a*), or the matters mentioned in paragraph (*b*)(i) to (iii), or the matter mentioned in paragraph (*c*), of section 18(2) of this Act; or

(*d*) the said section 25(1) does not apply to the relevant notice by virtue of section 29(4) of the Agriculture Act 1967 (which relates to notices to quit given by the Secretary of State or a Rural Development Board with a view to boundary adjustments or an amalgamation).

(2) Sub-section (1) of this section shall not apply in relation to the relevant notice where—

(*a*) the reasons given by the court for their decision to consent to the operation of the notice include the reason that they are satisfied as to the matter mentioned in section 26(1)(*e*) of the principal Scottish Act (which relates to the use of land for certain non-agricultural purposes); or

(*b*) the reasons so given consist of or include the reason

that the court are satisfied as to the matter mentioned in section 26(1)(*b*) of that Act or in paragraph (*a*) or paragraph (*c*) of section 18(2) of this Act but the court would have been satisfied also as to the matter mentioned in the said section 26(1)(*e*) if it had been specified in the application for consent,

and where the court would have been satisfied as mentioned in paragraph (*b*) of this sub-section they shall include a statement to that effect in their decision.

(3) In assessing the compensation payable to the tenant of an agricultural holding in consequence of the compulsory acquisition of his interest in the holding or part of it or the compulsory taking of possession of the holding or part of it, no account shall be taken of any benefit which might accrue to the tenant by virtue of section 9 of this Act.

(4) Any sum payable in pursuance of the said section 9 shall be so payable notwithstanding any agreement to the contrary.

(5) The following provisions of the principal Scottish Act shall apply to sums claimed or payable in pursuance of the said section 9 as they apply to compensation claimed or payable under section 35 of that Act, that is to say, sections 35(3), 61, 68 to 70, 75 to 78, 80, 82 to 84, 90, 93(6), 100 and Schedule 6.

(6) No sum shall be payable in pursuance of the said section 9 in consequence of—

(*a*) the termination of the tenancy of an agricultural holding or part of such a holding by virtue of a notice to quit unless the notice in consequence of which the termination occurs is served on the tenant after the initial date and the termination occurs after the date of the passing of this Act; or

(*b*) the resumption by the landlord of possession of part of the holding in pursuance of a provision in that behalf contained in the lease unless the resumption occurs after the date of the passing of this Act.

(7) No sum shall be payable in pursuance of section 9 of this Act in consequence of the termination of the tenancy of an agricultural holding or part of such a holding by virtue of a notice to quit where the relevant notice is given in pursuance of section 6(3) of the Agriculture Act 1958 (which relates to notice to

quit given to a tenant who has acquired right to the tenancy of the holding by virtue of section 16 of the Succession (Scotland) Act 1964 or as a legatee by virtue of section 20 of the principal Scottish Act) and—

(a) the landlord is terminating the tenancy for the purpose of using the land for agriculture only; and
(b) the notice contains a statement that the tenancy is being terminated for the said purpose:

Provided that if any question arises between the landlord and the tenant as to the purpose for which the tenancy is being terminated, the tenant shall, notwithstanding section 74 of the principal Scottish Act (matters to be referred to arbitration), refer the question to the Scottish Land Court for their determination.

(8) Section 73 of the principal Scottish Act (proceedings of the Land Court) shall apply for the purpose of the determination of any matter referred to the Scottish Land Court under subsection (7) of this section as it applies to any matter which they are required to determine under that Act.

(9) The provisions of Schedule 2 to this Act (which contains transitional provisions for certain cases) shall have effect for the purposes of this section in its application to Scotland.

(10) In this section—

(a) references to section 9 of this Act do not include references to it as applied by section 12 of this Act;
(b) 'the relevant notice' means a notice to quit given by the landlord of the agricultural holding in question in consequence of which compensation for disturbance becomes payable to the tenant of the holding as mentioned in the said section 9; and
(c) for the purposes of sub-section (1)(a), the purposes of the enactments relating to allotments shall be treated as excluded from the matters mentioned in section 26(1)(c) of the principal Scottish Act.

[1] See notes to Forms of Notice to Quit.

12.—(1) Where in pursuance of any enactment providing for the acquisition or taking of possession of land compulsorily by any person (hereafter in this Part of this Act referred to as an 'acquiring authority'), an acquiring authority acquire the interest in an agricultural holding or any part of it of the tenant of the holding or take possession of such a holding or any part of it, then, subject to the provisions of this Part of this Act, section 9 of this Act shall apply as if the acquiring authority were the landlord of the holding and compensation for disturbance in respect of the holding or part in question had become payable to the tenant as mentioned in sub-section (1) of that section on the date of the acquisition or taking of possession.

(2) No sum shall be payable by virtue of sub-section (1) of this section in respect of any agricultural holding held on a tenancy for a term of two years or upwards except in a case where the amount of compensation payable to the tenant of the holding by the acquiring authority in consequence of the acquisition or taking of possession in question is exceeded by the aggregate of the amounts which, if the tenancy had been from year to year, would have been so payable by way of compensation and by virtue of that sub-section; and in any such case the sum payable by virtue of that sub-section in consequence of the acquisition or taking of possession in question shall, subject to sections 13(3) and 14(3) of this Act, be of an amount equal to the excess.

(3) No sum shall be payable to the tenant of an agricultural holding by virtue of sub-section (1) of this section in consequence of such an acquiring of an interest or taking of possession as is there mentioned unless the date on which the acquisition or taking of possession occurs is later than the date of the passing of this Act and—

    (*a*) in the case of such an acquisition, unless the date on which notice to treat in respect of the interest to be acquired is served or treated as served on the tenant by the acquiring authority is after the initial date; and

    (*b*) where in the case of such a taking of possession prior notice of the taking of possession is by virtue of any enactment required to be served on the tenant by the acquiring authority, unless the date on which the notice is so served is after the initial date.

*   *   *   *   *   *

14.—(1) For the purposes of sub-section (1) of section 12 of this Act, a tenant of an agricultural holding shall be treated as not being a tenant of it in so far as, immediately before the acquiring of the interest or taking of possession mentioned in that sub-section, he was neither in possession nor entitled to take possession of any land comprised in the holding; and in determining for those purposes whether a tenant was so entitled, any such lease relating to the land as is mentioned in section 2(1) of the principal Scottish Act which has not taken effect as a lease of the land from year to year shall be disregarded.

(2) Section 12(1) of this Act shall not apply where the acquiring authority require the land comprised in the holding or part in question for the purposes of agricultural research or experiment or of demonstrating agricultural methods, or for the purposes of the enactments relating to smallholdings, nor where the Secretary of State acquires the land under section 57(1)(c) or section 64 of the Agriculture (Scotland) Act 1948; but where an acquiring authority exercise in relation to any land any power to acquire or take possession of land compulsorily which is conferred on the authority by virtue of sections 34, 35 or 39(3) of the Town and Country Planning (Scotland) Act 1947 or section 7 of the New Towns (Scotland) Act 1968, the authority shall be deemed for the purposes of this sub-section not to acquire the land for any of the purposes aforesaid.

(3) The provisions of Schedule 4 to this Act shall have effect for the purposes of section 12 of this Act in its application to Scotland.[1]

[1] Schedule 4 contains the important provision that the acquiring authority can (if it thinks too high a rent is being paid with a view to compensation) apply to an official arbiter (or Lands Tribunal for Scotland when it is set up) to fix the appropriate rent. There is no corresponding provision when the rent is too low.

15.[1]—(1) Except where compensation assessed in accordance with this sub-section would be less than if this sub-section were disregarded, in assessing the compensation payable by an acquiring authority to the tenant of an agricultural holding in connection with such an acquiring of an interest or taking of possession as is mentioned in section 12(1) of this Act, any provision in the lease entitling the landlord to resume land for building, planting, fencing, or other purposes (not being agricultural purposes) shall—

(a) in the case of an acquisition, be treated as if that provision authorised resumption of land for the purpose in question on the expiration of twelve months from the end of the year of the tenancy current when notice to treat in respect of the acquisition was served or treated as served on the tenant; and

(b) in the case of a taking of possession, be disregarded.

(2) [Not applicable to Scotland.]

(3) Where the landlord of an agricultural holding in Scotland resumes land in pursuance of such a provision in the lease as is mentioned in sub-section (1) of this section, compensation shall be payable by the landlord to the tenant, in addition to any other compensation so payable apart from this sub-section in respect of the land, of an amount which is equal to the value of the additional benefit (if any) which would have accrued to the tenant if the land had, instead of being so resumed, been resumed at the expiration of twelve months from the end of the year of tenancy current at a date two months before the date of resumption.

(4) Sub-sections (4) to (6) of section 10 of this Act shall apply to compensation claimed or payable under sub-section (2) of this section, and sub-sections (4) to (6) of section 11 of this Act shall apply to compensation claimed or payable under sub-section (3) of this section, as if for references to sums claimed or payable in pursuance of section 9 of this Act there were substituted references to compensation claimed or payable under the said sub-section (2) or sub-section (3), as the case may be; and section 12(3) of this Act shall apply to any increase of compensation in pursuance of sub-section (1) of this section as it applies to a sum payable by virtue of section 12(1) of this Act as if for references to the said section 12(1) there were substituted references to sub-section (1) of this section.

(5) For the purposes of sub-sections (1) to (3) of this section, the current year of a tenancy for a term of two years or upwards is the year beginning with such day in the period of twelve months ending—

(a) for the purposes of sub-section (1) or sub-section (2), with the date on which the notice mentioned in that sub-section is served; and

(b) for the purposes of sub-section (3), with a date two months before the resumption mentioned in that sub-section,

as corresponds to the day on which the term would expire by the effluxion of time.

[1] The words underlined are words substituted by sub-section 6 of this section in the application of the section to Scotland.

16. This Part of this Act shall apply to statutory small tenants as defined in the Small Landholders (Scotland) Act 1911 subject to the modifications set out in Schedule 5 to this Act.

Part II
Application of Part II to statutory small tenants in Scotland.
1911 c. 49.

17.—(1) In this Part of this Act—

Interpretation etc. of Part II.

'acquiring authority' has the meaning assigned to it by section 12(1) of this Act;

'the initial date' means 1st November 1967;

'possession' means actual possession;

'the principal Act' means the Agricultural Holdings Act 1948; and

1948 c. 63.

'the principal Scottish Act' means the Agricultural Holdings (Scotland) Act 1949;

1949 c. 75.

and unless the context otherwise requires expressions used in this Part of this Act and the principal Act or, as the case may be, the principal Scottish Act have the same meanings in this Part of this Act as in that Act.

(2) In this Part of this Act references to the termination of the tenancy of part of an agricultural holding are references to the resumption by the landlord of possession of that part of the holding and references to the acquisition of any property are references to the vesting of the property in the person acquiring it.

(3) Section 87(1) and (2) of the principal Act and section 86(1) and (2) of the principal Scottish Act (Crown land) shall have effect as if references to that Act included references to this Part of this Act.

(4) References in this section to this Part of this Act include references to Schedules 1 to 4 to this Act.

(5) In the application of this section to Scotland, in sub-section (2) the words from 'references to the termination' to 'and' shall be omitted.

PART III

TERMINATION OF TENANCIES OF AGRICULTURAL HOLDINGS IN SCOTLAND ACQUIRED BY SUCCESSION

18.[1]—(1) Section 6(3) of the Agriculture Act 1958 shall not apply to a notice to quit given to a tenant who has acquired right to the lease of an agricultural holding—

Termination in case of near relatives of deceased tenant.

(a) by virtue of section 16 of the Succession (Scotland) Act 1964, or

(b) as a legatee by virtue of section 20 of the principal Scottish Act,

where he is a near relative of the deceased tenant from whom he has acquired right to that lease; and accordingly section 25(1) of the principal Scottish Act shall, subject to the provisions of this section, apply to such a notice.

(2) Notwithstanding section 26(1) of the principal Scottish Act (which provides for the Scottish Land Court consenting to the operation of a notice to quit in certain circumstances, where the said section 6(3) would apart from the provisions of this section apply to the notice, the Scottish Land Court shall consent under the said section 25(1) to the operation of a notice to quit given to such a near relative as is mentioned in the foregoing sub-section—

(a) if they are satisfied that the near relative has neither sufficient training in agriculture nor sufficient experience in the farming of land to enable him to farm the holding to which the notice relates with reasonable efficiency, and if the notice contains a statement that it is given by reason of the matter aforesaid, or

(b) if they are satisfied—

(i) that the holding to which the notice relates, or where the holding forms only part of an agricultural unit, that unit, is not an agricultural unit which in the opinion of the Court is capable of providing full-time employment for an individual occupying it and for at least one other man,

(ii) that the notice is given in order to enable the landlord to use the holding for the purpose of effecting an amalgamation, and

(iii) that the amalgamation is proposed to be carried out within two years after the date of the termination of the tenancy specified in the notice,

and if the notice contains a statement that it is given in order to enable the landlord to use the holding for the purpose of effecting an amalgamation and specifies the land with which the holding is to be amalgamated, or

(c) if they are satisfied that the near relative is the occupier (either as owner or tenant) of agricultural land other than the holding to which the notice relates, or, where the holding forms only part of an agricultural unit, other than that unit, being agricultural land, which—

(i) has been occupied by him since a date prior to the death of the deceased tenant from whom he has acquired right to the lease of the said holding, and

(ii) is an agricultural unit which in the opinion of the Court is capable of providing full-time employment for an individual occupying it and for at least one other man,

and if the notice contains a statement that it is given by reason of the matter aforesaid and specifies the land:

Provided that, notwithstanding that they are satisfied as aforesaid, the Court shall withhold consent to the operation of the notice if in all the circumstances it appears to them that a fair and reasonable landlord would not insist on possession.

(3) For the purposes of paragraphs (*b*)(i) and (*c*)(ii) of the last foregoing sub-section, in assessing the capability of the unit of providing employment, it shall be assumed that the unit is farmed under reasonably skilled management, that a system of husbandry suitable for the district is followed and that the greater part of the feeding stuffs required by any livestock kept on the unit is grown there.

(4) The Court in giving consent to the operation of a notice to quit under the said section 25(1) as applied by this section on the grounds mentioned in sub-section (2)(*b*) of this section shall impose such conditions as appear to them requisite for securing—

(*a*) that the holding to which the notice relates will be used for the purpose of effecting an amalgamation with the land specified in the notice; and

(*b*) that the amalgamation will take place within two years after the date of the termination of the tenancy of the holding by reason of the notice;

and section 26(5) of the principal Scottish Act shall not apply to such a consent.

(5) Section 30 of the principal Scottish Act shall, with any necessary modifications, apply to a condition imposed under this section as it applies to a condition imposed under section 26 of that Act.

(6) This section shall apply to any notice to quit given to such a near relative after the passing of this Act.

(7) In this section, 'near relative' in relation to a deceased tenant of an agricultural holding means a surviving spouse, son or daughter, or adopted son or daughter, of that tenant.

(8) In the last foregoing sub-section, the reference to an adopted son or daughter of a deceased tenant shall be construed as a reference to a son or daughter adopted by him (whether alone or jointly with any other person) in pursuance of an adoption order within the meaning of section 23(5) of the Succession (Scotland) Act 1964.

1 See also Sec. 11 and Forms of Notice to Quit.

19.—(1) In the case of a notice to quit given by a landlord in pursuance of section 6(3) of the Agriculture Act 1958 to the tenant of an agricultural holding who is such a near relative of a deceased tenant as is mentioned in sub-section (1) of the last foregoing section, being a notice given between 26th January 1968 and the passing of this Act so as to have effect after the passing of this Act, the said section 6(3) shall not apply and section 25(1) of the principal Scottish Act shall, subject to the following provisions of this section, apply.

(2) In the case of such a notice to quit as is mentioned in the foregoing sub-section, the landlord may, within one month of the passing of this Act, notify the tenant in writing that the said section 6(3) no longer applies to the notice to quit but that, in the event of the tenant serving a notice on him under the next following sub-section, he will apply for the consent of the Scottish Land Court to the operation of the notice to quit on one or more of the following grounds, being a ground or grounds specified in the notification—

 (*a*) the matter mentioned in section 18(2)(*a*) of this Act;
 (*b*) that possession of the holding is required for the purpose of effecting an amalgamation with land specified in the notification;
 (*c*) the matter mentioned in section 18(2)(*c*) of this Act;
 (*d*) one or more of the matters set out in section 26(1) of the principal Scottish Act:

Provided that, if the landlord has not notified the tenant under this sub-section within the said month, the tenant shall be deemed to have served a counter-notice under the said section 25(1), and the period of one month referred to in section 27(1) of the principal Scottish Act shall be deemed to have expired.

(3) The tenant may, within one month of being notified in accordance with the last foregoing sub-section, serve a notice on the landlord requiring that the said section 25(1) shall apply to the notice to quit, and such a notice shall be deemed to be a counter-notice served under the said section 25(1) within the period mentioned therein.

(4) Notwithstanding section 26(1) of the principal Scottish Act, the Scottish Land Court shall consent under the said section 25(1) to the operation of such a notice to quit as is referred to in sub-section (1) of this section if they are satisfied with regard to—

    (a) the matter mentioned in paragraph (a), or

    (b) the matters mentioned in paragraph (b)(i) to (iii), or

    (c) the matter mentioned in paragraph (c),

of sub-section (2) of the last foregoing section:

Provided that, notwithstanding that they are satisfied as aforesaid, the Court shall withhold consent to the operation of the notice if in all the circumstances it appears to them that a fair and reasonable landlord would not insist on possession.

(5) Sub-sections (3) to (5) of section 18 of this Act shall apply to a consent given under the said section 25(1) as applied by this section on the grounds mentioned in sub-section (4)(b) of this section, as they apply to a consent given under the said section 25(1) as applied by the said section 18.

20.—(1) In this Part of this Act—

    'amalgamation' means a transaction for securing that agricultural land which is comprised in a holding to which a notice to quit relates and which together with some other agricultural land could form an agricultural unit, shall be owned and occupied with that other land; and

    'the principal Scottish Act' means the Agricultural Holdings (Scotland) Act 1949.

(2) Unless the context otherwise requires, expressions used in

this Part of this Act and the principal Scottish Act have the same meanings in this Part of this Act as in that Act.

. . . . .

Supplemental

50.—(1) Subject to sub-section (7) of section 45 of this Act, in this Act—

> 'the Minister' means, except in the application of this Act to Scotland, the Minister of Agriculture, Fisheries and Food and, in the application of this Act to Scotland, the Secretary of State;
>
> 'the Ministers' means the Minister of Agriculture, Fisheries and Food and the Secretary of State acting jointly; and
>
> 'notice' means notice in writing.

. . . . .

(3) Any reference in this Act to any enactment is a reference to it as amended, and includes a reference to it as applied, by or under any other enactment including an enactment in this Act.

(4) Where an offence under this Act committed by a body corporate is proved to have been committed with the consent or connivance of, or to be attributable to any neglect on the part of, any director, manager, secretary or other similar officer of the body corporate or any person who was purporting to act in any such capacity, he as well as the body corporate shall be guilty of that offence and shall be liable to be proceeded against and punished accordingly.

In this sub-section 'director', in relation to a body corporate established by or under any enactment for the purpose of carrying on under national ownership any industry or undertaking or part of an industry or undertaking, being a body corporate whose affairs are managed by its members, means a member of that body corporate.

51.—(1) Any power conferred by this Act to make regulations or a scheme or an order (other than an order under section 23(1)(*a*)) shall be exercisable by statutory instrument.

(2) Any statutory instrument containing an order or regulations made under any provision of this Act, other than an order under

section 23(1)(b) and regulations under section 2, shall be subject to annulment in pursuance of a resolution of either House of Parliament.

PART V

(3) No scheme shall be made under this Act unless a draft of the scheme has been approved by each House of Parliament.

(4) Any order or scheme made under any provision of this Act may be revoked or varied by a subsequent order or scheme made thereunder.

(5) Any order, scheme or regulations under this Act may make different provision for different circumstances; and nothing in any other provision of this Act authorising the making of different provision for such different cases as may be specified in that provision shall be construed as prejudicing the generality of the power conferred by this sub-section.

52. The enactments mentioned in Schedule 8 to this Act are hereby repealed to the extent specified in column 3 of that Schedule.

Repeals.

53. There shall be defrayed out of moneys provided by Parliament—

Financial provisions.

    (a) any expenses incurred by virtue of this Act by any Minister or government department (except the Postmaster General); and

    (b) any increase attributable to the provisions of this Act in the sums payable out of such moneys under any other Act;

and any sums received by the Minister or the Ministers by virtue of this Act shall be paid into the Exchequer.

54.—(1) This Act may be cited as the Agriculture (Miscellaneous Provisions) Act 1968.

Short title, commencement and extent, etc.

. . . .

(3) This Act, except Part IV and sections 10, 13, 47 and 48, extends to Scotland, and sections 11, 14 and 16 and Part III of this Act extend to Scotland only.

. . . .

336 AGRICULTURAL HOLDINGS (SCOTLAND) ACTS

SCHEDULES

. . . .

Section 11(9).

SCHEDULE 2

Transitional Provisions Relating to Payments Under
s. 9 in Scotland

1958 c. 71.

1. Where the relevant notice (not being a notice given in pursuance of section 6(3) of the Agriculture Act 1958) is served on the tenant after the initial date but not later than the commencement date and does not contain such a statement as is mentioned in section 25(2)(*c*) of the principal Scottish Act or section 11(1)(*a*) or (*b*) of this Act, then—

   (*a*) if an application for consent in respect of the relevant notice is made in pursuance of section 25(1) of the principal Scottish Act not later than the commencement date, any such statement as is mentioned in the said section 11(1)(*a*) which is included in the application shall be treated for the purposes of section 11 of this Act as included also in the relevant notice; and

   (*b*) if, in a case not falling within sub-paragraph (*a*) above, the landlord serves on the tenant, before or after the commencement date but before the expiration of the period of three months beginning with that date, a notice containing such a statement as is mentioned in the said section 11(1)(*a*) or (*b*) and indicating that the relevant notice is to be treated as having always included that statement, the relevant notice shall be so treated for the purposes of the said section 11.

2. Where the relevant notice is given in pursuance of section 6(3) of the Agriculture Act 1958, is served on the tenant after the initial date but not later than the commencement date and does not contain such a statement as is mentioned in section 11(7)(*b*) of this Act, then, if the landlord serves on the tenant, before or after the commencement date but before the expiration of the period of three months beginning with that date, a notice containing such a statement as is mentioned in the said section 11(7)(*b*) and indicating that the relevant notice is to be treated as having always

included that statement, the relevant notice shall be so treated for the purposes of section 11 of this Act:

Provided that this paragraph shall not have effect where the relevant notice is a notice to which, apart from the provisions of section 19 of this Act, the said section 6(3) would apply.

3. Where the relevant notice is a notice to which, apart from the provisions of the said section 19, the said section 6(3) would apply, and the landlord in a notification to the tenant under section 19(2) of this Act specifies a matter set out in section 26(1) of the principal Scottish Act, then—
   (a) in the case of a matter set out in paragraph (a), (b) or (c) of the said section 26(1), the relevant notice shall be treated for the purposes of section 11(1)(a) of this Act as if it had always contained a statement of that matter as a ground on which the carrying out of the purposes for which the landlord proposes to terminate the tenancy is desirable;
   (b) in the case of the matter set out in paragraph (d) of the said section 26(1), the relevant notice shall be treated for the purposes of section 11(1)(b) of this Act as if it had always contained a statement that the landlord would suffer hardship unless the notice had effect.

4. Where either—
   (a) the relevant notice is served on the tenant not later than the commencement date and contains such a statement as is mentioned in the said section 11(1)(a) or (b); or
   (b) a notice is served on the tenant under paragraph 1(b) of this Schedule,

and in either case no counter-notice in respect of the relevant notice in question has been served in pursuance of section 25(1) of the principal Scottish Act and the period during which such a counter-notice may be served has expired, the tenant may, within the period of three months beginning with the commencement date or, where the notice under paragraph 1(b) of this Schedule is served on a later date, beginning with the later date, make an application to the court for a determination that the court are satisfied that the relevant notice was given in order that the land in question may be used otherwise than for agriculture.

5. Where the court have, on or before the commencement date, given a decision consenting under section 25(1) of the

principal Scottish Act to the operation of the relevant notice and either—

(a) the reason given by the court for their decision is that they are satisfied as to the matter mentioned in section 26(1)(b) of that Act; or

(b) the reasons so given include that reason but not the reason that they are satisfied as to the matter mentioned in section 26(1)(e) of that Act,

the tenant may, at any time before the expiration of the period of three months beginning with the commencement date, make an application to the court for a determination that the reasons for their decision would have included the reason that they were satisfied as to the matter mentioned in the said section 26(1)(e) if that matter had been specified in the application for consent.

6. Where the court make a determination under paragraph 4 or paragraph 5 of this Schedule, section 11(1) of this Act shall not apply in relation to the relevant notice in question.

7. In this Schedule—

'the commencement date' means the date of the passing of this Act; and

'the court' and 'the relevant notice' have the same meanings as in section 11 of this Act.

. . . . .

Section 14(3).

SCHEDULE 4

Supplementary Provisions with Respect to Payments under s. 12(1) in Scotland

1. Subject to paragraph 4 of this Schedule, any dispute with respect to any sum which may be or become payable by virtue of section 12(1) of this Act shall be referred to and determined by the Lands Tribunal for Scotland.

2. If in any case the sum to be paid by virtue of the said section 12(1) to the tenant of an agricultural holding by an acquiring authority would, apart from this paragraph and paragraph 3 of this Schedule, fall to be ascertained in pursuance of section 9(2) of this Act by reference to the rent of the holding at a rate

which was not determined by arbitration under section 7 or section 8, or by the Scottish Land Court in pursuance of section 78, of the principal Scottish Act and which the authority consider is unduly high, the authority may make an application to the Lands Tribunal for Scotland for the rent to be considered by the Tribunal.

3. Where, on an application under paragraph 2 above, the tribunal are satisfied that—
- (a) the rent to which the application relates is not substantially higher than the rent which in their opinion would be determined for the holding in question on a reference to arbitration duly made in pursuance of section 7 of the principal Scottish Act (hereafter in this paragraph referred to as 'the appropriate rent'); or
- (b) the rent to which the application relates is substantially higher than the appropriate rent but was not fixed by the parties to the relevant lease with a view to increasing the amount of any compensation payable, or of any sum to be paid by virtue of the said section 12(1), in consequence of the compulsory acquisition or taking of possession of any land included in the holding,

they shall dismiss the application; and if the tribunal do not dismiss the application in pursuance of the foregoing provisions of this paragraph they shall determine that, in the case to which the application relates, the sum to be paid by virtue of the said section 12(1) shall be ascertained in pursuance of the said section 9(2) by reference to the appropriate rent instead of by reference to the rent to which the application relates.

For the purposes of sub-paragraph (a) of this paragraph, section 7(1) of the principal Scottish Act shall have effect as if for the reference to the next ensuing day there mentioned there were substituted a reference to the date of the application mentioned in the said sub-paragraph (a).

4. The enactments mentioned in paragraph 5 of this Schedule shall, subject to any necessary modifications, have effect in their application to such an acquiring of an interest or taking of possession as is mentioned in sub-section (1) of section 12 of this Act (hereafter in this paragraph referred to as 'the relevant event')
- (a) in so far as those enactments make provision for the doing, before the relevant event, of any thing connected

with compensation (including in particular provision for determining the amount of or the liability to pay compensation or for the deposit of it in a Scottish bank or otherwise), as if references to compensation, except compensation for damage or injurious affection, included references to any sum which will become payable by virtue of the said sub-section (1) in consequence of the relevant event; and

(b) subject to sub-paragraph (a) above, as if references to compensation (except as aforesaid) included references to sums payable or, as the context may require, to sums paid by virtue of the said sub-section (1) in consequence of the relevant event.

5. The enactments aforesaid are—

1845 c. 19. (a) the following provisions of the Lands Clauses (Scotland) Act 1845, that is to say, sections 56 to 60, 62, 63 to 65, 67 to 70, 72, 74 to 79, 83 to 87, 114, 115 and 117;

1947 c. 42. (b) paragraph 3 of Schedule 2 to the Acquisition of Land (Authorisation Procedure) (Scotland) Act, 1947;

1963 c. 51. (c) Parts I and II and section 40 of the Land Compensation (Scotland) Act 1963;

1968 c. 16. (d) paragraph 4 of Schedule 6 of the New Towns (Scotland) Act 1968;

(e) any provision in any local or private Act, in any instrument having effect by virtue of an enactment or in any order or scheme confirmed by Parliament or brought into operation in accordance with special parliamentary procedure, corresponding to a provision mentioned in sub-paragraph (a), (b) or (d) of this paragraph.

1949 c. 42. 6. Until sections 1 to 3 of the Lands Tribunal Act 1949 come into force as regards Scotland, this Schedule shall have effect as if for any reference to the Lands Tribunal for Scotland there were substituted a reference to an official arbiter appointed under Part I of the Land Compensation (Scotland) Act 1963; and sections 3 to 5 of that Act shall apply, subject to any necessary modifications, in relation to the determination of any question under this Schedule by an arbiter so appointed.

SCHEDULE 5

Sch. 5
Section 16.

Modifications of Part II for Statutory Small Tenants in Scotland

1. In section 9(3), at the end there shall be added the words 'and the reference to compensation for disturbance becoming payable to the tenant of an agricultural holding under the principal Scottish Act shall include a reference to the like compensation becoming payable to a statutory small tenant under section 13 of the Small Landholders and Agricultural Holdings (Scotland) Act 1931'.

1931 c. 44.

2. In section 11(5), after the words 'Scottish Act', there shall be inserted the words 'and that Act as read with section 32 of the Act of 1911' and for the words 'that Act' there shall be substituted the words 'the principal Scottish Act'.

3. In section 11(6), in paragraph (b) after the word 'lease', there shall be inserted the words ', or of the holding or part of the holding of a statutory small tenant on being so authorised by the Scottish Land Court under section 32(15) of the Act of 1911'.

4. In section 15(1), at the end there shall be added the following sub-section—

'(1A) Except where compensation assessed in accordance with this sub-section would be less than if this sub-section were disregarded, in assessing the compensation payable by an acquiring authority to a statutory small tenant as defined in the Act of 1911 in connection with such an acquiring of an interest or taking of possession as is mentioned in section 12(1) of this Act, any authorisation of resumption of the holding or part thereof by the Scottish Land Court under section 32(15) of the Act of 1911 for any purpose (not being an agricultural purpose) specified therein shall—

(a) in the case of an acquisition, be treated as if it became operative only on the expiration of twelve months from the end of the year of the tenancy current when notice to treat in respect of the acquisition was served or treated as served on the tenant; and

(b) in the case of a taking of possession, be disregarded.'

5. In section 15(3), after the word 'section' there shall be inserted the words 'or the landlord of the holding of a statutory small tenant resumes the holding or part thereof on being so authorised by the Scottish Land Court under section 32(15) of the Act of 1911'.

6. In Schedule 4, in paragraph 2 after the words 'Scottish Act', there shall be inserted the words 'or in the case of a statutory small tenant was not fixed by the Scottish Land Court in pursuance of sub-sections (7) and (8) of section 32 of the Act of 1911'.

7. In Schedule 4, in paragraph 3(*a*) after the word 'Act' there shall be inserted the words 'or in the case of a statutory small tenancy, the equitable rent which in their opinion would be fixed by the Scottish Land Court in pursuance of the said sub-sections (7) and (8)'.

8. In this Schedule, 'the Act of 1911' means the Small Landholders (Scotland) Act 1911.

INDEX

ACTS
amendments of other 202, 218 et seq., 302 et seq.
construction of references in other - - - - - 202
dates of operation - - - 2
objects of - - - - - 1
repeals, savings and transitional provisions 203, 220, 308, 335, 336
repealed, where counter-notice refers to - - - 139
savings, general - 204 et seq.
savings for other rights - - 206

ACTS (CITED)
Act of Regulations, 1695 - 85
Adoption Act, 1950 - - - 305
Acquisition of Land (Authorisation Procedure) (Scotland) Act, 1947 - - - 340
Agricultural Development Act, 1939 - - 175, 220
Agricultural Holdings (Scotland) Act, 1883 - 200, 201
Agricultural Holdings (Scotland) Act, 1908 46, 179, 203, 205
Agricultural Holdings (Scotland) Act, 1910 - - - 188
Agricultural Holdings (Scotland) Act, 1923 1, 2, 46, 48, 68, 76, 105, 114, 115, 116, 118, 157, 160, 164, 176, 196, 202, 204, 206, 220, 283, 287, 289, 300
Agricultural Holdings (Scotland) Act, 1931 40, 46, 68, 118, 164, 196, 202, 206, 283
Agricultural Marketing Act, 1931 - - - 112, 114
Agricultural Marketing Act, 1958 - - - - - 114

Agriculture Act, 1937 31, 162, 220
Agriculture Act, 1958 1, 2, 3, 8, 9, 16, 17, 22, 23, 28, 52, 105, 111, 113, 117, 130, 131, 137, 138, 139, 142, 145, 146, 168, 181, 209, 228, 241, 242, 302 et seq., 323
Agriculture Act, 1967 28, 139, 187, 323, 329
Agriculture (Amendment) Act, 1923 - - - - 220
Agriculture (Miscellaneous Provisions) Act, 1943 115, 206, 221
Agriculture (Miscellaneous Provisions) Act, 1963 3, 59, 63, 64, 217, 294
Agriculture (Miscellaneous Provisions) Act, 1968 1, 2, 3, 18, 26-29, 50, 58, 59, 130, 137, 142, 146, 152, 154, 175, 184, 185, 186, 190, 191, 192, 195, 202, 227, 228, 229, 304, 320, 322 et seq.
Agriculture (Miscellaneous War Provisions) Act, 1940 206
Agriculture (Safety, Health and Welfare Provisions), Act, 1956 - - - - 116
Agriculture (Scotland) Act, 1948 2, 4, 6, 22, 24, 46, 47, 48, 105, 115, 120, 125, 142, 148, 195, 196, 197, 204, 205, 221, 296, 300, 301, 309
Allotments (Scotland) Act, 1922 - - - - - 206
Arbitration (Scotland) Act, 1894 - - - - 94, 187
Bankruptcy (Scotland) Act, 1913 - - - - 134, 141

INDEX

Companies Act, 1948 - - 191
Conveyancing (Scotland) Act, 1924 - - - - 316
Crown Estate Act, 1961 - - 192
Crown Proceedings Act, 1947 65, 68
Crofters Holdings (Scotland) Act, 1886 - - 317, 318, 320
Crofters (Scotland) Act, 1955 320
Entail (Scotland) Act, 1914 - 89
Entail Acts - - - - - 190
Emergency Powers (Defence) Acts, 1939-1940 - - - 195
False Oaths (Scotland) Act, 1933 - - - - - - 76
Hill Farming Act, 1946 45, 85, 89, 98, 99-101, 114, 115, 219, 221, 288 et seq.
Housing (Scotland) Act, 1964 116, 167, 168, 210
Housing (Scotland) Act, 1966 116, 167, 168, 210
Hypothec Abolition (Scotland) Act, 1880 - - - 129
Income Tax Act, 1918 - 26, 151
Income Tax Act, 1952 - - 154
Interpretation Act, 1889 194, 205, 206, 227, 296
Land Compensation (Scotland) Act, 1963 - - - 340
Lands Clauses (Scotland) Act, 1845 - - - - - 340
Lands Tribunal Act, 1949 - 340
Limited Partnership Act, 1907 50
Livestock Rearing Act, 1951 - 116
Local Government (Scotland) Act, 1929 26, 112, 151, 220
New Towns (Scotland) Act, 1968 - - - - - - 340
Opencast Coal Act, 1958 112, 116, 123, 142, 165, 168, 169, 171, 172, 184, 190, 209, 210, 213
Recorded Delivery Service Act, 1962 - - - 136, 195
Registration of Leases (Scotland) Act, 1857 - - - 185
Removal Terms (Scotland) Act, 1886 - - - 19, 134
Reserve and Auxiliary Forces (Protection of Civil Interests) Act, 1951 - - 139, 314
Sheep Stocks Valuation (Scotland) Act, 1937 85, 89, 96, 285 et seq., 288, 289
Sheriff Courts (Scotland) Act, 1907 20, 134, 135, 136, 137, 227
Small Landholders (Scotland) Acts, 1886-1931 22, 53, 148, 154, 186, 329, 341, 342
Small Landholders and Agricultural Holdings (Scotland) Act, 1931 158, 160, 218, 286, 290, 320
Statute Law Revision Act, 1953 - - - - - - 175
Succession (Scotland) Act, 1964 1, 2, 3, 16, 17, 18, 130, 132, 243, 305, 306, 316 et seq., 325, 330
Town and Country Planning (Scotland) Act, 1947 - - 327
Tribunals and Inquiries Act, 1958 - - 8, 68, 79, 277
Valuation and Rating (Scotland) Act, 1956 - - 114, 154
Water (Scotland) Act, 1946 185, 191

ADDITIONAL PAYMENTS UNDER 1968 ACT
3, 25-26, 28, 58, 152, 228, 229, 322

AGENT OF LANDLORD OR TENANT - - - - - - 198

AGREEMENTS
Compensation under 91, 118, 160, 176, 238
Contracting out - 47, 49-51
fixed equipment, maintenance of - - - - - 109, 111
forms of - - 233, 238, 275
game damage, compensation for - - - 46-48, 125
implements etc., sold on quitting - - - - - 133
improvements executed under see Improvements.

INDEX

incoming tenant paying outgoing tenant's claims 9, 118, 164, 169
landlord and incoming tenant between - - - 91, 238
lease, where no written 5-6, 108, 234
letter of removal, as - - - 135
market gardens, compensation relating to, *see* Market Gardens.
notice of improvements, to dispense with - 35, 157, 167
notice to quit, holding agreed to be sold - - - 146, 233
rent, increase of - - - 114
settlement of claims on termination of tenancy - - 183
substituted compensation under, *see* "Substituted Compensation".
supplementary, to vary obligations - - - - 109, 275
tenancy, demand for execution of - - - - 108, 234

AGRICULTURAL EXECUTIVE COMMITTEES 2, 115, 116, 168, 181

AGRICULTURAL HOLDING, *see* HOLDING.

AGRICULTURAL LAND - - - 4

AGRICULTURAL LAND TRIBUNAL 303, 315

ALLOTMENTS - - - - - 148

APPEAL
from Sheriff Substitute to Sheriff incompetent - 78, 195
to Court of Session - 78, 85, 216

ARBITER
acceptance of office - - 7, 68
minute of - - - - - 268
appointment of - - 7, 214
  by agreement - - - - 66
  by Land Court - - 67, 189
  by Secretary of State 65, 66-67, 188, 214, 280
  forms - - - - 280

in valuations - - - 93-94
in writing - - - - 67
irrevocable except of consent - - - - - 68
minute of - - - 70, 253
joint intimation of completion - - - - 254
revocation of - - 68, 214, 255
Secretary of State, a party - 189
award, *see* Award
bias showing - - - - 88
counsel submitting questions of law to - - - 74, 102
crime, convicted of - - - 69
damages, power to assess - 102
death of or unable to act 68, 214
devolution by - - - - 101
minute of - - - - - 255
discretion of - - - 8, 70
disqualification and removal of - - - - - 69, 216
application to Sheriff for - 69
documents, power to call for - 75
duties, in arbitration for written lease - - - - - 5
duties in valuations - - - 101
duties in variation of rents - 7
expenses, power to award 80, 102, 216
incidental questions, power to deal with - - - - 89
interdict of 65, 68, 69, 70, 136
interest in dispute, having - 69
jurisdiction of - - - 66, 70
law agent, entitled to call in 74, 102
men of skill, assistance of 74, 102
misconduct or improper conduct of - - 69, 78, 216
named, agreement to appoint 66
objection to competency - 69
orders, incidental, forms of - 256
oversman, *see* Oversman
panel of arbiters - 67, 188, 214
personal knowledge of - - 74
proceed, compelled to 68, 101
proposed findings, note of 66, 79
qualifications of - - - 54
remuneration of - 68, 81, 188
resignation of - - - - 68
state a case, order for 77, 216, 261 *et seq.*
state a case, refusal to 77, 87, 218

INDEX

ARBITRATION
  See also under matters referred to arbitration.
  arbiter, see Arbiter
  claims, see Claims
  clerk in, see Clerk
  competency - - - 66-67
  death of party - - - - 68
  documents called for 72, 214, 256
  evidence - - - 73-76, 215
  expediting, regulations for
    187, 217
  expenses of - - - 80, 188, 216
  expert opinion or advice 74, 102
  forms - - - - - 216, 253
  hearing and inspection - - 72
  incidental questions in 89, 256
  irregularities in - - - - 87
  jurisdiction of arbiter - 66, 70
  Land Court, references to,
    see Land Court
  matters to be referred to 54, 186
  outside the Act, see VALUATION
  procedure in - 55, 70 et seq.
  proof - - - - - 72, 73
  prorogation, minute of - 269, 270
  provisions as to - - - 187, 214
  reference under S.4 or S.5(5)
    of 1949 Act - - - 4, 235
  reference under S.7 of 1949
    Act - - - - - - 6
  relevancy, questions of 53, 64-66
  stated case - 54, 76 et seq., 89
    216, 261
  statement of case and particulars of claims
    48, 56, 59, 63, 64, 214, 215, 249
  submission, minute of, see
    Submission
  suspension and interdict - 54
  tenancy, matters arising out
    of - - - 58, 63, 186
  tenants, outgoing and incoming questions between
    54, 55
  terms of lease adjustment of - 61
  'valuation' and 'arbitration'
    distinction between - - 89

ARBITRATIONS AND VALUATIONS
  (outside Act) - 89 et seq., 215

AWARD - - - - 82 et seq.
  alteration or cancellation of - 84

  alternative - - - - - 83
  amounts for claims to be
    stated separately - - 83, 215
  common law submission, in - 267
  expenses to be included in
    80, 102, 216
  form of - - 83, 215, 267, 278
  interim - - - 83, 215, 217
  mistakes in - - 84, 95, 215
  not probative - - - - 84
  partial - - - - - - 83
  proposed findings - 66, 79, 258
  reasons for - - - - 8, 68, 79
  recording of - - - - 84, 185
  reduction of, see Reduction
    of Award
  setting aside, see Reduction
  stamping of - - - - 84
  supplement to - - - - 84
  time for making of 82, 101, 215
    extension by parties
      101, 215, 255, 257
    extension by Secretary of
      State - 82, 215, 277 et seq.
  time for payment of sums
    awarded
      84, 184, 215, 277 et seq.
  valuations in - - - 101, 102

BAD HUSBANDRY
  see Certificate of Bad Husbandry

BANKRUPTCY OF TENANT
  62, 134, 137, 139, 141, 200

BENEFIT
  feeding stuffs, re - - - 38
  improvements - 31, 32, 38, 39, 43, 49, 161, 166
  land resumed - - - 149, 174
  temporary pasture - - - 39
  variation of rent - - 112, 114

BEQUEST OF LEASE
  16-19, 129, 303, 320
  contracting out of - - 49, 130
  intimation of - - - 2, 130
  legatee, see Legatee
  to whom allowed - 16, 129, 131

BRACKEN, etc.
  33, 36, 38, 209, 211, 212

BREACH OF CONTRACT - - 40, 181

INDEX

BREACH OF LEASE
  see Lease

BRIDGES, MAKING OF
  34, 208, 210, 212

BUILDINGS AND FIXTURES
  arbitrations - - - 60, 108
  common law fixtures - - 13
  compensation for - - 52, 60
  contracting out of the provisions, re - - - - - - 49
  definitions, see Words and Phrases
  erection of buildings 208, 212, 213
  fire insurance premium, re 5, 109
  fixed equipment, see Fixed Equipment
  fixtures, intention to remove 12, 60, 238
    dilapidation or deterioration of - - - - - 112
    landlord electing to purchase - - 12, 60, 123
    purchased by landlord, payment for - - - 62, 123
    repair of - - - - 211, 213
    tenant's right to remove 12, 60, 123
    valuation of - 12, 60, 123
  improvements which require consent 32, 56, 57, 210, 212
    which require notice 34, 159, 208, 212
  liability for maintenance 62, 108 et seq.
    contracting out - - - 109
    provision limiting liability of tenant - - - - 5
    tenant's failure - - - 141
    where transferred 5, 61, 110
  market garden improvements 52, 177, 213
  record of - - - - 58, 108
  renewal of - - - - - 109
  repairs, agreed compensation for - - - - - - 36
    compensation for 36, 41, 59, 159, 209, 211, 213
    consent to - - - 32, 36
    extraordinary - - - - 41
    notice of - 56, 57, 160, 209
    replacement of - - - 109

CANAL, MAKING OF - - - 148

CERTIFICATE OF BAD HUSBANDRY
  application for - - 23, 145, 312
  prevention of deterioration after grant of - - - 145, 312

CHALKING OF LANDS 209, 211, 212

CHARGE ON HOLDING
  51, 185, 190, 191

CLAIMS
  See also under the particular subjects for which claims are made
  adjustments, extension of time for - - - - 252, 257
  arbiter, order by, for - - 256
  holding vested in more than one person - - - - 174
  landlord by - 63 et seq., 266
  notice, see Notice
  forms of - - - 247, 248
  particulars of
    48, 56, 59, 63, 64, 214
  receiving claims and objections - - - 54, 165, 266
  relevancy and competency of 64 et seq.
  settlement of, on termination of tenancy - - 183 et seq.
    application to extend time for - - - - - 184
    by arbitration - - - 183
    statement of case and particulars of - 56, 64, 249
  tenant by - 55 et seq., 259, 266

CLAY BURNING - - - 209, 211

CLAYING OF LAND - 209, 211, 212

CLERK IN ARBITRATION
  appointment of - - 66, 71
  forms - - - - 260, 268
  fees, etc. of - - - 71, 82, 216
  layman as - - - - - 74

CLOVER - - - 66, 211, 213

COAL, WORKING OF - - - 148

COMMISSIONER OF CROWN LANDS - - - - - 192

COMMON LAW
  arbitration or valuation, sub-

348 INDEX

jects of - - 54, 89 et seq.
submission, form of - - - 263
submission, award, form of - 267

COMPANY, AS TENANT - - - 50

COMPENSATION
See also under the particular subjects in respect of which claims are made.
additional payments under 1968 Act, see Additional Payments under 1968 Act.
adjustment of, in respect of ploughing grants - - - 175
agreed scale - - - 36, 45
agreements, under 91, 118, 160, 176, 238
assessment of - - - 46, 55
bankruptcy of tenant - - 62
change of tenancy, after - - 163
charging of estate with see Charge on Holding
claims for see Claims Contracting out - - 49, 126
fair and reasonable - 45, 160
holding vested in more than one person - - - - 174
not payable for anything done in compliance with the Act 175
part of holding, in respect of 173
payable by purchaser - - 199
procedure for ascertainment and recovery - - - 46
recovery of - - - 46, 184
rights to under 1949 Act in lieu of rights under earlier Acts - - - - - 48, 206
substituted see Substituted Compensation
tables prepared by Secretary of State - - - - - 37
tenant gives notice to quit where - - - 53, 180
tenant quitting before commencement of the Act - 203
tenant, removed for non-payment of rent, where 16, 129
tenant paying to outgoing tenant - -118, 164, 169, 238

COMPULSORY ACQUISITION
325 et seq.

CONSENT OF LAND COURT
conditions accompanying 23, 142
notice to quit, to 18, 21, 22, 27, 62, 138, 139, 141 et seq., 153, 312, 323

penalty for breach of conditions - - - - 23, 145

CONSENT OF LANDLORD
agreements between incoming and outgoing tenants - - 117
application for - - - - 243
compensation for improvements paid by incoming tenant - - - 117, 164
improvements, form of consent - - - - - - 244
to, as condition of compensation - - 156, 166
to which consent is required 32, 56, 57, 166, 208, 210, 219
to which consent is not required 35 et seq., 209, 211, 212
limited owner's powers - - 190
market garden improvements 51, 178

CONSENTS, VALIDITY OF - - 192

CONSUMING VALUE- - - - 95

CONTRACTING OUT 49 et seq., 126
(and see Agreements)

CORN, CONSUMPTION OF 36, 209, 211, 213

COTTAGES, ERECTION OF - - 147

COURT OF SESSION
appeal to 78, 85, 186, 189, 216, 218
power to compel arbiter to proceed - - - - 68, 101

CROP, AWAY-GOING VALUATIONS 90, 93, 95, 103, 187

CROPPING
custom of - - 31, 118, 120

INDEX 349

freedom of   11, 31, 39, 118 *et seq.*, 162, 172
  let for four years - - 198
  restrictions - - - - 162
  system of, meaning - - 120

CROWN LAND
  application of Act to - - 192

CUMULATIVE FERTILITY - 42, 170

CUSTOM
  cropping, of - - 31, 118, 120
  Evesham custom   53, 179, 181
  fodder, consumption of - - 91
  varying rights of tenants at away-going - - 91, 92, 93

DAIRYING PLANT, FIXED   34, 208

DAMAGES
  arbiter, assessment by   54, 90, 121
  deterioration for - 12, 54, 119
  liquidated - - - 14, 126
  measure of - - - 121, 126
  temporary pasture, in respect of - - - - - 161

DEER *see* Game

DEFENCE REGULATIONS 115, 116, 159, 195
DEFINITIONS, *see* WORDS AND PHRASES

DEPARTMENT OF AGRICULTURE
  appointment of arbiter by   67
  appropriate authority, the   159
  landlord, as - - 67, 189

DE-RATING - - - 112, 114

DETERIORATION OF HOLDING
  arbitration, claim referred to   54
  compensation for   31, 63, 163, 171
    conditions for payment of   63
  damages for and interdict against - - - 12, 54, 119
  notice of claim, *see* Notice by landlord
  powers of supervision etc., repeal of - - - - - 302
  prevention of, after grant of certificate - - 145, 212
  record of holding, required in claim - - - 128, 172
  rent, arbitration effect - - 112

DILAPIDATIONS   15, 31, 112, 170

DIPPERS - - - - - - 208

DISTURBANCE
  compensation for 16, 17, 23, 24, 25, 26, 51, 52, 59, 132, 149, 150 *et seq.*, 229, 322
  amount of - 24, 25, 26, 150
  charge on holding, in respect of - - - - 190
  conditions for payment of 58, 150 *et seq.*
  contracting out - - - 50
  expense of preparing claim   150
  expense of removal - - 150
  holding vested in more than one person - - - 174
  notice to quit, *see* Notice to Quit
  part of holding - - 151, 154
  sub-tenant, compensation to 26, 151
  sum for reorganisation of tenant's affairs   26 *et seq.*, 152, 322

DOCUMENTS
  arbiter's order, form of - 256
  production of - 72, 75, 215
  for purposes of sheep stock valuation - - - - 289

DRAINAGE
  improvements 33 *et seq.*, 158, 208, 211
  substituted compensation - 45

DUNG
  benefit - - - - - - 38
  compensation for - - 38, 44
  incoming tenant, bound to take 91 *et seq.*
  left, steelbow - - - - 38
  succession in question of - 38
  valuing 44, 90, 91 *et seq.*, 103, 187

EARLY RESUMPTION, COMPENSATION FOR - - 28, 59, 327

ELECTRICAL EQUIPMENT 34, 208, 212
EMBANKMENTS   33, 34, 208, 211, 212

INDEX

ENTAILED ESTATES
  power to apply money for improvements - - - - 190
  sheep stock taken over by 89, 199

ESTATE MANAGEMENT
  good - - - - 197, 214
  sound, distinguished from good - - - 142, 143

EVESHAM CUSTOM - 53, 179, 181

EVIDENCE
  competency of - - - - 95
  oath on - - - 74, 75
  refusal to listen to by arbiter 73
  shorthand notes of - - - 74
  witnesses, citation of - - 76
    exclusion of - - - - 76
    expense of - - - - 81
    perjury of - - - - 76
    skilled - - - - - 74
    subject to any legal objection 74
  valuations in - - - 90, 95

EXPENSES
  agent and client basis - - 81
  arbiter, against the - - - 82
  arbitration - 80 *et seq.*, 216
  award by arbiter - 80, 102, 215
  claim for compensation 80, 133
  clerk, of - - - 80, 82, 216
  discretion of arbiter - - 81, 216
  disturbance claim - - - 80
  extrajudicial type - - - 80
  Secretary of State, incurred by 193
  stated case of - 79, 82, 219
  taxation, subject to - 81, 216
  time for payment of - 84, 215
  witnesses of - - - 81, 216

FALLOW GROUND
  incoming tenant bound to take bare - - - - 91
  valuation of - - - 90, 187

FEEDING STUFFS
  consumption on holding 36, 37 *et seq.*, 43 *et seq.*
  unexhausted value - - 37, 91

FENCES
  in bad order - - - - 200
  making of - - 208, 210

removal of - - 208, 210
valuation of 90, 95, 103, 187

FEUING - - - - - 21

FIARS' PRICES - - - 103, 104

FIRE, DAMAGE BY - - - - 214

FIXED EQUIPMENT
  definition of - - - - 196
  fire insurance of - - - - 109
  fixtures distinguished from - 123
  maintenance of 5, 108 *et seq.*, 235
  record of - - 14, 108 *et seq.*
  repair of - - - 34, 109
  works of this nature, executed by landlord - - - - 116

FIXTURES, MACHINERY AND BUILDINGS - - - - 60

FIXTURES, *see* Buildings and Fixtures

FODDER, VALUATION OF - 91, 95

FOLDS - - - - - - 210

FORDS - - - - - - 208

FORMS
  Agricultural Holdings (Scotland) Forms 1912, 1926, 1934 278
  Agricultural Holdings (Specification of Forms) (Scotland) Instrument 1960 - 277
  list of - - - - 224 *et seq.*

FREEDOM OF CROPPING
  see Cropping

FRUIT BUSHES - 210, 212, 213

FRUIT TREES
  planting of - - - 33, 52, 213
  protection of - 209, 210, 212
  removal of - - - 52, 177

GAME, DAMAGE BY
  compensation for 46 *et seq.*, 52
  conditions of payment of 60, 125
  contracting out 47, 50, 126
  landlord's right to indemnity 47, 125
  notice of claim for 47, 48, 125
    forms - - - 239, 240
  where right to take game not vested in landlord - 47, 126

INDEX

GARDENS 33, 47, 147, 208, 210, 212

GRAIN, Valuation of - - 103, 104

GRANT COMMITTEE ON SHERIFF
COURTS - - - - - 103

GRANTS, STATE 7, 10, 45, 115, 175

GRASS - - - - - 39, 91

GRAZING - - 4, 106, 138

GROUSE *see* Game

GUARDIAN
appointment of to landlord or tenant - - - - - 191

HAULAGE - - - - 33, 208

HAY SHEDS - - - - - 34

HEIR OF ENTAIL - - - - 190

HIGH FARMING
37, 42 *et seq.*, 128, 170, 247

HERITAGE IN SUCCESSION - - 316

HOLDING
affected by Act - - - - 4
charge on 51, 185, 190, 191
definition of - - 155, 198
deterioration of, *see* Deterioration
entry and inspection
72 *et seq.*, 128
improvements, *see* Improvements
market garden, to be treated as 62
particulars of, to be included in written lease - - - 213
protection of - - - - 11
record of, *see* Record of Holding
resumption of, *see* Resumption of Holding
sale of, *see* Sale of Holding
succession to, *see* Succession
tenant's obligations - - - 11
vested in more than one person - - - - - 174

HOP GARDENS - - - - 211

HUSBANDRY
bad, *see* Certificate of Bad Husbandry
deterioration of holding, *see* Deterioration of Holding
fertility, improvement made to restore - - - - 55
freedom of cropping, *see* Cropping
high farming
37, 42 *et seq.*, 128, 170, 247
good - - 141, 143
rules of good, *see* Rules of Good Husbandry
special standard - - - 169
claim for - - - - 170

IMPLEMENTS
payment for - 62, 93, 133
sheds for - - - - - 208

IMPROVEMENTS
buildings to, *see* Buildings and Fixtures
compensation
1, 3, 16, 29 *et seq.*, 52
agreed for new - 34, 166
agreed for old 157, 158, 160
agreed for temporary 36, 117
allowed by Act - 29, 33
"benefit" taken into consideration - 31, 32, 38
conditions for payment of
34, 35, 42
customary - - - - 156
fair and reasonable - - 45
general condition of holding - - - 30, 172
leases entered into on or after 1st January, 1921 - - 30
may depend on time begun 30
measure of - 29, 166
notice of claim - - - 42
payable in spite of contravention of lease - - 36
reduced or excluded 31, 161
substituted, *see* Substituted Compensation
where made in previous tenancy - - - - 30

INDEX

where there is sub-tenant - 30
where tenant removed for non-payment of rent - 16
where tenant undertook execution - - 30, 32
compensation for new
    29, 34, 36, 42, 45, 165 et seq., 208 et seq.
  agreed - - - 34, 166
  amount of - - - - 166
  change of tenancy - - 168
  charge on holding, in respect of - - - - - - 190
  conditional on approval by Land Court - - - 167
  conditional on consent of landlord - - 32, 166
  conditional on notice to landlord - - 166, 208
  right of tenant paying to outgoing tenant 91, 169
compensation for old
    29, 34, 36, 42, 43, 155 et seq., 210 et seq.
  amount of - - - - 156
  conditional on consent of landlord - 156, 210, 212
  conditional on notice to landlord - 157, 211, 212
  right of tenant paying to outgoing tenant - - 164
continuous adoption of a special standard and system of farming
    3, 14, 32, 42 et seq., 48, 58, 113, 169, 209, 247
  (see also High Farming)
consent to, by Land Court, see Consent of Land Court
contracted for - - 3, 32
entailed moneys, power to apply - - - - - 190
fixed equipment, see Fixed Equipment
increase of rent, due to 10, 112
Land Court, certain questions referred to - - 35, 54
landlord, undertaken by 34, 160
landlord's failure to carry out
    35, 56, 57, 58, 168
landlord's objections to tenant's - - - 34, 167
manurial improvements - 38

market garden improvements, see Market Garden
1923 Act Improvement
  29, 34, 56, 155, 156, 157, 197
1931 Act Improvement
  29, 34, 40, 56, 155, 156, 157
notice by landlord - 34, 35
notice of, agreement to dispense with - 35, 157, 167
old, carried out before 1909
    46, 203
pasture, see Pasture
permanent - - 32 et seq.
previous tenant's - - - 30
record of holding in - - 14, 15
repealed statutes, made under 46
scheduled
    31 et seq., 43 et seq., 56 et seq., 208 et seq.
scheme for - - - - - 219
State grant for - - 10, 112
sub-tenant, effected by - 30
temporary 35 et seq., 208
unexhausted value of 37, 164
value of - 30, 43 et seq.
variation of rent, on- - - 10

INSPECTION OF HOLDING
by arbiter - - 9, 72, 88
form of intimation - - 257
landlord's - - - - - 128
on authority of Secretary of State - - - - - 193
provisions as to - - - 193

INSURANCE
liability for premiums 5, 108
sums recovered under fire policies - - - - - 133

INTERDICT
arbiter, of 65, 68, 69, 70, 136
landlord's remedy of
    12, 54, 119, 121
Secretary of State appointing arbiter - - - - - 65

INTERPRETATION CLAUSE 195 et seq.

IRRIGATION 33, 208, 210, 212

IRRITANCY OF LEASE 14, 134, 137, 141

INDEX

LAND COURT
acquirer of lease, in case of objection to - - 17, 132
applications to, forms of 225, 253
appointment of arbiter by 67, 189
certificate of bad husbandry 312
consent of, *see* Consent of Land Court
definition - - 196, 307
improvements, approval of 34, 58, 167
legatee, in case of objections to 16, 132
"market garden direction" 53, 62, 180, 181
proceedings of - - - - 186
record of holding, differences in - - - 15, 127
references to, in lieu of arbitration 27, 54, 97, 189
not subject to stated case or appeal - - - - - 189
rent, variation of, by - - 9
Rules of - - - 55, 189
sheep stock valuations 89, 90, 97, 99, 285, 288
substituted for Secretary of State - - - - - 192

LAND IMPROVEMENT COMPANIES 191

LANDLORD
agent of - - - - - 198
agreement with incoming tenant - - - 91, 238
change of, notice of 194, 249
charge on holding, by - - 190
claims by - - 63 *et seq.*, 113
consent of, *see* Consent of Landlord
Crown Lands in relation to - 192
definition of 3, 4, 129, 194, 196, 199
Department of Agriculture as 67
guardian to, appointment of 191
inspection, in absence of - 73
limited owners' power of consent - - - - - 190
miscellaneous provisions 11 *et seq.*
need not be infeft - - - 113
notices by, *see* Notice by Landlord

objection to acquirer of lease, by - - - - - 131
objection to legatee by - - 130
obligation to incoming tenant 42
partner in tenant firm - - 50
power of entry on holding - 128
pro indiviso proprietors - 137
rights and claims against outgoing tenant, assignation of - - - - - 217
savings for other rights - - 206
Secretary of State 55, 192
shareholder in tenant company - - - - - 50
undertaking to reinstate building - - - - - 214

LAND TRIBUNAL FOR SCOTLAND 327, 338

LAW AGENT
arbiter entitled to call in 74, 102

LEASE
acquisition of 131 *et seq.*, 242
abbreviated form of - - - 272
"agreement" distinguished from - - - - - 50
bequest of, *see* Bequest of Lease
breach of conditions of 139, 141
continuance of - 11, 117
custom of country - - - 198
definition of 50, 105, 106, 135, 196, 198
irritancy of - - 14, 134, 137, 141
ish - - 135, 137
intestacy of tenant on - - 17
lease entered into before 1921 49
lease entered into after 31st July 1931 - - - - 63
lease entered into after 1st Nov. 1948 63, 105, 107, 134
minimum term of - - - 4
mowing or grazing for 4, 106
penalties in - - 120, 126
provisions as to 105 *et seq.*, 213
renunciation by tenant - - 200
resumption under 21, 61, 135
revision of - - - - - 107
Secretary of State, where approved by - - - 106
succession, in - - 16, 317

M

INDEX

supplementary agreements to
vary obligations - - 109, 275
tacit relocation, held under
4, 5, 20, 61, 107, 134
tenancy agreements, demand
for execution of - - 107, 234
term of, variation of - - 11
termination of, claims arising
out of - - - - - 64
termination of, distinguished
from termination of tenancy - - - - - 137
termination of, by landlord on
tenant's death - - - 2
terms, adjustments of 61 etc., 107
terms as to permanent pasture,
variation of
10, 49, 55, 116 et seq.
terms implied by Act in new 5, 62
written, provisions for
5, 107 et seq., 213
year for less than - - - 61
year, for single - - - 198
year to year, less than from 4, 106

LEGATEE
application to be declared
tenant - - 17, 130, 241
bequest of lease, does not
accept - - 17, 130
must intimate - 16, 130, 241
express exclusion in lease,
where - - - - - 200
has possession pending proceedings - - - - 130
landlord objecting to 17, 130, 242
notice to quit - 17, 305, 329
termination of tenancy where
legatee has acquired - - 329

LIMING OF LAND
36, 162, 209, 211, 212

LIMITED OWNERS
power to give consent etc. - 190

LIQUIDATED DAMAGES 14, 126

LIVESTOCK - - - 116, 197

MACHINERY 12, 34, 60, 123, 210

MANURE
application of, as improvement
36, 37 et seq., 161,
162, 209, 211, 213
compensation, ascertaining - 43
crops removed during last 2
years - - - - - 44
conditions attending freedom
of cropping - - - - 11
dung, see Dung
full equivalent manurial value
11, 12, 119, 120, 214
outgoing tenant required to
offer at market value - - 55
particulars required from tenant - - - - - 35
purchased (including artificial)
application of
36, 37 et seq., 161,
162, 209, 211, 213
removal, after notice to terminate tenancy - - - 122
residual value - - - - 37
sub-tenant, improvements
made by - - - - 30
succession in question of - 38
unexhausted values - 37, 90
valuation of - - 37, 90
valuation of, principles of - 95

MANURIAL BENEFIT - - - 95

MANURIAL IMPROVEMENTS 30, 38

MANURIAL VALUE - 38, 95

MARKET GARDENS
compensation for improvements 32, 36, 51 et seq., 165, 213
agreements as to 32, 36, 182
definition of - 178, 197, 199
Evesham Custom 53, 179, 181
holding, treated as
53, 62, 177 et seq.
improvements - - 62, 213
Land Court, direction by
53, 62, 180, 181
lease current 1898 - - 178, 179
special provisions as to 51 et seq.
substituted compensation
32, 36, 45, 52

MARKET VALUE 89, 95, 97 et seq.

INDEX

MARLING OF LAND - 209, 211, 212

MARTINMAS 19, 92 *et seq.*, 197, 201

MINERALS, WORKING OF - - 148

MOVEABLES IN SUCCESSION - - 316

MOWING LETS - - 4, 106

NEGATIVE PRESCRIPTION - - 101

NOTICE BY LANDLORD
acquirer of lease objecting to 131, 243
breach of conditions of tenancy to remedy - 138, 231
buildings etc., electing to purchase - - - 123, 239
change of landlord 194, 248, 249
claim of - - - - - 63
contract for sale of farm, of 146, 233
deterioration, claim for compensation for 47, 48, 62, 172, 247
fixtures, electing to purchase 12, 123, 239
improvements, intending to execute - - - 34, 246
completion of - - - 61
objections to 34, 167, 245
lease, adjustment of terms of 62, 107
legatee, objecting to 17, 130, 242
notice to quit, *see* Notice to Quit
pasture permanent, variation of terms as to - - - 116
record to be made, requiring 240
removal of tenant from part of holding, for 147, 232
rent, for payment of - - 231
increase of, due to improvements 34, 61, 115, 237
resumption of holding 21, 135
revision of lease for securing 107
terminate tenancy, to 52, 107, 122, 227, 228, 232
termination of tenancy, claims - - - 64, 183
written lease for securing - 107

NOTICE BY LEGATEE 16, 130, 131, 241

NOTICE BY TENANT
arbitration requiring 111, 230, 236, 311
buildings and fixtures, intention to remove 12, 60, 123, 238
claims, of 42, 46, 48, 55, 58, 59, 69, 183, 247, 248
counter-notice 62, 138, 139, 143, 145, 148, 150, 151, 230, 232
game damage, claim for 47, 60, 124, 239
holding vested in more than one person, where - - 174
improvements, for approval of - - - 34, 166, 243
dispensed with - 35, 58
intention to make 34, 56, 57, 157, 166, 245
which do not require 32, 35 *et seq.*, 209, 211, 212
which require 35, 56, 57, 208, 211, 212, 219
intention to claim compensation for special system of farming - - - 169, 247
intention to flit, of - - - 137
lease, adjustment of terms of - - - 61, 107
market garden improvements 51, 62, 165, 182, 213
notice to quit, counter notice to 17, 18, 21, 23, 24, 62, 138, 230
permanent pasture, variation of terms as to - 116, 309
quit, intention to 21, 53, 134, 180
form of written notice - 21
record to be made, requiring 240
rent, arbitration as to 61, 236
variation of - - 61, 111
repairs, intention to execute 57, 160, 209
sale of implements, stock, etc 58
sale of holdings, notice to quit to remain in force following - - - 146, 234
sub-tenant, to 22, 51, 143, 144, 312
terminate tenancy, to 107, 122
termination of tenancy claims 183

INDEX

written lease for securing - 107

NOTICE TO QUIT 19 *et seq.*
  acquirer of lease to 110, 228
  agreement as to validity 146, 233
  amendments as to (1958) - 303
  arbitration, questions referred
    to 21, 143, 230, 320
  contracting out of, provisions
    for - - 20, 50, 134
  consent to - - 137 *et seq.*, 303
  counter notice by tenant, *see*
    Notice by Tenant
  date of, at the 25, 138, 140
  form of - - 21, 134, 140
  forms, statutory 21, 223 *et seq.*
  freedom of cropping restrict-
    ing - - - 11, 119
  given before November 1948 203
  given after 1st November,
    1967 - 59, 322 *et seq.*
  grounds for to be stated
    18, 19, 25, 28, 138, 139
  incontestable 2, 18, 305, 329
  landlord by, *see* Notice by
    Landlord
  Land Court, consent of, *see*
    Consent
  near relative successor to
    2, 18, 23, 27, 228, 306, 329
  part of holding
    21, 23, 58, 147, 149, 173
    form - - - - - 232
    right to treat as notice to
      quit whole
      21, 58, 145, 148, 233
  penalty for breach of con-
    ditions accompanying 23, 313
  period of notice required
    20, 121, 134, 319
  postponed by Land Court - 144
  procedure to be followed by
    tenant - - - - - 23
  pro-indiviso proprietors - 137
  provisions as to giving 134 *et seq.*
  restrictions on operation of
    21, 62, 137 *et seq.*
  sale of holding, provisions in
    case of - - - 136, 146
    agreement on - - - - 233
    to remain in force following 234
  service of - 48, 134, 136, 194
  statutory provisions, exclusion

  of, incompetent - - 20, 135
  sub-tenant to 22, 23, 144, 145, 312
  suspended, pending arbitra-
    tion on decision of Land
    Court - - - 144, 311
  tenant by, *see* Notice by Tenant
  validity of 136, 137, 140
  where breach not remedied - 138
  where holding agreed to be
    sold - - - 146, 233
  where holding vests in more
    than one person - - - 175
  where land required for a use
    other than agriculture - 138
  where landlord not legal
    owner - - - - - 199
  where landlord prejudiced by
    breach - - - - - 139
  where relates to permanent
    pasture - - - - - 138
  where rent unpaid - - - 138
  where rules of good husbandry
    not complied with - - - 138
  where tenant notour bank-
    rupt - - - 134, 139
  withdrawal of - - - 74, 140

NOTICES
  forms of - - - 227 *et seq.*
  service of 48, 134, 136, 194

NOTOUR BANKRUPTCY
  134, 139, 141, 180

OBSTACLES TO CULTIVATION
  33, 36, 209, 211, 213

ORCHARDS 33, 208, 210, 212

OSIER BEDS 33, 208, 210, 212

OVERSMAN
  appointment of - - 94, 268
  minute of acceptance 94, 269
  arbiter obstructive, where
    death of - - - 101, 102
  decision of, is final - - - 94
  devolution by arbiter on
    101, 102, 255
  duties of - - - 101 *et seq.*
  valuations in - - 93, 286

PANEL OF ARBITERS, *see* Arbiter

PARTNERSHIP, AS TENANT - 50

INDEX

PARTRIDGES, see Game

PASTURE
  meadow includes - - - 197
  permanent, definition of - 39
    breaking up of, benefit - 163
    improvement of 34, 45, 209
    laying down 33, 208, 210, 212
    notice to quit - 138, 140
    ploughing up - - - 176
    reduction of - - - 309
    restoration of, compensation excluded - - - 10
    variation of terms as to 10, 49, 55, 116 *et seq.*, 309
    written agreements, in connection with - - - 45
  temporary 39 *et seq.*, 209, 211, 213
    basis of value - - - - 40
    "benefit" in - - - - 39
    compensation in respect of 36, 161, 168, 209
    definition of - - - 39
    in England - - - - 39
    laying down 39, 209, 211, 213
    laid down during previous lease - - - - - 40
    second year's grass - - 209
    to be left by tenant - - 117
    tenant's pasture - - - 176

PAYING GUESTS - - - - 198

PENAL RENTS - - - 14, 126

PENS - - - - - 208

PERMANENT PASTURE, see Pasture

PHEASANTS, see Game

PIER, MAKING OF - - - 148

PLOUGHING GRANTS - - - 175

POULTRY FOLDED ON LAND - 36

PROCEDURE IN VALUATIONS, see VALUATIONS

PRODUCE
  compensation for failing to replace - - - - - 120
  definition of - 120, 197

    disposal of - - - - 118
  landlord agreeing to purchase - - - - 62, 133
  payment for, on termination of tenancy - - 62, 133

PROROGATION, MINUTE OF 269, 270

QUITTING, see REMOVING

RAILWAY, MAKING OF - - - 148

RATES RELIEF - - - 112, 114

RECLAMATION OF WASTE LAND 34, 208, 210, 212

RECORD OF HOLDING 14, 127
  buildings and fixtures 14, 15, 108, 127, 123
  claim for deterioration, in - 173
  cost of making - 15, 127
  form of - - 14, 127, 297
  fixed equipment and alteration of holding - 58, 63, 170
  improvements, in 14, 15, 42, 127, 165
  Land Court to determine differences - - 15, 127
  landlord or tenant may require 14, 127, 240
  where essential - - - - 128

REDUCTION OF AWARD 85 *et seq.*
  award ad factum praestandum 86
  bad award - - - - 85
  bias shown by arbiter - - 88
  bribery and corruption - - 85
  challenge, grounds of - - 85
  Common Law grounds - - 85
  evidence heard in absence of other party - - - - 88
  failure to deal with claim - 88
  failure to issue within time limit - - - - - 86
  failure to show basis of valuation - - - - - 287
  falsehood - - - - - 85
  fraud - - - - - 87
  misconduct of arbiter 85, 86, 87, 216
  mistake admitted - - 86
  reference not exhausted 86, 87
  refusal to admit material

evidence - - - - 87
refusal to state a case - - 87
required details omitted - - 85
wrongful admission of evidence - - - - - 87

REGULATIONS AND ORDERS
Agricultural Holdings (Servicemen) Scotland Regulations 1952 - - - - 139
Agricultural Holdings (Scotland) Regulations 1950 - 111
Agricultural Holdings (Specification of Forms) (Scotland) Rules 1960 - - - 277
Agricultural Records (Scotland) Regulations 1948 - 296
Defence Regulations 195, 208
revocation and variation of - - - - - 195
Rules of Court 78, 218
under former Acts, general savings - - - - 204

REMOVING
expenses and loss attributed to - - 26, 62, 133, 150
letter of removal, an agreement - - - 51, 135
non payment of rent, for 15, 128, 134, 270
notice by tenant of intention to quit 19 *et seq.*, 53, 137, 180
Sheriff Courts (Scotland) Act 1907, provisions of 20, 137
suitable successor to be found 53, 180
where interest under lease vests in executor - - - 319

RENEWALS, *see* Buildings

RENT
abatement of - - - - 154
absolute right to increase 10, 114
arbitration, demand by landlord for 6, 111, 236, 302
arbitration demand by tenant for - - 60, 236, 302
basis of fixing new rent 7 *et seq.*
disturbance claims, in 23 *et seq.*, 151, 154

"economic" rent - 1, 8
fixing, amendments as to (1958) 302
increase of, effect from - - 182
improvements by landlord 10, 34, 61, 114 *et seq.*
landlord, claims by - - - 64
lease, to be included in - - 213
non-payment, removal for 15, 128, 134, 270
notices requiring payment of 237, 140
payment of, not referred to arbitration - - - - 54
penal rents - - - 14, 126
reduction of where partly dispossessed - 61, 149
variation of 6 *et seq.*, 111 *et seq.*
conditions for - - - 60
deterioration, in - 7, 112
improvements 7, 111, 114 *et seq.*
method of serving reference 6

REORGANISATION OF TENANT'S AFFAIRS, SUM FOR 26 *et seq.*

REPAIRS, *see* Buildings

RESERVOIRS - - 21, 33, 148

RESUMPTION OF HOLDING
compensation for 28, 137, 174, 327
notice of - - - - 135, 137
part of holding, compensation for - 61, 149, 173, 324
reduction of rent where tenant dispossessed of part 61, 149
under lease - - - 21, 61, 135

ROADS, MAKING OF 34, 148, 208, 210, 212

ROOTS - - - - - 122

ROTATION OF CROPS - - - 140
see also Cropping

RULES OF GOOD HUSBANDRY 24, 140, 197

SALE OF HOLDING
notice by landlord to tenant 146, 232
notice to quit, agreement as to 146, 233

INDEX

SECRETARY OF STATE
  applications to - - - - 225
  approving letting - - - 106
  arbiter, appointment of
      65, 66 et seq., 188,
      189, 214, 235, 281
  Secretary of State, a party - 189
  as landlord or tenant 53, 192
  award, application for extension of time of making
      82, 215, 283
  consent of, for case under one year - - - - - 61, 106
  direction of, to carry out improvements - - - - 114
  expenses and receipts - - 193
  former power to make market garden direction - - - 181
  power to make rules as to arbitration expenses etc. - 187
  power to extend time for settlement of claims on termination of tenancy - - 183
  power to revoke or vary order 195
  power to create a charge on holding - - - - 185
  power to vary schedules 1 and 4 - - - - - 189
  tables prepared by - - - 37

SECURITY OF TENURE 1, 21 et seq.
  contracting out of - - - 50
  for near relative successor - 18
  restrictions on operation of notice to quit - - - 21

SERVICE OF NOTICES
      48, 134, 136, 194

SEWAGE - - - 34, 209

SHEAF SHEDS - - - 34, 208

SHEEP DIPPING ACCOMMODATION
      34, 211, 212

SHEEP PENS, ETC. - - - - 208

SHEEP STOCKS
  classes, designations of - - 294
  entailed estates, taken over by 89
  Hill Farming Act 1946
      89, 114, 115, 288 et seq.

Sheep Stocks Valuation (Scotland) Act 1937
      89, 96, 99, 285 et seq.
  valuation of
      65, 90, 93, 95 et seq.,
      187, 285, 288 et seq.

SHERIFF
  appeal to on remuneration of arbiter - - - - - 188
  arbiter, removal of application for - - - 216, 262
  no appeal from Sheriff Substitute - - - - - 195
  order to state a case application for - - - 216, 262
  review of taxation of expenses
      128, 216
  stated case for opinion of 76, 214

SILAGE AND SILOS
      34, 93, 208, 210, 211

SLUICES 33, 208, 210, 212

SMALLHOLDINGS, PROVISION OF 148

STAMP DUTY
  minute of submission, on - 70

STATED CASE - - 76 et seq.
  appeal to Court of Session 170, 186
  application to Sheriff for order on arbiter 77, 87, 262
  expenses of - - 79, 82, 218
  for opinion of Sheriff 76, 218, 286
  remit for amendment - - 78
  refusal to state a case 77, 87
  revisal by parties - - - 77

STATUTORY INSTRUMENTS, see Regulations and Orders

STEELBOW 38, 44, 92, 103

STIPEND - - - - 26

STRAW, VALUATION OF
      90, 92, 103, 104, 210

STRAWBERRY PLANTS - 52, 213

INDEX

SUBMISSION
  award in Common Law submission - 90, 267
  duration of - - - - 101
  form of - - - - - 263
  minute of - 70, 87, 94, 263
  prorogation, minute of 269, 270

SUBSTITUTED COMPENSATION
  market garden improvements 45, 52, 165, 182
  under agreements 3, 32, 36, 45, 46, 52, 56, 57, 157, 159, 160, 166

SUB-TENANT
  disturbance, compensation for 26, 151
  improvements by - - - 30
  notice to quit to 22, 51, 143, 144
  period of let to - - 3, 4
  right to claims competent to principal tenant - - - 3

SUCCESSION
  dung-hills or manure in questions of succession - 38
  holdings to, 17 *et seq.*, 129 *et seq.*, 304
  where lease vests in executor of tenant - - - 320
  where successor a near relative 17, 18, 27, 28, 306
  where successor not a near relative - 17, 18, 228
  provisions as to 2, 304 316 *et seq.*
  where notice to quit is given by tenant - - - - 58

TABLES PREPARED BY SECRETARY OF STATE - - - - - 37

TACIT RELOCATION
  4, 51, 107, 117, 121, 134, 135, 136, 137
  contracting out of - 50, 107

TAXATION BY AUDITOR
  arbitration expenses subject to 81 *et seq.*, 216
  record expenses in making 15, 128

TEMPORARY PASTURE, *see* Pasture

TENANCY
  breach of conditions, notice to remedy - - 138, 231
  change of - 164, 168
  "commencement of the tenancy" - - - 41, 114
  joint-tenancy - - - - 51
  new tenancy - - 1, 181
  questions arising out of 59, 63, 89
  security of tenure, provisions as to - - - 21, 50
  termination of, claims arising out of 7, 62, 183 *et seq.*, 271, 324
  definition of - 124, 197
  devices to simplify - - 50
  notice of - - - 20, 52, 121
  part of holding occupied after - - - - - 184
  purpose of - - 138, 141
  settlement of claims on 183, 252, 283
  termination of lease, distinguished from - - - 137
  time for notice of - - 20
  where tenant near relative successor - - 18, 228, 329
  terms of, *see* Lease

TENANT
  agent of - - - - - 198
  away-going - - - - 128
  bankruptcy of 62, 134, 139, 141, 180, 200
  claims between outgoing and incoming - 54, 55, 90, 91
  claims by 55 *et seq.*, 113, 259, 266
  company as - - - - 50
  death of - - - - - 16
  definition of 3, 4, 164, 197, 199
  entry at Martinmas - - - 92
  entry at Whitsunday - - 92
  executor of - - - 17, 317
  guardian to, appointment of - 191
  miscellaneous provisions - 11
  removal of, *see* Removing
  out-going - - - - - 188
  partnership as - - - - 50
  quitting where notice given before 1st November 1948 - 203
  savings for other rights etc. - 206

INDEX

sum for reorganisation of affairs - - - 26 et seq.
trust deed for creditors - 139, 180

TERMS
  removal terms - - - - 19

THREE-YEAR AVERAGE PRICE
                          100, 292

THRESHING MILLS   34, 92, 93, 208

TIME
  for adjusting claims extension of - - - - - 252
  improvement, commencement of - - - - - 29
  notice to terminate tenancy 20, 52, 121
  payment of sum awarded, for 84
  pronouncing award, for   82, 255

TRAMWAY, MAKING OF   - 148

TREE ROOTS -   36, 209, 211, 213

TREES, PLANTING OF   21, 148, 213

TRUST DEED - - -   139, 180

VALUATION
  arbiter and oversman, appointment of - - - 93
  arbitrations etc. outside Act 65, 89 et seq., 187
  away-going 65, 89, 95, 102 et seq., 187
  common law, subjects of - 89
  crops, of, see Crop
  dung, of, see Dung
  fallow ground, of, see Fallow Ground
  fences, of, see Fences
  incoming tenant's obligation to take over at - 44, 92
  Martinmas entry   92, 99, 100
  procedure and principles of 94, 100
  procedure in making   102 et seq.
  questions excluded under the Act - - - 65, 90
  sheep stock, see Sheep Stocks
  silage, see Silage

statements etc. to be submitted - - - - - 90
straw of, see Straw
turnips of - - - - - 95
valuation and arbitration, distinction between - - - 89

VEGETABLE CROPS   -   52, 213

WARPING OF LAND   33, 208, 211

WATER COURSES
           34, 148, 208, 210, 212

WATER MEADOWS
               33, 208, 210, 212

WATER POWER AND SUPPLY
           33, 114, 208, 210, 212

WEIRING OF LAND
               33, 208, 211, 212

WHARF, MAKING OF - - - 148

WHITSUNDAY
       19 et seq., 92, 98, 99, 100, 197, 201

WIREWORK IN HOP GARDENS - 211

WITNESSES, see Evidence

WORDS AND PHRASES
  absolute owner - - - 195
  Act, the - - - - - 277
  Act of 1883, the - - - 30
  Act of 1886, the - - - 320
  Act of 1900, the - - - 30
  Act of 1931, the - - - 320
  Act of 1947, the - - - 307
  Act of 1948, the - - - 307
  Act of 1949, the - - - 320
  acquirer - - - - - 132
  acquiring authority - - - 329
  adopted - - - - - 305
  agreement in writing - - 160
  agricultural holding 4, 49, 105, 174, 195, 286, 307
  agricultural land -   4, 105
  agricultural lease - - - 320
  agricultural unit - - - 195
  agriculture - -   4, 105, 195

INDEX

all compensation	182
amalgamation	333
any benefit	162
appointed day	307
appropriate authority	159
arable land	11, 119
arbiter	286
as soon as is reasonably possible thereafter	109
assignee	199
at my away-going	201
at the date of the giving of the notice	25, 140
basic ewe value	100, 292
basic lamb value	100, 292
building	13, 195
commencement of the tenancy	41, 114
contract of tenancy	307
cumulative or accumulated fertility	170
custom of the country	198
de minimis non curat lex	141
Defence Regulations	195
definite and limited period	140
delivery	218
determination of Tenancy	137, 200
directly attributable	153
director	334
during the tenancy	170
economic rents	1, 8
eild ewe	294
eild sheep	294
either party	107, 135
equivalent allowance or benefit	114
erection or enlargement of buildings for the purpose of trade or business of market gardener	181
ewe	294
ewe hogg	294
expenses reasonably incurred	153
expiration of the stipulated duration of any lease	120, 121
fair and reasonable compensation	16
fair and reasonable landlord	142
fair rent	8
fair value to an incoming tenant	9, 123
farming of land	198
fiars prices	103
fixed equipment	123, 196
fixtures	12, 13, 123
fodder	93
for the protection of the landlord	182
former enactment relating to agricultural holdings	196
full equivalent manurial value	12, 119
game	125
good husbandry	11, 22, 141, 143
greater hardship	22, 142, 143
hefting	97
high farming	170
holding	155, 198
initial date	329
ish	20
lamb	294
Land Court	9, 196, 307
landlord	4, 156, 196, 199, 307
lease	196, 290, 307, 320
lease in writing	108
liming of land	36
livestock	197
livestock rearing	116
manuring	36, 43, 162
market value	89, 95, 97
Martinmas	20, 197, 201
member of his family	131, 304
Minister, the	307, 334
muirburn	26, 290
near relative	17, 332
near relative successor	18
new improvement	197
new tenancy	53
1923 Act improvement	155, 197
1931 Act improvement	155, 197
non-statutory land	174
not being agricultural purposes	137
notice	334
notice to quit	152
old improvement	155, 197
open market value, *see* market value	
other purposes	137
other than for agriculture	142, 143
outgoing tenant	188
parties claiming through them respectively	75
pasture	10, 197
permanent pasture	10, 33, 39
possession	329

INDEX

prescribed - - - - - 197
prescribed period - - - 111
present rent - - - - 154
principal act - - - - 329
principal Scottish Act 329, 333
proof from the stack - - 104
produce - - - 120, 197
proper rent - - - - - 8
quitting - - - - - 155
reasonable opportunity 122, 126
reasonable time - 159, 181
rent - - - 153, 154
roots - - - - - 122
rules of good husbandry 24, 143
Scottish Act of 1948 - - 307
Scottish Act of 1949 - - 307
shall be taken into account - 162
shott - - - - - 294
sheep—eild ewe, eild sheep, eild gimmer, ewe, ewe hogg, gimmer, lamb, shott, tup - 294
sound estate management - 22
standard or system of farming 170
statement of case - - - 64
steelbow - - - - - 38
stipulated - - - 107, 135
substantial and otherwise suitable person - - - 180
suitable and adequate provisions - - - - - 120
surface damages - - - 149
system of cropping - - - 120
tacit relocation - - 20, 107
temporary pasture 39 *et seq.*, 210
tenant 4, 197, 199, 290, 307
tenant's pasture - - - - 176
termination of the lease or tenancy - - 137, 197, 201
three-year average price - 100
threshing out - - - - 104
to an incoming tenant - - 156
took effect - - - - 114
trade or business - 4, 105
treated as a separate unit - 143
tup - - - - - 294
two years' rent of the holding 154
unavoidable cause - - - 131
unavoidably incurred - - 153
value to an incoming tenant - 170
warping or weiring of land - 33
Whitsunday - 20, 197, 201
works for the application of water power or for the supply of water - - - 33
year before the expiration of the lease - - - 119, 121
year before the tenant quits - 122
year's rents - - - - 154

The Sheriff Clerk,
Sheriff Court House,
Hope Street,
LANARK.